Advanced Technologies for Biomass

Advanced Technologies for Biomass

Editors

Andrea Di Carlo
Elisa Savuto

MDPI • Basel • Beijing • Wuhan • Barcelona • Belgrade • Manchester • Tokyo • Cluj • Tianjin

Editors
Andrea Di Carlo
University of L'Aquila
Italy

Elisa Savuto
University of L'Aquila
Italy

Editorial Office
MDPI
St. Alban-Anlage 66
4052 Basel, Switzerland

This is a reprint of articles from the Special Issue published online in the open access journal *Energies* (ISSN 1996-1073) (available at: https://www.mdpi.com/journal/energies/special_issues/Advanced_Technologies_Biomass_Production).

For citation purposes, cite each article independently as indicated on the article page online and as indicated below:

LastName, A.A.; LastName, B.B.; LastName, C.C. Article Title. *Journal Name* **Year**, *Volume Number*, Page Range.

ISBN 978-3-0365-6153-0 (Hbk)
ISBN 978-3-0365-6154-7 (PDF)

Contents

Xiaodan Liu, Xuping Feng, Lingxia Huang and Yong He
Rapid Determination of Wood and Rice Husk Pellets' Proximate Analysis and Heating Value
Reprinted from: *Energies* **2020**, *13*, 3741, doi:10.3390/en13143741 1

**Grigorios Rekleitis, Katherine-Joanne Haralambous, Maria Loizidou
and Konstantinos Aravossis**
Utilization of Agricultural and Livestock Waste in Anaerobic Digestion (A.D): Applying the
Biorefinery Concept in a Circular Economy
Reprinted from: *Energies* **2020**, *13*, 4428, doi:10.3390/en13174428 13

Barbara Malsegna, Andrea Di Giuliano and Katia Gallucci
Experimental Study of Absorbent Hygiene Product Devolatilization in a Bubbling
Fluidized Bed
Reprinted from: *Energies* **2021**, *14*, 2399, doi:10.3390/en14092399 27

Falko Marx, Paul Dieringer, Jochen Ströhle and Bernd Epple
Design of a 1 MW$_{th}$ Pilot Plant for Chemical Looping Gasification of Biogenic Residues
Reprinted from: *Energies* **2021**, *14*, 2581, doi:10.3390/en14092581 49

**Douglas Alberto Rocha de Castro, Haroldo Jorge da Silva Ribeiro, Lauro Henrique Hamoy
Guerreiro, Lucas Pinto Bernar, Sami Jonatan Bremer, Marcelo Costa Santo, et al.**
Production of Fuel-Like Fractions by Fractional Distillation of Bio-Oil from Açaí (*Euterpe oleracea*
Mart.) Seeds Pyrolysis
Reprinted from: *Energies* **2021**, *14*, 3713, doi:10.3390/en14133713 75

**Sunyong Park, Hui-Rim Jeong, Yun-A Shin, Seok-Jun Kim, Young-Min Ju,
Kwang-Cheol Oh, et al.**
Performance Optimisation of Fuel Pellets Comprising Pepper Stem and Coffee Grounds
through Mixing Ratios and Torrefaction
Reprinted from: *Energies* **2021**, *14*, 4667, doi:10.3390/en14154667 103

**Conceição de Maria Sales da Silva, Douglas Alberto Rocha de Castro, Marcelo Costa Santos,
Hélio da Silva Almeida, Maja Schultze, Ulf Lüder, et al.**
Process Analysis of Main Organic Compounds Dissolved in Aqueous Phase by Hydrothermal
Processing of Açaí (*Euterpe oleraceae*, Mart.) Seeds: Influence of Process Temperature,
Biomass-to-Water Ratio, and Production Scales
Reprinted from: *Energies* **2021**, *14*, 5608, doi:10.3390/en14185608 119

**Maria Elizabeth Gemaque Costa, Fernanda Paula da Costa Assunção, Tiago Teribele,
Lia Martins Pereira, Douglas Alberto Rocha de Castro, Marcelo Costa Santo, et al.**
Characterization of Bio-Adsorbents Produced by Hydrothermal Carbonization of Corn Stover:
Application on the Adsorption of Acetic Acid from Aqueous Solutions
Reprinted from: *Energies* **2021**, *14*, 8154, doi:10.3390/en14238154 143

Rachele Foffi, Elisa Savuto, Matteo Stante, Roberta Mancini and Katia Gallucci
Study of Energy Valorization of Disposable Masks via Thermochemical Processes:
Devolatilization Tests and Simulation Approach
Reprinted from: *Energies* **2022**, *15*, 2103, doi:10.3390/en15062103 165

Nianze Zhang, Chunyan Tian, Peng Fu, Qiaoxia Yuan, Yuchun Zhang, Zhiyu Li and Weiming Yi
The Fractionation of Corn Stalk Components by Hydrothermal Treatment Followed by Ultrasonic Ethanol Extraction
Reprinted from: *Energies* **2022**, *15*, 2616, doi:10.3390/en15072616 . . . . **189**

Andrea Di Carlo, Elisa Savuto, Pier Ugo Foscolo, Alessandro Antonio Papa, Alessandra Tacconi, Luca Del Zotto, et al.
Preliminary Results of Biomass Gasification Obtained at Pilot Scale with an Innovative 100 kWth Dual Bubbling Fluidized Bed Gasifier
Reprinted from: *Energies* **2022**, *15*, 4369, doi:10.3390/en15124369 . . . . **205**

Article

Rapid Determination of Wood and Rice Husk Pellets' Proximate Analysis and Heating Value

Xiaodan Liu [1,2], Xuping Feng [1,2], Lingxia Huang [3,*] and Yong He [1,2,4,*]

[1]   College of Biosystems Engineering and Food Science, Zhejiang University, 866 Yuhangtang Road, Hangzhou 310058, China; xdlww@zju.edu.cn (X.L.); pimmmx@163.com (X.F.)
[2]   Key Laboratory of Spectroscopy Sensing, Ministry of Agriculture, Hangzhou 310058, China
[3]   The Rural Development Academy, Zhejiang University, 866 Yuhangtang Road, Hangzhou 310058, China
[4]   State Key Laboratory of Modern Optical Instrumentation, College of Optical Science and Engineering, Zhejiang University, 866 Yuhangtang Road, Hangzhou 310058, China
*   Correspondence: lxhuang@zju.edu.cn (L.H.); yhe@zju.edu.cn (Y.H.); Tel.: +86-571-889-82143 (Y.H.)

Received: 24 June 2020; Accepted: 15 July 2020; Published: 20 July 2020

Abstract: Biomass pellets are a potential renewable and clean energy source. With the advantages of perfect combustion performance and easy storage and transport, biomass pellets have gradually replaced fossil fuels and become widely used. Rapid and accurate determination of biomass pellets' quality is critical to efficient energy use. Laser-induced breakdown spectroscopy (LIBS) combined with chemometric methods were utilized. The gross calorific value (CV) and ash content (Ash), volatile matter (VM) and fixed carbon (FC) were firstly measured and analyzed. LIBS spectra and their corresponding elements of biomass pellet samples were analyzed. Three quantitative analysis models for quality indexes including partial least-squares regression (PLSR), least squares-support vector machines (LS-SVM), extreme learning machines (ELM) were further built. All models performed well, especially the LS-SVM model which obtained the best determination results, with all R^2 values over 0.95. Concurrently, the modeling performance of ash was slightly better than that of the other three quality indexes, which further confirmed the feasibility of using relevant elements to predict biomass quality indexes. The overall results indicated that LIBS coupled with suitable chemometrics could be an alternative promising method to determine quality indexes of biomass pellets and further improve energy utilization by using biomass materials with better quality.

Keywords: biomass pellet; laser-induced breakdown spectroscopy; chemometrics; quality indexes

---

1. Introduction

The energy crisis is one of the greatest challenges that our society is facing nowadays, so it is essential to develop sustainable new energy resources to reduce the use of fossil fuels. In recent years, biomass energy has drawn wide attention around the world as a clean and renewable energy source [1,2]. Especially for biomass pellets there is a growing market demand [3], since have the advantages of a high combustion efficiency, convenient storage and transport, environmental friendliness and low cost [4,5]. In order to make full use of this biomass energy source, it is important to analyze the combustion performance of the biomass pellets. Industrial analysis components including carbon content, volatile matter content and ash content are all important factors affecting biomass combustion performance [6–9]. The calorific value is a direct indicator of biomass combustion performance [10,11]. Thus, accurate and fast determination of all these quality indexes are of great importance in the efficient utilization of energy. Traditional detection methods are mainly industrial technology and chemical analysis, and detection methods of different indexes are different. It is time, cost, and labor-consuming to measure all these indexes simultaneously [12]. Therefore, it is particularly necessary to develop newly rapid quantitative analysis techniques.

Laser-induced breakdown spectroscopy (LIBS) is a recently emerged spectroscopic technique based on elemental analysis [13]. It provides a reliable alternative for qualitative and quantitative analysis of the composition of samples [14,15]. With the merits of fast detection, little sample or no pretreatment, multi-elemental simultaneous detection capability, LIBS has been widely applied to sample analysis in many fields ranging from agriculture and biological to industry [16–19]. In recent years, there have been many researches on energy using LIBS. Li et al. [20] built a PLS quantitative prediction model for the calorific value of coal based on LIBS combined with different spectral pre-processing methods. The best result was achieved when using the 11 points smoothing combined with the second-order derivation, with both the correlation coefficient of calibration set and correlation coefficient above 99%. Yao et al. [21] used artificial neural networks to quantitatively analyze the ash, volatile matter, fixed carbon, and gross calorific value of coal, and a reliable and accurate result was achieved, with the correlation coefficients of all prediction models above 97%. Dong et al. [22] employed LIBS and multiple linear regression combined with the partial least squares regression (PLSR) and support vector regression (SVR) to analyzed the carbon contents in coal samples. The best results of carbon calibration were obtained using LIBS coupled with an MLR model-added SVR correction. The determination coefficient (R^2), root mean standard error of cross validation (RMSECV) and root mean standard error of prediction (RMSEP) were 0.99%, 0.00039%, and 1.43%, respectively. AINTS et al. [23] focused on the quantitative assessment of the calorific value of Estonian oil shale using LIBS. The ordinary multivariate regression analysis was applied, and the correlation coefficients of the prediction model for the calorific value and moisture content were above 90% and 75%, respectively. Lu et al. [24] investigated the feasibility of using LIBS to determine gross calorific value, carbon content, volatile matter content and ash content. Several pre-processing methods combined with PLS model were used, and it was turned out that the PLS model based on spectra that combined baseline correction with Z-score standardization performed best, with the coefficient of determination of all prediction models above 96%. To the best of our knowledge, there are no studies focused on quantitative prediction of biomass fuels properties using LIBS combined with different chemometrics.

Based on the abovementioned situation, the objective of this study was: (1) to explore the feasibility and accuracy of using LIBS to analyze the properties of biomass pellets; (2) to select important spectral lines for building the relationships between the important elemental emission lines and quality indexes; and (3) to build accurate and reliable prediction models for the quality indexes of biomass pellets.

2. Materials and Methods

2.1. Sample Preparation

The biomass pellets were provided by different bioenergy companies, and they are representative of the most commonly used varieties found on the biofuel market in China. They mainly included rice husks, pine wood, rubber wood, and red wood. The distance between the sample surface and the lens for laser focusing will affect the intensity of the spectral signals. Besides, uneven samples will also cause interference in the spectral signals. Thus, a uniform and regularly shaped sample is of great importance. In order to obtain accurate and reliable experimental data, the biomass pellets were first milled and then compressed into a circle tablet using a pressing machine (FY-24, SCJS, Tianjin, China) for 30 s and under a pressure of 10 MPa. The resulting side length and thickness were 10 mm and 2 mm, respectively. The tablets were then dried at 60 °C for 6 h in an oven. A total of 128 tablets samples were prepared. They were divided into calibration and prediction sets in a ratio of 3:1 by joint x–y distances (SPXY) methods [25]. Thus, 96 samples and 32 samples were obtained for calibration and prediction, respectively. Besides, the four kinds of biomass pellets were distributed in both calibration and prediction sets to make the analyses more reliable.

2.2. Determination of Quality Indexes

In this study, industrial analysis components were also determined according to the traditional methods. Ash content (Ash) and volatile matter (VM) were measured according to ASTM D1102-84 and ASTM E872-82 [26]. The gross calorific value (CV) (MJ·kg^{-1}) was determined in accordance with the China national standards GB/T30727-2014 using an oxygen bomb calorimeter (5E-AC, Chansha Kaiyuan Instrument Co., Ltd., Changsha, China). Fixed carbon (FC) was calculated after determining the Ash and VM content according to the following equation:

$$FC\% = 100\% - Ash\% - VM\% \tag{1}$$

The values of Ash, VM, and FC were dry basis.

2.3. Experimental Apparatus and LIBS Measurement

A self-built LIBS system (Figure 1) was used in this work. It mainly comprised of a Q-switched Nd:YAG pulsed laser (Vlite-200, Beamtech Optronics, Beijing, China) with the maximal energy of 200 mJ @532 nm for generating laser pulse to ablate samples, an Echelle spectrograph (ME5000, Andor Technology, Belfast, UK) equipped with an (intensified charge coupled device) ICCD camera (iStar DH340T, Andor Technology, Belfast, UK) for collecting spectral signals, and a digital delay pulse generator (DG645, Stanford Research Systems, San Jose, CA, USA) for controlling the delay time between the laser and the ICCD camera. In addition, lens, mirrors and X-Y-Z stage were used for laser transmission and sample placement. A Hg: Ar lamp (HG-1, Ocean Optics, Winter Park, FL, USA) and a Deuterium Halogen light source (DH-2000-BAL-CAL, Ocean Optics, Winter Park, FL, USA) were used for the calibration of the wavelength and intensity. Experimental parameters were optimized by the response surface methodology (RSM) to obtain a best signal to noise ratio (SNR) and relative standard deviation (RSD). Based on the optimization, the energy of laser pulse was 60 mJ with 1 Hz repetition rate, the delay time was 1.5 µs, and the gate width of ICCD was 10 µs. Besides, the distance between the lens and sample surface was optimized to 98 mm when a good spectral signal was obtained.

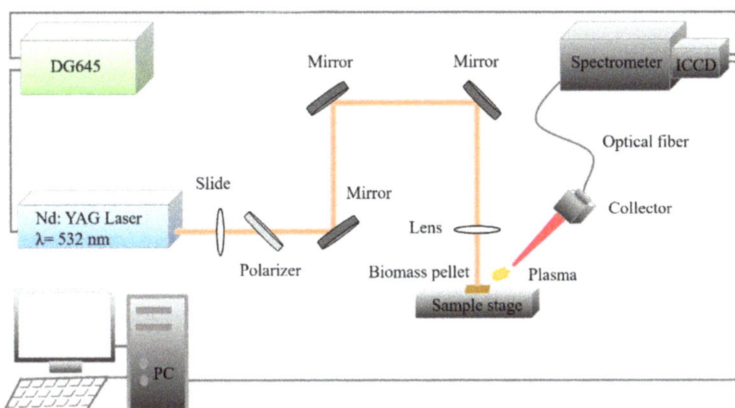

Figure 1. Schematic diagram of the LIBS system.

LIBS measurements were conducted after setting the parameters. First, the laser beam was adjusted to be focused 2 mm below the sample surface to avoid any interference caused by an uneven sample surface. The signal detector was then adjusted to ensure that the spectral signal could be collected. After that, a total of 80 spectra of each sample were obtained by controlling the moving path of the X-Y-Z stage. 16 position of each tablet sample were ablated with five successive accumulations. At last, the average of 80 spectra was used to be the spectrum of the sample.

2.4. Data Pretreatment and Analyze

In order to eliminate the impact caused by experimental instruments and radiation generated by plasma formation, and further to improve the precision and repeatability of LIBS detection, several pre-processing methods were introduced, including baseline correction, wavelet transform, normalization, and outlier discarding. Baseline correction could eliminate background signals, and wavelet transform could reduce any noise. Normalization could effectively reduce the fluctuation of each shot and improve signal stability. Detailed information of these methods were mentioned in our previous work [27]. Outlier discarding could improve the reliability and accuracy of the data. A self-developed outlier removing algorithm based on median absolute deviation (MAD) was used in this work. In order to meet our particular demands, a relatively stable spectral line of CN 388.29 nm was applied to identify outliers. Once the difference between the intensity and median of CN 388.29 nm exceeded 2.5 times the MAD, the spectrum would be considered as an outlier and be removed. Outliers removing was repeated until there were no outliers or a maximum of 25% of the original spectra were removed.

2.5. Chemometrics for Data Analyze

Several chemometric methods were conducted after data pretreatment, including partial least squares regression (PLSR), least squares-support vector machines (LS-SVM), extreme learning machines (ELM). The specific procedure of LIBS data analysis is shown in Figure 2.

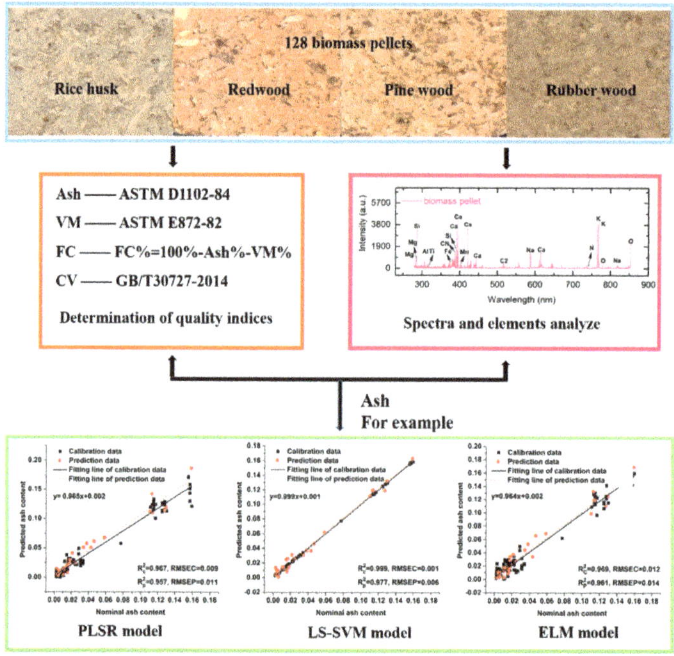

Figure 2. Flowchart of data analysis for quality indexes prediction of biomass pellets.

Partial least-squares regression (PLSR) is a widely used linear analysis algorithm for the quantitative analysis of large datasets, especially for spectra [28–31]. The main principle of this method is to convert the large linear original data into independent latent variables which still contain most of the information of the original variables, and the first few important latent variables will be used for further analysis [32,33]. In this work, PLSR models were applied based on the value of the biomass pellets' quality indexes and the intensities of LIBS spectrum. Full cross-validation was used to reduce the possibility of overfitting of models. Besides, the number of latent variables is also important for modeling, and it was determined when the minimal mean squared error was obtained [34].

Different from PLSR, the least-squares support vector machine (LS-SVM) is a commonly used non-linear machine learning method that can be used to deal with both linear and non-linear problems [35–38]. It has been turned out to be a good choice for quantitative analyses for a small dataset [39,40]. According to the previous study, kernel function and its parameters including sig2 and gam have a great influence on the performance of the LS-SVM model. Kernel function is the formula that maps complex data to high-dimensional space for linear calculation. The parameter sig2 reflects the complexity of the sample distribution in the feature space. The parameter gam was used to control the fitting error of the function. These two parameters together determine the generalization ability and training error of the function [41]. In this case, radial basis function (RBF) was adopted as a kernel function due to the advantage of good generalization ability. Then we optimized sig2 and gam when the minimum value of RMSECV was obtained in the grid-search within the range of 10^3–10^{10}. Besides, full cross-validation was also used as the same as that in PLSR models. The whole process was conducted in MatLab.

Extreme learning machine (ELM) is a recently developed machine learning algorithm based on single-hidden layer feedforward neural networks (SLFNN), and it runs in a simpler and easier way than traditional neural network methods [42,43]. Output weights can be obtained by randomly initializing input weights and hidden layer biases in global optimization [44,45]. In particular the number of hidden layer biases is obtained by continuous optimization in a predefined range which is usually less than the number of samples for modeling. This property makes ELM runs significantly faster with guaranteed learning accuracy. Besides, it has the advantage of good generalization performance [46] which makes it suitable for both quantitative and qualitative analysis. In this case, we tried to explore the feasibility of quantitative prediction for the quality indexes of the biomass pellet.

In addition, performances of all the abovementioned models were evaluated by several indicators, including the coefficient of determination in calibration and prediction sets (R^2_C and R^2_P), root mean square error (RMSEC and RMSEP). R^2 reflects the relationship between the predicted value of the quality indexes of biomass pellet and their reference value, RMSE reflect the prediction error of the model. The closer the R^2 to 1, the smaller the RMSE, the better the model will perform.

3. Results and Discussion

3.1. Quality Indexes Statistics

The descriptive statistics of the quality indexes for all samples are listed in Table 1. Due to the effect of types and processing methods, the quality indexes of the biomass pellets are totally different. It is obvious that VM took up a large proportion in all components, which varied from 74.96% to 88.99%. FC followed, varying from 10.39% to 21.22%. The Ash contents were in the range of 0.26% to 15.94%. CV varied in the range of 15.1 to 19.54 (MJ·kg^{-1}). These ranges provide the foundation for developing reliable quantitative prediction models of the quality indexes.

Table 1. Statistical description of four quality indexes for all samples.

Quality Indexes	Maximum	Minimum	Mean	Standard Deviation
Ash content (%)	15.94	0.26	4.43	5.10
Volatile matter (%)	88.99	74.96	80.22	2.71
Fixed carbon (%)	21.22	10.39	15.83	2.63
Calorific value	19.54	15.10	17.71	1.10

In order to more comprehensively evaluate the quality of biomass pellets, we have explored the internal relationships of these quality indexes. Table 2 shows the Pearson's correlation coefficients among the quality indexes of 128 biomass pellets. It can be clearly seen that CV was significantly positively related to VM, with correlation coefficients of 0.751. It also had a significant negative correlation with Ash, and the correlation coefficients was −0.903. Besides, Ash and VM were significantly negatively correlated. FC has a positive correlation with Ash and a negative correlation with VM, but there is no significant correlation with CV. These results were also roughly consistent with previous research [12,47]. These results also indicated that the higher the value of VM is, the lower the value of Ash is, and the higher the value of CV will be, which further lead to a better combustion performance of the biomass pellets.

Table 2. Pearson's correlation coefficients of four quality indexes for all samples.

Quality Indexes	Ash	VM	FC	CV
Ash	1	−0.923 [**]	0.292 [**]	−0.903 [**]
Volatile matter	−0.923 [**]	1	−0.636 [**]	0.751 [**]
Fixed carbon	0.292 [**]	−0.636 [**]	1	−0.063 [ns]
Calorific value	−0.903 [**]	0.751 [**]	−0.063 [ns]	1

Significance: [**] $p < 0.01$; [ns] not significant.

3.2. Spectral Analysis

The spectra of the four different types of biomass pellets (Figure 3) were used as examples. To eliminate the effects of noise, only spectra of 280.25–854.12 nm (18354 variables) were used for the analyses. Based on previous research and the National Institute of Standards and Technology (NIST) database, the specific elements and molecular bands corresponding to the spectral lines were given in Figure 3. These emissions included O (777.19 nm, 844.63 nm), N (742.36 nm), K (766.49 nm, 769.89 nm), Ca (393.39 nm, 422.67 nm, 445.47 nm, 616.21 nm), Na (588.99 nm, 819.48 nm), Mg (280.27 nm, 285.21 nm), Al (309.27 nm), Fe (373.48 nm, 374.94 nm), Mn (403.07 nm), Si (288.15 nm, 390.5 nm), Ti (319.19 nm), C2 (516.52 nm), CN (388.32 nm) etc. According to previous research, these elements have direct effects on the quality indexes. Metal oxides formed from the elements K, Ca, Na, Mg, Al, Fe, Mn, Si, Ti, are the main constituents of ash [48–50]. The main constituents of VM including CO_2, NO_2, CO, = etc were also formed from elements C, O and N [51–53]. FC was also directly related to the element C and molecular bands of C2 and CN [24,54]. Besides, all elements affected each other due to the ionization and the combination of elements in the process of compound formation [55]. Therefore, there should have a good correspondence between these quality indexes and the elements. In addition, CV was highly affected by Ash and VM as mentioned before. Thus, the value of CV could also predicted by the content of these elements. It could also be clearly observed from Figure 3 that the elements in different samples are consistent, but the contents of the elements were different. This further indicated the possibility of predicting these the quality indexes using the contents of the elements. Therefore, chemometric models based on the spectral lines covered these elements were further built to predict the value of the quality indexes quantitatively.

3.3. Prediction of Quality Indexes

PLSR, LS-SVM and ELM were applied to build the quantitative models for accurate prediction. The quantitative modeling results were stated in Table 3.

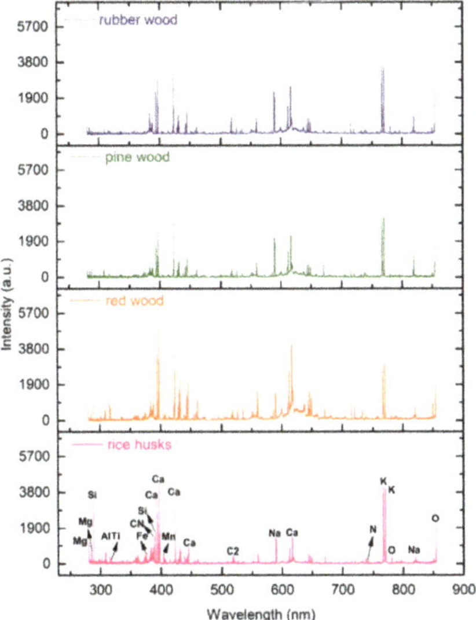

Figure 3. Spectra of the four categories of biomass pellets in 280.25–854.12 nm.

Table 3. Prediction results of chemometric models.

Quality Indexes	Model	Parameter	R^2_C	RMSEC	R^2_P	RMSEP
Ash (wt.%)	PLSR	6	0.967	0.009	0.957	0.011
	LS-SVM	$(4.377 \times 10^5, 1 \times 10^5)$	0.999	0.001	0.977	0.006
	ELM	15	0.969	0.012	0.961	0.014
VM (wt.%)	PLSR	7	0.943	0.006	0.924	0.007
	LS-SVM	$(3.375 \times 10^4, 206)$	0.999	0.001	0.974	0.006
	ELM	43	0.952	0.006	0.932	0.007
FC (wt.%)	PLSR	11	0.947	0.006	0.937	0.007
	LS-SVM	$(2.133 \times 10^4, 3.589 \times 10^3)$	0.979	0.001	0.953	0.009
	ELM	32	0.874	0.009	0.836	0.015
CV (MJ·kg^{-1})	PLSR	4	0.946	0.159	0.931	0.274
	LS-SVM	$(6.656 \times 10^7, 6.349 \times 10^4)$	0.993	0.103	0.971	0.149
	ELM	18	0.958	0.140	0.944	0.241

Parameters of different models; the optimal number of LVs for partial least-squares regression (PLSR), different sig2 and gam values for the least-squares support vector machine (LSSVM), the number of hidden layer nodes for extreme learning machine (ELM). VM for volatile matter, FC for fixed carbon, CV for Calorific value.

Ash is the residue left after the complete combustion of all combustibles in the biomass and the minerals undergo a series of complex reactions such as decomposition and combination. Its melting characteristics have a great influence on the combustion performance of biomass. Previous research revealed that biomass with low ash content has a better combustion performance [56,57]. Figure 4 shows the fitting results of models for Ash. It's clear that all models performed well, with all R^2 above 0.95. Especially LS-SVM gave the best result, with R^2 in calibration and prediction sets reaching 0.999 and 0.977, respectively. It could be concluded that LIBS combined with chemometrics is a promising way to predict ash content and further to provide reference for evaluating the combustion performance of biomass materials.

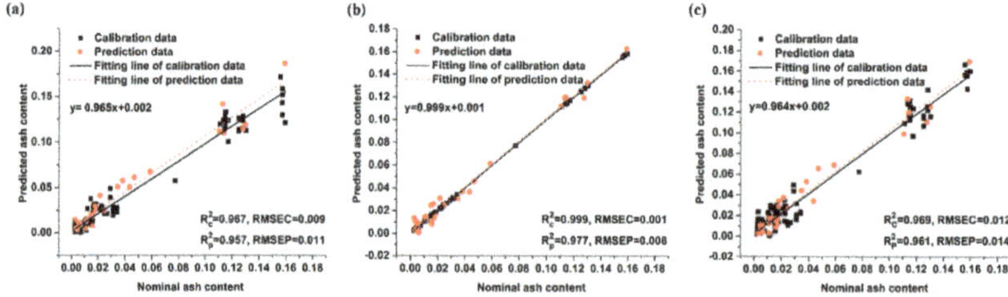

Figure 4. Model fitting results for Ash. (**a**) PLSR (partial least-squares regression); (**b**) LS-SVM (the least-squares support vector machine); (**c**) ELM (extreme learning machine).

VM refers to the gaseous product produced by the decomposition of biomass organic matter. It has a great effect on the ignition and combustion of biomass and can reflect the quality of biomass. In contrast to ash, the higher the VM content is, the better the combustion performance of the biomass will be. In this study, calibration and prediction results of models for VM are provided in Figure 5. It could be seen that all data fitted well in both calibration and prediction set. In particular, LS-SVM obtained more accurate results, with R^2 in both calibration and prediction sets over 0.97. On the whole, the modeling results showed that LIBS combined with LS-SVM could be a good tool for selecting biomass with higher VM content, which might be used as a good fuel.

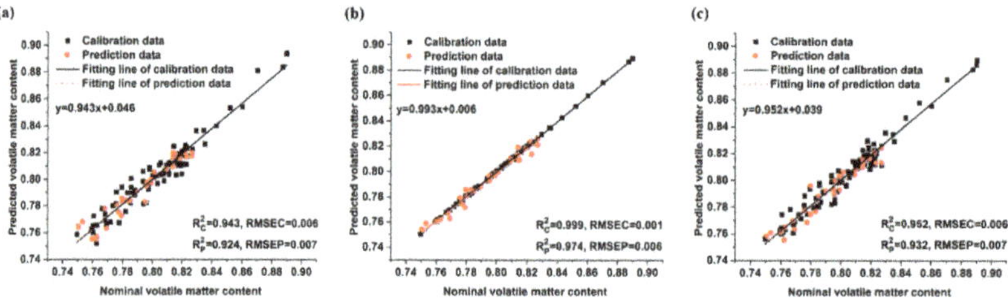

Figure 5. Model fitting results for VM. (**a**) PLSR (partial least-squares regression); (**b**) LS-SVM (the least-squares support vector machine); (**c**) ELM (extreme learning machine).

Fixed carbon is the residue remaining after removing moisture, Ash, and VM from biomass. The ignition point of fixed carbon is very high [58]. Generally, the higher the fixed carbon content is, the greater the temperature for ignition and combustion is [59]. Figure 6 shows the model fitting results for FC. As seen, LS-SVM performed perfect, and R^2 in both calibration and prediction sets were over 0.95. ELM gave the worst results which was still acceptable, with R^2 in calibration and prediction sets were 0.874 and 0.836. Thus, using LIBS and suitable chemometrics to predict the content of FC could provide a reference for selecting better biomass fuels.

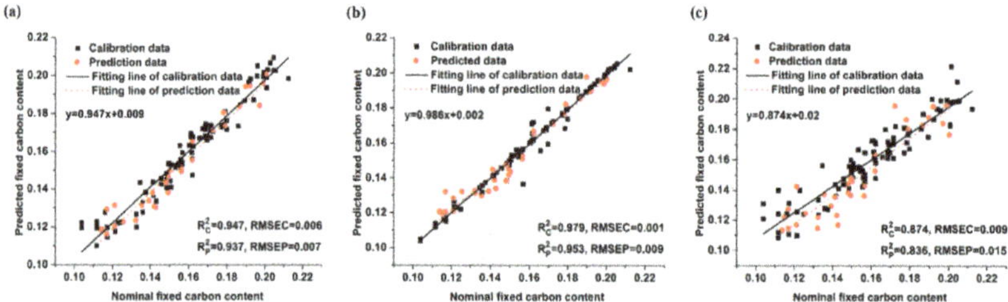

Figure 6. Model fitting results for FC. (**a**) PLSR (partial least-squares regression); (**b**) LS-SVM (the least-squares support vector machine); (**c**) ELM (extreme learning machine).

CV refers to the amount of heat released per unit mass of biomass fuels when it is completely combusted. It is an important indicator for evaluating the combustion performance of biomass. Figure 7 exhibits the calibration and prediction results of CV. It can be observed that all models gave a satisfactory result, and the best model for CV determination was obtained using LSSVM (R^2_C and R^2_P were 0.993 and 0.971, respectively). Thus, fast and accurately predicting of CV by using LIBS and LS-SVM model could be conducive to industrial applications.

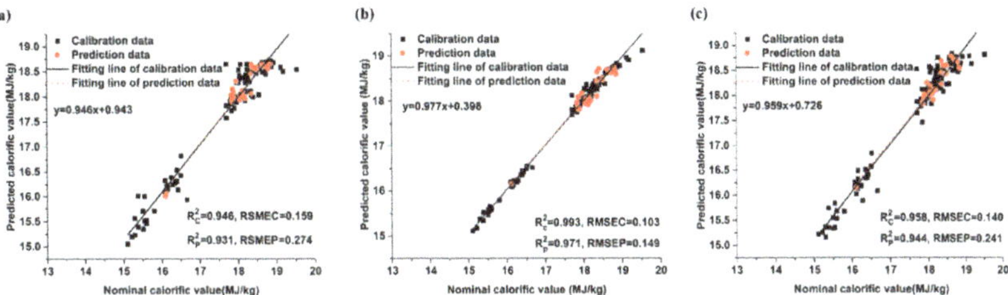

Figure 7. Model fitting results for CV. (**a**) PLSR (partial least-squares regression); (**b**) LS-SVM (the least-squares support vector machine); (**c**) ELM (extreme learning machine).

As a group, all models performed well for predicting these quality indexes. LS-SVM model performed best. As for the four indices, the prediction results for FC are relatively poor, it might be caused by the moisture. The biomass is extremely hydrophobic and it might quickly reabsorb some moisture although we have dried the samples, thus there might be some deviations in the determination of nominal fixed carbon content according to Equation (1). Besides, the prediction results of ash were slightly better that the other three indexes. This might be credited to the fact that that Ash was more directly related to elements, and this further indicated the capability of using LIBS to predict biomass quality.

4. Conclusions

In the present study, LIBS technology combined with three chemometric methods was adopted to determine the four quality indexes of biomass pellets. Statistics of CV and industrial analysis components including Ash, VM, FC were firstly determined using traditional methods. The internal relationships of these quality indexes were then analyzed based on these statistics. CV and VM were both negatively related to Ash and FC, and the relationships between them were significant, except for that between CV and FC. Besides, there was a significantly positive relationship between CV and VM, Ash and FC. LIBS spectra of four different kinds of samples were further analyzed, and this showed that the spectral lines corresponded to the elements that were related to the four indexes. PLSR, LS-SVM and ELM were applied to build quantitative analysis models for these quality indexes. All models showed a satisfactory prediction ability, especially LS-SVM models, with all R^2 values above 0.95. Among the four indexes, the prediction accuracy of Ash was relatively better. On the whole, LIBS technology coupled with LS-SVM methods exhibited great potential in determining the quality indexes of biomass pellets. It might be developed as a good tool for fast industrial detection of biomass fuels, further improving the efficiency of energy use.

Author Contributions: X.L. and X.F. performed the measurements; X.L. wrote the manuscript; X.L., X.F., Y.H and L.H. designed the experiment; Y.H. and X.F. reviewed the initial design of the experiments and provided guidance for the writing of the manuscript; All authors have read and agreed to the published version of the manuscript.

Funding: This work was supported by the Planned Science and Technology Project of Guangdong Province, China (no. 2019B020216001), and the Project of Guangdong Province Universities and Colleges Pearl River Scholar Funded Scheme, China (no. 2016).

Conflicts of Interest: The authors declare no conflict of interest.

References

1. Keles, S.; Kar, T.; Bahadır, A.; Kaygusuz, K. Renewable energy from woody biomass in Turkey. *J. Eng. Res. Appl. Sci.* **2017**, *6*, 652–661.
2. David, E.; Kopac, J.; Armeanu, A.; Niculescu, V.; Sandru, C.; Badescu, V. Biomass—Alternative renewable energy source and its conversion for hydrogen rich gas production. *E3S Web Conf.* **2019**, *122*, 01001. [CrossRef]
3. Kaygusuz, K.; Toksoy, D.; Bayramoğlu, M.M. Global utilization of wood pellet for residential heating. *J. Eng. Res. Appl. Sci.* **2017**, *6*, 688–697.
4. Kiss, I.; Alexa, V.; Sárosi, J. Biomass from Wood Processing Industries as an Economically Viable and Environmentally Friendly Solution. *Analecta Tech. Szeged.* **2016**, *10*, 1–6. [CrossRef]
5. Feng, X.; Yu, C.; Liu, X.; Chen, Y.; Zhen, H.; Sheng, K.; He, Y. Nondestructive and rapid determination of lignocellulose components of biofuel pellet using online hyperspectral imaging system. *Biotechnol. Biofuels* **2018**, *11*, 88. [CrossRef]
6. Nazari, M.M.; San, C.P.; Atan, N.A. Combustion Performance of Biomass Composite Briquette from Rice Husk and Banana Residue. *Int. J. Adv. Sci. Eng. Inf. Technol.* **2019**, *9*, 455–460. [CrossRef]
7. Parascanu, M.M.; Sandoval-Salas, F.; Soreanu, G.; Valverde, J.L.; Silva, M.L.S. Valorization of Mexican biomasses through pyrolysis, combustion and gasification processes. *Renew. Sustain. Energy Rev.* **2017**, *71*, 509–522. [CrossRef]
8. Li, J.; Paul, M.C.; Younger, P.; Watson, I.; Hossain, M.; Welch, S. Prediction of high-temperature rapid combustion behaviour of woody biomass particles. *Fuel* **2016**, *165*, 205–214. [CrossRef]
9. Oladejo, J.; Adegbite, S.; Gao, X.; Liu, H.; Wu, T. Catalytic and non-catalytic synergistic effects and their individual contributions to improved combustion performance of coal/biomass blends. *Appl. Energy* **2018**, *211*, 334–345. [CrossRef]
10. Sarikaya, A.C.; Acma, H.H.; Yaman, S.; Sarikaya, A.C.; Haykiri-Acma, H.; Serdar, Y. Synergistic Interactions During Cocombustion of Lignite, Biomass, and Their Chars. *J. Energy Resour. Technol.* **2019**, *141*, 12. [CrossRef]
11. Estiati, I.; Freire, F.B.; Freire, J.T.; Aguado, R.; Olazar, M. Fitting performance of artificial neural networks and empirical correlations to estimate higher heating values of biomass. *Fuel* **2016**, *180*, 377–383. [CrossRef]
12. Feng, X.; Yu, C.; Shu, Z.; Liu, X.; Yan, W.; Zheng, Q.; Sheng, K.; He, Y. Rapid and non-destructive measurement of biofuel pellet quality indices based on two-dimensional near infrared spectroscopic imaging. *Fuel* **2018**, *228*, 197–205. [CrossRef]
13. De Oliveira, D.M.; Fontes, L.M.; Pasquini, C. Comparing laser induced breakdown spectroscopy, near infrared spectroscopy, and their integration for simultaneous multi-elemental determination of micro- and macronutrients in vegetable samples. *Anal. Chim. Acta* **2019**, *1062*, 28–36. [CrossRef] [PubMed]
14. Moncayo, S.; Manzoor, S.; Rosales, J.; Anzano, J.; Cáceres, J. Qualitative and quantitative analysis of milk for the detection of adulteration by Laser Induced Breakdown Spectroscopy (LIBS). *Food Chem.* **2017**, *232*, 322–328. [CrossRef] [PubMed]
15. Gondal, M.; Habibullah, Y.; Baig, U.; Oloore, L. Direct spectral analysis of tea samples using 266 nm UV pulsed laser-induced breakdown spectroscopy and cross validation of LIBS results with ICP-MS. *Talanta* **2016**, *152*, 341–352. [CrossRef]
16. Peng, J.; He, Y.; Jiang, J.; Zhao, Z.; Zhou, F.; Liu, F. High-accuracy and fast determination of chromium content in rice leaves based on collinear dual-pulse laser-induced breakdown spectroscopy and chemometric methods. *Food Chem.* **2019**, *295*, 327–333. [CrossRef]
17. Sun, C.; Tian, Y.; Gao, L.; Niu, Y.-S.; Zhang, T.; Li, H.; Zhang, Y.; Yue, Z.; Delepine-Gilon, N.; Yu, J. Machine Learning Allows Calibration Models to Predict Trace Element Concentration in Soils with Generalized LIBS Spectra. *Sci. Rep.* **2019**, *9*, 1–18. [CrossRef]
18. Bonta, M.; Quarles, C.D.; Russo, R.; Gonzalez, J.J.; Hegedus, B.; Limbeck, A. Elemental mapping of biological samples by the combined use of LIBS and LA-ICP-MS. *J. Anal. At. Spectrom.* **2016**, *31*, 252–258. [CrossRef]
19. Cui, M.; Deguchi, Y.; Wang, Z.; Tanaka, S.; Fujita, Y.; Zhao, S. Improved Analysis of Manganese in Steel Samples Using Collinear Long-Short Double Pulse Laser-Induced Breakdown Spectroscopy (LIBS). *Appl. Spectrosc.* **2018**, *73*, 152–162. [CrossRef]

20. Li, W.; Lu, J.; Dong, M.; Lu, S.; Yu, J.; Li, S.; Huang, J.; Liu, J. Quantitative Analysis of Calorific Value of Coal Based on Spectral Preprocessing by Laser-Induced Breakdown Spectroscopy (LIBS). *Energy Fuels* **2017**, *32*, 24–32. [CrossRef]

21. Yao, S.; Mo, J.; Zhao, J.; Li, Y.; Zhang, X.; Lu, W.; Lu, Z. Development of a Rapid Coal Analyzer Using Laser-Induced Breakdown Spectroscopy (LIBS). *Appl. Spectrosc.* **2018**, *72*, 1225–1233. [CrossRef] [PubMed]

22. Dong, M.; Wei, L.; Lu, J.; Li, W.; Lu, S.; Li, S.; Liu, C.; Yoo, J.H.; Li, W. A comparative model combining carbon atomic and molecular emissions based on partial least squares and support vector regression correction for carbon analysis in coal using LIBS. *J. Anal. At. Spectrom.* **2019**, *34*, 480–488. [CrossRef]

23. Aints, M.; Paris, P.; Tufail, I.; Jõgi, I.; Aosaar, H.; Riisalu, H.; Laan, M. Determination of the calorific value and moisture content of crushed oil shale by libs. *Oil Shale* **2018**, *35*, 339. [CrossRef]

24. Lu, Z.; Chen, X.; Yao, S.; Qin, H.; Zhang, L.; Yao, X.; Yu, Z.; Lu, J. Feasibility study of gross calorific value, carbon content, volatile matter content and ash content of solid biomass fuel using laser-induced breakdown spectroscopy. *Fuel* **2019**, *258*, 116150. [CrossRef]

25. Galvão, R.K.H.; De Araújo, M.C.U.; José, G.E.; Pontes, M.J.; Silva, E.C.; Saldanha, T.C.B. A method for calibration and validation subset partitioning. *Talanta* **2005**, *67*, 736–740. [CrossRef]

26. Yan, W.; Perez, S.; Sheng, K. Upgrading fuel quality of moso bamboo via low temperature thermochemical treatments: Dry torrefaction and hydrothermal carbonization. *Fuel* **2017**, *196*, 473–480. [CrossRef]

27. Liu, X.; Feng, X.; Liu, F.; Peng, J.; He, Y. Rapid Identification of Genetically Modified Maize Using Laser-Induced Breakdown Spectroscopy. *Food Bioprocess Technol.* **2018**, *12*, 347–357. [CrossRef]

28. Guo, G.; Niu, G.; Shi, Q.; Lin, Q.; Tian, D.; Duan, Y. Multi-element quantitative analysis of soils by laser induced breakdown spectroscopy (LIBS) coupled with univariate and multivariate regression methods. *Anal. Methods* **2019**, *11*, 3006–3013. [CrossRef]

29. Xiao, S.; He, Y. Application of Near-infrared Spectroscopy and Multiple Spectral Algorithms to Explore the Effect of Soil Particle Sizes on Soil Nitrogen Detection. *Molecules* **2019**, *24*, 2486. [CrossRef]

30. Zhang, C.; Jiang, H.; Liu, F.; He, Y. Application of Near-Infrared Hyperspectral Imaging with Variable Selection Methods to Determine and Visualize Caffeine Content of Coffee Beans. *Food Bioprocess Technol.* **2016**, *10*, 213–221. [CrossRef]

31. Wold, S.; Sjöström, M.; Eriksson, L. PLS-regression: A basic tool of chemometrics. *Chemom. Intell. Lab. Syst.* **2001**, *58*, 109–130. [CrossRef]

32. Martín, M.; Hernández, O.; Jiménez, A.; Arias, J.; Jiménez, F. Partial least-squares method in analysis by differential pulse polarography. Simultaneous determination of amiloride and hydrochlorothiazide in pharmaceutical preparations. *Anal. Chim. Acta* **1999**, *381*, 247–256. [CrossRef]

33. Sjöström, M.; Wold, S.; Lindberg, W.; Persson, J.-Å.; Martens, H. A multivariate calibration problem in analytical chemistry solved by partial least-squares models in latent variables. *Anal. Chim. Acta* **1983**, *150*, 61–70. [CrossRef]

34. Rapid Identification of Varieties of Walnut Powder Based on Laser-Induced Breakdown Spectroscopy. *Trans. ASABE* **2017**, *60*, 19–28. [CrossRef]

35. Morellos, A.; Pantazi, X.E.; Moshou, D.; Alexandridis, T.; Whetton, R.; Tziotzios, G.; Wiebensohn, J.; Bill, R.; Mouazen, A.M. Machine learning based prediction of soil total nitrogen, organic carbon and moisture content by using VIS-NIR spectroscopy. *Biosyst. Eng.* **2016**, *152*, 104–116. [CrossRef]

36. Yang, C.; Yang, J.; Ma, J. Sparse Least Squares Support Vector Machine With Adaptive Kernel Parameters. *Int. J. Comput. Intell. Syst.* **2020**, *13*, 212–222. [CrossRef]

37. Bao, Y.; Liu, F.; Kong, W.; Sun, D.; He, Y.; Qiu, Z. Measurement of Soluble Solid Contents and pH of White Vinegars Using VIS/NIR Spectroscopy and Least Squares Support Vector Machine. *Food Bioprocess Technol.* **2013**, *7*, 54–61. [CrossRef]

38. Wang, H.-Q.; Sun, F.; Cai, Y.-N.; Ding, L.-G.; Chen, N. An unbiased LSSVM model for classification and regression. *Soft Comput.* **2009**, *14*, 171–180. [CrossRef]

39. He, Y.; Liu, X.; Lv, Y.; Liu, F.; Peng, J.; Shen, T.; Zhao, Y.; Tang, Y.; Luo, S. Quantitative Analysis of Nutrient Elements in Soil Using Single and Double-Pulse Laser-Induced Breakdown Spectroscopy. *Sensors* **2018**, *18*, 1526. [CrossRef]

40. Wang, L.S.; Wang, R.J.; Lu, C.P.; Wang, J.; Huang, W.; Jian, Q.; Wang, Y.B.; Lin, L.Z.; Song, L.T. Quantitative Analysis of Total Nitrogen Content in Monoammonium Phosphate Fertilizer Using Visible-Near Infrared Spectroscopy and Least Squares Support Vector Machine. *J. Appl. Spectrosc.* **2019**, *86*, 465–469. [CrossRef]

41. Boukhari, Y.; Boucherit, M.N.; Zaabat, M.; Amzert, S.; Brahimi, K. Optimization of learning algorithms in the prediction of pitting corrosion. *J. Eng. Sci. Technol.* **2018**, *13*, 1153–1164.
42. Liu, F.; Wang, W.; Shen, T.; Peng, J.; Kong, W. Rapid Identification of Kudzu Powder of Different Origins Using Laser-Induced Breakdown Spectroscopy. *Sensors* **2019**, *19*, 1453. [CrossRef] [PubMed]
43. Huang, G.-B.; Zhou, H.; Ding, X.; Zhang, R. Extreme Learning Machine for Regression and Multiclass Classification. *IEEE Trans. Syst. Man Cybern. Part B* **2011**, *42*, 513–529. [CrossRef] [PubMed]
44. Ding, Y.; Yan, F.; Yang, G.; Chen, H.; Song, Z. Quantitative analysis of sinters using laser-induced breakdown spectroscopy (LIBS) coupled with kernel-based extreme learning machine (K-ELM). *Anal. Methods* **2018**, *10*, 1074–1079. [CrossRef]
45. Bian, X.; Fan, M.-R.; Guo, Y.; Wang, J.-J.; Li, S.-J.; Chang, N. Spectral quantitative analysis of complex samples based on the extreme learning machine. *Anal. Methods* **2016**, *8*, 4674–4679. [CrossRef]
46. Abuassba, A.O.M.; Zhang, D.; Luo, X.; Shaheryar, A.; Ali, H. Improving Classification Performance through an Advanced Ensemble Based Heterogeneous Extreme Learning Machines. *Comput. Intell. Neurosci.* **2017**, *2017*, 1–11. [CrossRef]
47. Gillespie, G.; Everard, C.D.; McDonnell, K. Prediction of biomass pellet quality indices using near infrared spectroscopy. *Energy* **2015**, *80*, 582–588. [CrossRef]
48. Singh, S.; Ram, L.C.; Masto, R.E.; Verma, S.K. A comparative evaluation of minerals and trace elements in the ashes from lignite, coal refuse, and biomass fired power plants. *Int. J. Coal Geol.* **2011**, *87*, 112–120. [CrossRef]
49. Vassilev, S.V.; Vassileva, C.G.; Song, Y.-C.; Li, W.-Y.; Feng, J. Ash contents and ash-forming elements of biomass and their significance for solid biofuel combustion. *Fuel* **2017**, *208*, 377–409. [CrossRef]
50. Mlonka-Medrala, A.; Magdziarz, A.; Gajek, M.; Nowińska, K.; Nowak, W. Alkali metals association in biomass and their impact on ash melting behaviour. *Fuel* **2020**, *261*, 116421. [CrossRef]
51. Ren, X.Y.; Cai, H.Z.; Chang, J.M.; Fan, Y.M. TG-FTIR Study on the Pyrolysis Properties of Lignin from Different Kinds of Woody Biomass. *Paper Biomater.* **2018**, *3*, 1–7.
52. Li, R.; Konnov, A.A.; He, G.; Qin, F.; Zhang, D. Chemical mechanism development and reduction for combustion of NH3/H2/CH4 mixtures. *Fuel* **2019**, *257*, 116059. [CrossRef]
53. Vassilev, S.V.; Vassileva, C.G.; Vassilev, V.S. Advantages and disadvantages of composition and properties of biomass in comparison with coal: An overview. *Fuel* **2015**, *158*, 330–350. [CrossRef]
54. Zuo, Z.; Yu, Q.; Xie, H.; Duan, W.; Liu, S.; Qin, Q. Thermogravimetric analysis of the biomass pyrolysis with copper slag as heat carrier. *J. Therm. Anal. Calorim.* **2017**, *129*, 1233–1241. [CrossRef]
55. Yao, S.; Zhang, L.; Xu, J.; Yu, Z.; Lu, Z. Data Processing Method for the Measurement of Unburned Carbon in Fly Ash by PF-SIBS. *Energy Fuels* **2017**, *31*, 12093–12099. [CrossRef]
56. Atan, N.A.; Nazari, M.M.; Azizan, F.A. Effect of torrefaction pre-treatment on physical and combustion characteristics of biomass composite briquette from rice husk and banana residue. *MATEC Web Conf.* **2018**, *150*, 06011. [CrossRef]
57. Fang, J.; Zhan, L.; Ok, Y.S.; Gao, B. Minireview of potential applications of hydrochar derived from hydrothermal carbonization of biomass. *J. Ind. Eng. Chem.* **2018**, *57*, 15–21. [CrossRef]
58. Kok, M.V.; Ozgur, E. Thermal analysis and kinetics of biomass samples. *Fuel Process. Technol.* **2013**, *106*, 739–743. [CrossRef]
59. Dalólio, F.S.; Da Silva, J.N.; De Oliveira, A.C.C.; Ferreira-Tinôco, I.D.F.; Barbosa, R.C.; Resende, M.D.O.; Albino, L.F.T.; Coelho, S. Poultry litter as biomass energy: A review and future perspectives. *Renew. Sustain. Energy Rev.* **2017**, *76*, 941–949. [CrossRef]

Article

Utilization of Agricultural and Livestock Waste in Anaerobic Digestion (A.D): Applying the Biorefinery Concept in a Circular Economy

Grigorios Rekleitis [1,*], Katherine-Joanne Haralambous [2], Maria Loizidou [2] and Konstantinos Aravossis [3]

[1] Department of Chemical Sciences, School of Chemical Engineering, National Technical University of Athens (NTUA), 15780 Athens, Greece
[2] Unit of Environmental Science and Technology, Department of Chemical Sciences, School of Chemical Engineering, National Technical University of Athens (NTUA), 15780 Athens, Greece; harjo@chemeng.ntua.gr (K.-J.H.); mloiz@chemeng.ntua.gr (M.L.)
[3] School of Mechanical Engineering, National Technical University of Athens (NTUA), 15780 Athens, Greece; arvis@mail.ntua.gr
* Correspondence: greklitis@gmail.com; Tel.: +30-69-4207-8027

Received: 26 July 2020; Accepted: 26 August 2020; Published: 27 August 2020

Abstract: There has been intense debate over the depletion of fossil fuel reserves in recent decades as well as the greenhouse gas emissions that are causing climate change. At the same time, new legislation in Greece, national policies, European policies, and realistic needs, need effective waste management and the protection of national resources. As a result, it seems a necessity to exploit waste treatments, while expanding the use of renewable energy sources. In this study, an attempt is made to focus our interest and research on a specific biomass waste stream, namely the waste biomass from the agricultural-livestock sector. The possibility of processing these wastes through the technology of biomass biorefinery with anaerobic digestion as its central process will be studied. The technology of anaerobic digestion is a process widely used for the treatment of agricultural residues and livestock waste as well as for the exploitation of energy crops (energy development, soil enhancement) mainly in countries in Europe and globally. This study reviews the biorefinery biomass technology, the energy production technology, production of biofuels, and new materials from waste biomass at the behest of the circular economy and bioeconomy. Additionally, this research will be an introduction in maximizing the potential of the full utilization of agricultural and livestock waste, and the by-products that can be produced from these processes.

Keywords: biogas; circular economy; bioeconomy; wastes; energy; R.E.S; biomass; anaerobic; digestion; agriculture; livestock; Greece; biorefinery

---

1. Introduction

Effective energy output and reduced consumption have been our primary objectives in achieving a secure global supply of energy. Today, many countries in Europe and around the world are encouraging the use of renewable energy sources with assured return prices or carbon trading schemes [1–4]. Biomass, which has great potential for exploitation, is a very important substrate for the supply of renewable energy. Anaerobic digestion is used as an effective method for converting biomass into energy [5].

Anaerobic digestion (AD) is a process of decomposing organic matter in an oxygen-free environment through a microbial composition. It is a cycle found in many inhospitable natural habitats such as aquifers, sediments, water-soluble soils, and in mammals' intestines. It can also be used on a wide variety of raw materials including industrial, agricultural, municipal, food wastes, and bio-chemical wastes. Biogas is provided by the biodegradation of organic materials and the lack of oxygen. In nature, it is the result of anaerobic fermentation of biodegradable materials and is a form of biofuel [5]. Biogas technology is an option with major benefits, providing eco-friendly energy and at the same time addressing the waste management crisis. An estimation of 1,000,000 tons of waste will produce 1 MW of electrical energy (biogas) for around ten years annually [5]. Biogas is derived primarily from the anaerobic digestion of animal waste (wastewater produced from pig farms, dairy farms e.tc), industrial waste, wastewater, and municipal waste. Biogas is composed primarily of methane and carbon dioxide, which can be used for generating heat and electricity, but also as an internal combustion engine fuel [5]. Biogas production from a biological process was first recorded in the United Kingdom in 1895 [6]. The method has since been developed and widely implemented for the treatment of wastewater and also for the stabilization of sludges. The oil crisis of the 1970s brought about new ideas of using alternative energy sources including anaerobic digestion (AD). Global efforts have been made to displace fossil fuels for the production of energy and also to find environmentally sustainable alternatives for the management and recycling of organic wastes. Some of AD's most significant applications today are treating agricultural substrates through biogas plants. The highest percentage of installed agricultural biogas plants are in Europe, and especially in North America [5,7,8].

Initially, the National Renewable Energy Action Plan (NREAP) targets for Greece are presented, showing that new installations of biomass projects are a necessity for Greece, and the country must move quickly to achieve these targets. Then, the definitions of biomass, bioeconomy, and circular economy are presented, focusing on the utilization of agricultural and livestock waste. The important problem of waste biomass management is presented and examined through the investigation of available technologies through the international literature. The central pillar in this work is the process of anaerobic digestion.

This study essentially presents the operation of an anaerobic biorefinery where the anaerobic digester is a central element for the biotransformation of raw materials into various intermediate and high-end products. Biorefinery seems to be the optimal technology for the utilization of agricultural and livestock waste, contributing to a triple benefit: (a) energy production; (b) environmental protection; and (c) production of new materials/products. A flow diagram of these contributing processes is shown schematically that can be used in the biogas plants and also in the biorefinery plant (laboratory, pilot, large scale). Finally, the advantages of an installation of a biorefinery plant are presented, and the benefits to the environment, the local community, and the country.

2. Renewable Energy Sources (RES) Production in Greece

The Greek power sector relies heavily on fossil fuels, the majority of which are imported. Approximately 54 percent of its energy needs are met by petroleum products, compared to an EU level average of 33.4 percent. In addition to being used in the transport sector, these petroleum products are often converted into energy in significant quantities. The non-connected Greek Islands, in particular, get their electricity primarily from expensive and inefficient diesel power stations. Nearly 61% of Greece's primary energy needs are met through imports with the remaining 39% covered by domestic energy sources, mostly lignite 77 percent and RES 22 percent [9].

The EU criteria laid down in Directive 2009/28/EC (EEL 140/2009) (Promoting renewable energy use and amending and subsequently repealing Directives 2001/77/EC and 2003/30/EC) provide that the contribution of renewable energy sources (RES) to total final energy consumption for Greece by the end of 2020 must be 18% [10,11].

As provided from Greek Law 3851/2010 [12], the national RES goals are set as follows:

a. Contribution of the RES produced energy to the total final gross energy consumption in 2020: 20%.
b. Contribution of the electrical energy of RESs to total electrical consumption: at least 40%.
c. Contribution of the energy generated by RES for the heating and cooling energy consumption: at least 20%.

d.   Contribution of electrical energy generated by RES to total consumption of electrical energy for transportation: at least 10%.

Table 1 shows the RES electricity installed capacity per technology, and the National Renewable Energy Action Plan targets for Greece.

Table 1. RES electricity installed capacity per technology (September 2017) [10].

Technology	Electricity Installed Capacity in September 2017 (MW)	National Renewable Energy Action Plan (NREAP) Target/2017 (MW)	National Renewable Energy Action Plan (NREAP) Target/2020 (MW)
Wind plants	2451.00	5430.00	7500.00
Solar Photovoltaic systems	2604.00	1456.00	2200.00
Small hydroplants	231	233	350
Large hydroplants	3173.00	3396.00	4300.00
Biomass-Biofuels	61	160	350
Concentrated solar power (CSP)	0	140	250
Geothermal plants	0	20	120
Total	8520.00	10,835.00	15,070.00

Biomass, which has great potential for exploitation, is an important raw material for the supply of renewable energy.

In Greece, only a few biomass energy projects have been developed, mainly for municipal solid waste management. The installed biomass power capacity currently stands at 58 MW for a total of 12 individual projects [9], while the biomass target for Greece for the year 2020 is 350 MW.

3. Biomass

3.1. General

Biomass is the organic matter of waste and residues from agriculture, livestock sector, forestry, and other industries including waste fisheries and the organic matter of industrial and municipal wastes.

Biomass, with proper treatment can be converted into fuel as well as energy. Below are briefly presented the methods of biomass conversion, the products that are being produced, and the types and applications of biofuels in Greece [5].

3.2. Biomass Conversion Methods

Biomass can be converted into energy and utilized products with the following conversion methods. Table 2 shows these biomass conversion methods [13–15].

Table 2. Biomass conversion methods [13–15].

Conversion Methods		
Chemical Conversion Methods	**Thermochemical Conversion Methods**	**Biochemical Conversion Methods**
1. Hydrolysis (acid, alkaline, self-hydrolysis)	1. Burning	1. Enzymatic hydrolysis
2. Ozonolysis	2. Gasification	2. Fermentation
3. Meteorization	3. Pyrolysis 4. Liquidation	3. Anaerobic digestion

There is great interest from the international scientific community for further utilization of waste biomass through new innovative methods of conversion. Processing of biomass, with the conversion methods above-mentioned, can produce a number of products that have a high added value. Many of these products are presented in Table 3 [15–22].

Table 3. High added value products [15–22].

Products with High Value Added Value	
Biofuels	**High Value Added Products**
1. Biodiesel mainly FAME (fatty acid methyl esters) and RME (rapeseed methyl esters)	1. Alcohols (xylitol, glycerol) 2. Organic acids (Citric, gluconic acid)
2. Alcohols (bioethanol, biomethanol)	3. Sweeteners (fructose)
3. Biogas	4. Vitamins (C, B12)
4. Syngas synthetic gas	5. Antibiotics (streptomycin, cephalosporins)
5. Hydrogen	6. Proteins and amino acids
6. Solid fuels (briquettes/pellets)	7. Bio-fertilizers 8. Bio-insect repellents

3.3. Applications of Biofuels in Greece

Biofuels have a variety of applications in Greece and globally. Their main applications in brief, are presented as following [15–18]:

- Heating of buildings from biomass combustion in individual/central boilers.
- Energy production in agricultural industries.
- Energy production in wood industries.
- Energy production in wastewater treatment plants (mainly from biological treatment of sludges) and sanitary landfills (mainly from capture and combustion of gas).
- Liquid biofuels (e.g., biofuels for transportation).

4. Biomass, Bioeconomy, and Circular Economy

From the time of the Industrial Revolution onward, our planet has lived in a linear economy. Our modern society is based mainly on the consumer model. This model is based on the following process: "use of natural resources, produce, consume, dispose". Natural resources provide inputs to factories, which are then used to create goods and eventually, will be disposed of in the environment. This linear model of economy of mass production and mass consumption is testing the natural limits of the planet and threatens the stability of our future.

Today, we consume our natural resources faster than we can replace them. By 2030, if we continue at the same pace, there will be a lack of natural resources [23]. At the same time, the global population will grow, which will increase the demand for advanced industrial products.

A circular economy is the answer to the ambition for sustainable development, taking into account the continuing concern about the ongoing reduction of environmental resources and the depletion of the natural environment. The circular economy is a centered economic model based mainly on reducing the resources used in the manufacturing process. A circular economy is also based on stressing the use of renewable energy sources, the full treatment of agricultural and livestock waste, the reuse of products and waste, and the generation of energy from waste. A circular economy is the answer to sustainable development, taking into account the growing concerns regarding natural resource constraints and environmental degradation.

4.1. Preserving a Sustainable Model Should Be a Top Policy Priority

Environmental initiatives of the United Nations (UN) embody those concerns and recommend effective measures to maximize the benefits of preserving a green planet. The Ellen MacArthur Foundation is a famous organization founded in 2010 that has made major contributions to form European policies for a sustainable environment. The organization stresses [24] that a viable circular economy will help Europe revitalize its competitiveness by three percentage points per year. This would offer a significant gain of up to 0.6 trillion EURO by 2030, increasing the gross domestic product (GDP) by seven percentage points at the EU level compared to current growth.

4.2. Circular Economy and Waste

Research by the Ellen MacArthur Foundation focuses on planning based on circular economic principles and is trying to help businesses turn a linear production line into a circular economy model as circular economists cite waste as a useful commodity. The Ellen MacArthur Foundation [24] defines and classifies waste into four different categories:

- Exhausted resources: materials and energy that cannot be constantly regenerated (e.g., fuel).
- Products with recurring life cycles: products that have an artificially short lifespan (e.g., smartphones).
- Products that become useless when their lifespan is ended, (e.g., cars that remain unused for 90% of their lifespan).
- Waste with built-in value: these are materials and energy that are not recovered from discarded products and can be reused (e.g., non-reusable fabrics).

4.3. Biomass and Bioeconomy

The key aim of a bioeconomy is to manage various agricultural, industrial, and forestry activities in a sustainable manner. Therefore, the main effort to ensure a successful transition to bioeconomics is to achieve a sustainable design that combines sustainability and profitability. This is the big challenge for bioeconomics for the next few decades. Science and technology play a key role in the transition to sustainable bioeconomics.

Defining bioeconomics is a critical step. The concept of bioeconomics is to deal with the global population growth, the depletion of resources, climate change, and other environmental concerns. The concept itself is very basic. In fact, it is about satisfying the needs of the present generation without undermining the potential of the next generation to fulfill their own needs.

In terms of bioeconomics, many international institutions, governments, and scientific communities have put forward their own interpretations in terms of bioeconomics. The problem here is that bioeconomics is a multidimensional concept and its meaning depends largely on who defines the term. Economists, industrialists, economists, governments, scientists, and so on can use various concepts of bioeconomy.

We present here the well-known definition (which is still widely accepted) regarding bioeconomics. The bio-economy, according to the European Union, "includes the production and conversion of renewable biological resources into food, feed, organic products and bioenergy. It also includes food, forestry, agriculture, fisheries, pulp and paper production" [25].

4.4. Bioeconomics, Sustainability, and Agriculture

Agriculture is the foundation of any economic model. Agriculture is the key-role to the concept of bioeconomics and has a privileged position in bioeconomics. As an economic model, bioeconomics focuses primarily on meeting the food needs due to the human population growth. Agricultural sustainability means meeting people's needs for feed, food, crops, fuel, and energy, while at the same time protecting and enhancing the quality of the environment and preserving natural resources for the needs of future generations. The goal of a bioeconomy is very simple when compared with previous and existing economic systems, which is to use all available resources (water, land and energy) to feed the human population growth.

4.5. Bioeconomy in Greece

Agricultural sector in Greece contributed with 27 billion euros in 2017, almost 4.1 percent to gross domestic product (GDP) of Greece. Seventy percent of this contribution comes from agricultural products, and the remaining 30 percent from animal products. In addition, there is a lot of unexploited waste in the agricultural and livestock sector in Greece. Through research [26], the annual waste generation in Greece from the agricultural and livestock sector has been calculated to be around 57,983,751 tn/y. A total of 53 percent of these wastes comes from the agricultural sector and 47 percent from the livestock sector.

5. Biogas Plant from Anaerobic Digestion of Agricultural and Livestock Wastes

Biogas production technology from waste biomass is an effective solution with many major advantages. Biogas delivers environmentally sustainable energy and also contributes to a comprehensive and integrated treatment of agricultural and livestock waste, reducing pollution discharge to the atmosphere by more than 55 per cent [5,7]. The production of biogas consists mainly of carbon dioxide and methane. Anaerobic digestion is the decomposition of organic matter without oxygen and is an effective method for converting biomass into energy. It can also be used on a wide variety of raw materials including industrial, urban waste, agricultural, municipal, food waste, and biochemical waste.

5.1. Agricultural and Livestock Waste Biomass

Biomass is the necessary raw material for the anaerobic digestion process. Agricultural waste (such as residues from wheat, sorghum, rice, etc.), livestock wastes (such as animal manures, animal bedding straw) and waste from the food industries are an abundant and often unlimited source of biomass. The amount of methane produced from these materials is usually very high, although pretreatment methods are usually required for maximum methanogenic efficiency [20].

5.2. Operation of a Biogas Plant

The following flowchart (Figure 1) shows how a biogas plant operates:

The process flow below examines the various steps taken at a biogas facility. Originally, the first stage in the unit input is the collection and storage of raw materials for anaerobic digestion. This is followed by the improvement of the substrate material, consisting of the stages of chopping, filtering, preparation mixing, pasteurization, etc. for the following anaerobic digestion cycle. In the stage of anaerobic digestion, the already formed substrate is led to primary (primary digestion) and then secondary digesters (secondary digestion) to conduct and complete the anaerobic digestion cycle. Biogas is derived from the anaerobic digestion process and is subjected to a series of treatments (desulfurization, drying, etc.) to make it suitable for use. Finally, the processed biogas is burnt in an ICE (internal combustion engine) unit for electrical and thermal energy production. The residue from the anaerobic digestion process (digestate) is processed and then available for the fields as liquid fertilizer (agriculture production). The general structure and technique of the biogas plant process is intended for the digestion and anaerobic treatment of a complete mix of farms and residues such as cow manure, pig manure (livestock production), and agricultural residues.

Through the scientific literature, a flowchart can be created that shows the processes that take place in a biogas plant and how the energy is produced [5,7,8].

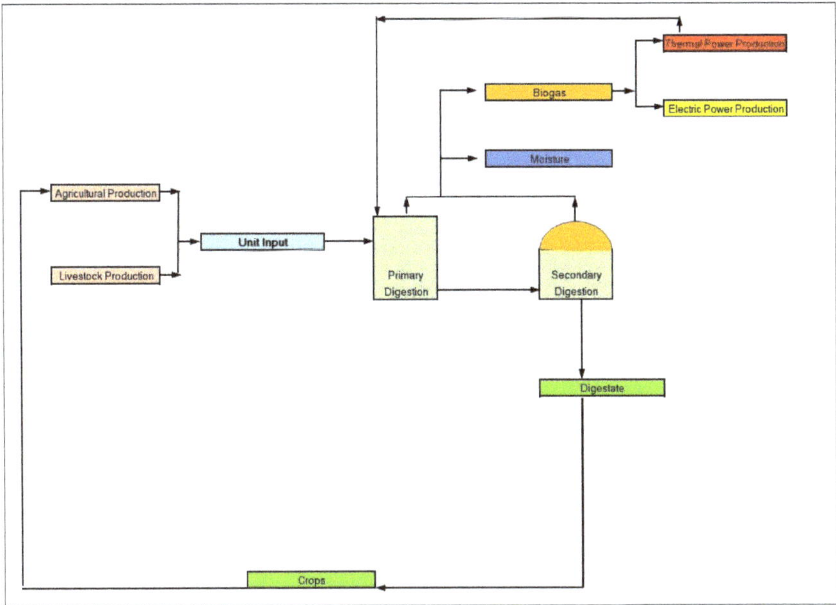

Figure 1. Process flowchart of a typical biogas plant from agricultural and livestock production.

5.3. Digestate as a Fertilizer

Digestate, which originates from the anaerobic digestion of agricultural and livestock organic waste, contains nutrients such as phosphorus, nitrogen, and potassium that may be beneficial as fertilizer for application to soil [27,28]. Therefore, the ammonia in the digested product may differ according to the substrate.

Parameters such as salinity, pH, heavy metals, and nutrients in digested products are important chemical properties to consider. The common residual pH is on a scale of 6.7 to 8.4 [29,30] and is generally slightly alkaline due to the degradation in the production of volatile fatty acids (VFA) and ammonia during the anaerobic digestion process.

5.4. Advantages of a Biogas Plant

A major advantage of biogas plants is that these systems can be easily extended with the installation of modern internal combustion engines (ICE) and can be easily transported without much difficulty or modification.

Additionally, a significant advantage of this technology is that the lifespan of these plants can exceed 25 years with low maintenance and operating costs.

Furthermore, these systems are ideal for central and distributed energy production as they can make a major contribution to the 'distributed power generation' of the area that they are being installed.

Biogas plants offer diversification of energy output provided by internal combustion engines and can create conditions for economic growth in the area that they are being installed.

Finally, the anaerobic digestion of agricultural and livestock waste in biogas plants has significant environmental and agricultural benefits such as [31]:

a.   Low emissions of greenhouse gases (CO_2, CH_4).
b.   Cost savings by the farmer.
c.   Improved soil fertility.
d.   Economic and ecological recycling of waste and wastewater.

e. Reduced discomfort related to odor removal and insects (e.g., bugs, mosquitoes).
f. Reduction of pathogens in the digestate.

In addition, and in terms of fertility production, the environmental benefits of applying anaerobic digestion technologies to both agricultural and livestock waste are shown below.

a. **Advantages for Farmers** [5,7]:

- Improving the quality of organic fertilizers/reducing inorganic fertilizers.
- Reduction of phytotoxic substances and odors.
- Reduction of weeds, etc.
- Stabilization and improvement of soil fertility.

b. **Advantages for Local Community** [5,7]:

- Reducing pollutants and odors.
- Positive effect on the protection of water resources.
- Positive effect on climate protection.
- Compared to other fuels, biogas shows positive behavior.

5.5. Biogas Potential in Greece

In Greece, in most cases, the local authorities and the regional and national bodies are responsible for policy-making, but also for the collection, treatment, and final disposal of municipal waste. In these cases, the availability of waste is stable (with the exception of small seasonal fluctuations due to tourism). In contrast, the management of agricultural-livestock waste is a major problem that needs to be solved due to the high potential, but also because of its spatial dispersion throughout the country. In some cases, there is a lack of knowledge about the potential of waste and the method of treatment of these wastes [32].

In areas such as Greece, the seasonal production of waste (e.g., waste from fruits, potato processing, etc.) is an important factor for the successful implementation of a biogas project. Biogas can be produced from almost all organic wastes. Today in Europe, there are relatively limited volumes of biogas coming from wastewater plants, sanitary landfills, and industrial facilities. The large volume of biogas in 2030 is projected to be produced from large central combustion plants, livestock facilities, and food processing industries [32].

In Greece, biogas is produced mainly from wastewater plants, sanitary landfills, and industrial facilities. According to estimates of the Center for Renewable Energy Sources and Saving of Greece (CRES) based on conservative scenarios, it is estimated that anaerobic digestion from only livestock waste could supply cogeneration biogas units with a total capacity of 350 MW of installed power [32].

6. Biorefinery

6.1. Definition of Biorefinery

There are many different definitions that explain the term "biorefinery". Some examples are:

- A biorefinery is an installation that incorporates biomass conversion processes and equipment.
- A biorefinery produces fuels, energy, and high-added value chemicals. The concept is similar to that of an oil refinery, which produces multiple fuels and petroleum products [33].
- A biorefinery is a complex of industries based on organic raw materials and produce chemicals, fuels, energy, and other products and materials [34].
- A biorefinery is a promising plant, where biomass is converted to a range of valuable products [35].
- A biorefinery is the integrated bio-industry that uses various technologies to produce chemicals, biofuels, food, biomaterials (including fiber), and electricity [36].

- A biorefinery is a comprehensive bio-industry complex, using various technologies for the production of chemicals, food ingredients, biofuels, and biomass energy [37].

6.2. Biocellulose Biorefinery Biomass Plants, the Future of the Chemical Industry

An essential component of society is the sustainable development of the energy and chemical industries. Chemical and energy industries, however, are heavily dependent on non-renewable resources like oil, coal, and natural gas. The shortages of raw materials and the effects of climate change pose a great threat to any type of sustainable development.

At this point, lignocellulosic biomass is one of the most abundant renewable raw material. A lignocellulosic biorefinery (lignocellulosic feedstock biorefinery) is a promising model for the full utilization of lignocellulosic biomass, with a major role in the development of the energy and chemical industries [20].

Lignocellulosic biomass is a suitable raw material for the chemical and energy industries and is extremely necessary for the sustainable development of the energy and chemical industries. This choice is always influenced by many factors such as the availability of raw materials, the technical/technological level required, various economic criteria, and environmental and ecological implications.

6.3. Flowchart of Biorefinery Plant

Through the scientific literature [38–42], a flowchart (Figure 2) was created that shows the processes that take place in a biorefinery plant and also the energy, intermediate products, and high value products that are produced.

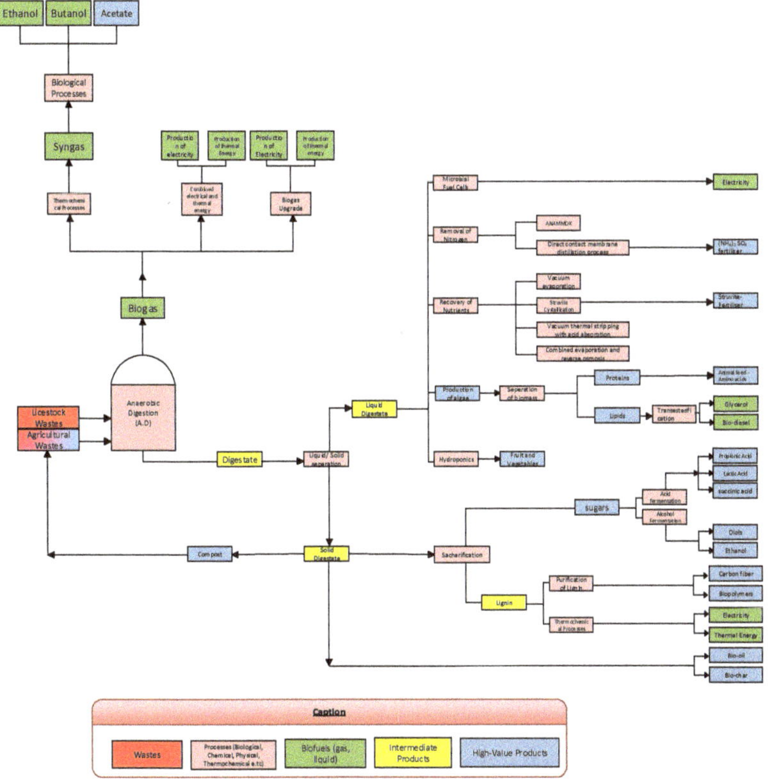

Figure 2. Process flowchart of a biorefinery plant [39–43].

6.4. Energy, Biofuels, and High Value Products

Figure 2 shows the processes that take place inside a biorefinery plant, with anaerobic digestion as its central process. Table 4 shows the incoming agricultural and livestock waste and the intermediate products, biofuels, and high value products that can be produced through the treatment of these wastes [38–42].

Table 4. Waste, intermediate products, biofuels, and high value products in a biorefinery [38–42].

Description	Type (Waste/Intermediate Products/Biofuels/High Value Products)
Incoming livestock waste	Waste
Incoming agricultural waste	Waste
Biogas produced by anaerobic digestion	Biofuel
Electricity generation from combined production of electric power and heat	High Value Product
Thermal energy production from combined production of electric power and heat	High Value Product
Electricity generation from combined production of electric power and heat from biogas upgrade	High Value Product
Thermal energy production from combined production of electric power and heat from biogas upgrade	High Value Product
Ethanol	Biofuel
Butanol	Biofuel
Acetate	High Value Product
Liquid digestate	Intermediate Product
Solid digestate	Intermediate Product
Production of electricity from combined production of electric power and heat from microbial fuel cells	High Value Product
$(NH_4)_2SO_4$ fertilizer	High Value Product
Struvite-Fertilizer	High Value Product
Proteins	High Value Product
Lipids	High Value Product
Animal feed-amino acids	High Value Product
Glycerol	High Value Product
Biodiesel	Biofuel
Sugars	Intermediate Product
Propionic acid	High Value Product
Lactic acid	High Value Product
Succinic acid	High Value Product
Diols	High Value Product
Ethanol from solid digestion	Biofuel
Lignin	Intermediate Product
Carbon fiber	High Value Product
Biopolymers	High Value Product
Electricity from thermochemical treatment of lignin	High Value Product
Thermal Energy from thermochemical treatment of lignin	High Value Product
Bio oil	High Value Product
Bio-char	High Value Product
Compost	High Value Product

7. Conclusions

This research aims to study, investigate, and evaluate the feasibility of the full utilization of organic waste, particularly waste from agricultural and livestock units. In this work, we showed that the implementation of biorefinery has several great benefits as it provides environmental-friendly energy and, at same time, exploits energy efficiency and contributes to the comprehensive and holistic waste management of livestock and agricultural wastes.

Additionally, this study is an introduction to an anaerobic biorefinery as promising technology where anaerobic digestion serves as a hub for the conversion of waste to energy, intermediate products, and many high value products.

The installation of a biorefinery plant in a region where agricultural and livestock wastes are produced will have the following benefits:

a.   A high biomass capacity that can be extracted from agricultural waste will lead to a triple benefit: (a) energy production, (b) production of high-value products, and (c) environmental protection.

b.   The development of a central biorefinery plant from livestock and agricultural wastes will help to centralize organic waste management and protect vulnerable habitats in a broader region [21].

c.   The operation of a biorefinery plant will contribute to Greece's target goals for RES installations [10].

d.   The anaerobic digestion process is extremely beneficial for organic farming, particularly for soils that require nutrients. The organic matter in the digested soil can create humus in the soil and this is a unique advantage for low-carbon content soils [5,7,29].

e.   It will improve the energy autonomy and energy security of the wider installation area by improving the flexibility of Greece's energy network, and with the result of leading to the self-sufficiency of electricity imports during peak times [10].

f.   It will help to centralize the treatment of organic waste and to protect fragile habitats in the installation region [8,9].

g.   Large amounts of nitrogen, ammonia, hydrogen sulfide, oxides, and volatile compounds will not be emitted into the air [43].

h.   Large amounts of methane will not be released into the atmosphere, which is the main greenhouse gas that induces climate change [43].

i.   Chemical fertilizers that cause ongoing regional soil, surface water, and underground water pressures will be replaced with organic fertilizer from biorefineries at the behest of the circular economy [44–46].

j.   This will lead to an increase in the productivity of agricultural products as organic fertilizers will improve annual production quantitatively and qualitatively [44–46].

k.   It will help create new jobs [5].

l.   It will effectively deal with and eliminate the waste produced [5,7,8].

m.   It will be a central lever of research and innovation with an emphasis on environmental protection technologies [5].

n.   It will be a model of sustainable development, bioeconomy, and circular economy [24].

o.   It will provide significant economic benefits to the local community [5,7,8].

In this work, a biorefinery plant has a special position as one of the largest facilities for the processing of biomass into high value products in the circular economy. This design presents a dynamic culture of innovation, where new materials that have significantly higher value or energy content than biogas or the digestate material can be produced.

Author Contributions: Conceptualization, G.R., K.-J.H., M.L. and K.A.; Supervision, K.-J.H.; Writing—review & editing, G.R. All authors have read and agreed to the published version of the manuscript.

Funding: This research received no external funding.

Conflicts of Interest: The authors declare no conflict of interest.

References

1.   European Climate Adaptation Platform Climate-ADAPT. *Climate-ADAPT Strategy 2019-2021*; European Climate Adaptation Platform Climate-ADAPT: Luxembourg, January 2019.

2.   Brussels European Council Conclusions of 8/9 March 2007 (Document Reference: 7224/1/07 REV 1 dated 2 May 2007), European Council. Available online: http://www.consilium.europa.eu/uedocs/cms_data/docs/pressdata/en/ec/93135.pdf (accessed on 14 August 2020).

3. European Climate Summit 2014, Chair's Summary: "Leaders Committed to Limit Global Temperature Rise to Less than 2 Degrees Celsius from Pre-Industrial Levels.", European Council. Available online: http://www.un.org/climatechange/summit/2014/09/2014-climate-change-summary-chairs-summary/ (accessed on 14 August 2020).

4. Intended Nationally Determined Contribution of the EU and Its Member States Submission by Latvia and the European Commission on Behalf of the European Union and Its Member States, Dated 6 March 2015, European Council. Available online: http://ec.europa.eu/clima/news/docs/2015030601_eu_indc_en.pdf (accessed on 14 August 2020).

5. Al Seadi, T.; Rutz, D.; Prassl, H.; Köttner, M.; Finsterwalder, T.; Volk, S.; Janssen, R. *Biogas Handbook*; Niels Bohrs, V., Ed.; University of Southern Denmark: Esbjerg, Denmark, 2003; pp. 9–10. ISBN 9788799296200.

6. Metcalf, L.; Eddy, H.P. *Wastewater Engineering: Treatment Disposal Reuse*, 3rd ed.; Revised by George Tchobanoglous, Frank Burton; McGraw-Hill: New York, NY, USA, 1991.

7. Al Seadi, T. *Good Practice in Quality Management of AD Residues from Biogas Production*, Report Made for the International Energy Agency, Task 24-Energy from Biological Conversion of Organic Waste; IEA Bioenergy and AEA Technology Environment: Oxfordshire, UK, 2001.

8. Al Seadi, T.; Holm Nielsen, J. *Utilisation of Waste from Food and Agriculture: Solid Waste: Assesment, Monitoring and Remediation*; 2004 Waste Management Series 4; Elsevier: Amsterdam, The Netherlands, 2004; pp. 735–754. ISBN 0080443214.

9. Greece Energy Situation. Available online: https://energypedia.info/wiki/Main_Page (accessed on 17 April 2020).

10. European Parliament. *Directive 2009/28/EC of the European Parliament and of the Council of 23 April 2009 'on the Promotion of the Use of Energy from Renewable Sources and Amending and Subsequently Repealing Directives 2001/77/EC and 2003/30/EC'*; European Council: Brussels, Belgium, 2009.

11. Ministry of Development, Greece (2020). Available online: http://www.mindev.gov.gr/ (accessed on 17 April 2020).

12. Greek Law. *Nr.3851/2010: Accelerating the Development of Renewable Energy Sources to Deal with Climate Change and other Regulations Addressing Issues under the Authority of the Ministry of Environment*; Energy and Climate Change: Athens, Greece, 2010.

13. U.S Department of Energy (DOE)—Energy Efficiency and Renewable Energy: Biomass Conversion, DOE/EE-1436 June 2016. Available online: https://www.energy.gov/sites/prod/files/2016/07/f33/conversion_factsheet.pdf (accessed on 14 August 2020).

14. Nautiyal, O.P.; Dugesh, P. *Biomass to Fuel: Conversion Techniques. Energy Resources: Development, Harvesting and Management*, 1st ed.; Uttarakhand Science Education and Research Centre (USERC), Government of Uttrakhand: Dehradun, Uttarakhand, India, 2016; pp. 155–194.

15. Grassi, G.; Collina, A.; Zibetta, H. *Biomass for Energy, Industry and Environment 6th E.C Conference, Athens, Greece, 22–26 April 1991*; Taylor & Francis: Boca Raton, FL, USA, 1992; ISBN 185166730X.

16. Mousdale, D.M. *Biofuels-Biotechnology, Chemistry and Sustainable Development*; CRC Press: Boca Raton, FL, USA, 2008.

17. Drapcho, C.M.; Nhuan, N.P.; Walker, T.H. *Biofuels Engineering Process Technology*; McGraw-Hill Education: New York, NY, USA, 2008.

18. Demirbas, A. *Biofuels-Securing the Planet's Future Energy Needs*; Springer: Berlin/Heidelberg, Germany, 2009; ISBN 9781848820111.

19. Scragg, A.H. *Biofuels: Production, Application and Development*; CABI Publishing: Wallingford, UK, 2009; pp. 1–237.

20. Petersson, A.; Thomsen, M.H.; Hauggaard-Nielsen, H.; Thomsen, A.B. Potential bioethanol and biogas production using lignocellulosic biomass from winter rye, oilseed rape and faba bean. *Biomass Bioenerg.* **2007**, *31*, 812–819. [CrossRef]

21. Sawatdeenarunat, C.; Surendra, K.C.; Takara, D.; Oechsner, H.; Kumar Khanal, S. Anaerobic digestion of lignocellulosic biomass: Challenges and opportunities. *Bioresour. Technol.* **2014**, *178*, 178–186. [CrossRef] [PubMed]

22. Yang, L.; Xu, F.; Ge, X.; Li, Y. Challenges and strategies for solid-state anaerobic digestion of lignocellulosic biomass. *Renew. Sustain. Energy Rev.* **2015**, *44*, 824–834. [CrossRef]

23. Hult International Business School, Cambridge, MA, USA. Available online: https://www.hult.edu/blog/tag/sustainability/ (accessed on 17 April 2020).
24. Mac Arthur, E. Towards a Circular Economy. *J. Ind. Ecol.* **2013**, *1*, 23–44. Available online: https://mvonederland.nl/system/files/media/towards-the-circular-economy.pdf (accessed on 15 June 2020).
25. European Association for Chemicals. *Biotechnology and Energy*; European Association for Bioindustries (EuropaBio): Brussels, Belgium, 2011.
26. EUBIONETIII Project Partners. Available online: https://www.researchgate.net/figure/257547972_fig8_Fig-8-Amounts-of-unexploited-agro-industrial-residues-in-selected-EU-member-countries (accessed on 15 June 2020).
27. Mishima, D.; Tateda, M.; Ike, M.; Fujita, M. Comparative study on chemical pretreatments to accelerate enzymatic hydrolysis of aquatic macrophyte biomass used in water purification processes. *Bioresour. Technol.* **2006**, *97*, 2166–2172. [CrossRef] [PubMed]
28. Lissens, G.; Thomsen, A.B.; De Baere, L.; Verstraete, W.; Ahring, B.K. Thermal wet oxidation improves anaerobic biodegradability of raw and digested biowaste. *Environ. Sci. Technol.* **2004**, *38*, 3418–3424. [CrossRef] [PubMed]
29. Madsen, M.; Holm-Nielsen, J.B.; Esbensen, K.H. Monitoring of anaerobic digestion processes: A review perspective. *Renew. Sustain. Energy Rev.* **2011**, *15*, 3141–3155. [CrossRef]
30. Tampio, E.; Ervasti, S.; Rintala, J. Characteristics and agronomic usability of digestates from laboratory digesters treating food waste and autoclaved food waste. *J. Clean. Prod.* **2015**, *94*, 86–92. [CrossRef]
31. IEA Bioenergy. Task 37. In *The Role of Anaerobic Digestion and Biogas in the Circular Economy*; IEA Bioenergy: Paris, France, 2018.
32. Centre for Renewable Energy Sources and Saving of Greece (CRES). *Energy Policy & Planning Directorate*; Database: Athens, Greece, 2017.
33. National Renewable Energy Laboratory. Available online: www.nrel.gov (accessed on 15 June 2020).
34. Kamm, B.; Hüttl, R.; Grünewald, H. Lignocellulosic Feedstock Biorefinery—Combination of technologies of agroforestry and a biobased substance and energy economy. *Forum Forsch.* **2006**, *19*, 53–62.
35. U.S. Department of Energy (DOE). Available online: https://www.energy.gov/ (accessed on 15 June 2020).
36. European Commission Current Situation and Potential of the Bio-Refinery Concept in the EU: Strategic Framework and Guidelines for Its Development. Available online: https://cordis.europa.eu/project/id/44275/reporting (accessed on 15 June 2020).
37. European Association for Bioindustries (EuropaBio). *Building a Bio-based Economy for Europe in 2020*; European Association for Bioindustries: Brussels, Belgium, 2011.
38. Choi, S.; Song, C.W.; Shin, J.H.; Lee, S.Y. Biorefineries for the production of top building block chemicals and their derivatives. *Metab. Eng.* **2015**, *28*, 223–239. [CrossRef] [PubMed]
39. Elbersen, H.W.; Bindraban, P.S.; Blaauw, R.; Jongman, R. Biodiesel from Brazil. In *The Biobased Economy—Biofuels, Materials and Chemicals in the Post-Oil Era*; Langeveld, H., Sanders, J., Meeusen, M., Eds.; Earthscan: London, UK, 2010; pp. 283–301.
40. Bozell, J.J.; Petersen, G.R. Technology Development for the Production of Biobased Products from Biorefinery Carbohydrates—The US Department of Energy's "Top 10" Revisited. *Green Chem.* **2010**, *12*, 539–554. [CrossRef]
41. Sheets, J.P.; Yang, L.; Ge, X.; Wang, Z.; Li, Y. Beyond land application: Emerging technologies for the treatment and reuse of anaerobically digested agricultural and food waste. *Waste Manag.* **2015**, *44*, 94–115. [CrossRef] [PubMed]
42. Vaneeckhaute, C. Nutrient Recovery from Bio-Digestion Waste: From Field Experimentation to Model-based Optimization. Ph.D. Thesis, Ghent University, Ghent, Belgium, Faculte' des Sciences et deGe'nie, Universite', Laval, QC, Canada, 2015.
43. Balde, H.; VanderZaag, A.C.; Burtt, S.D.; Wagner-Riddle, C.; Crolla, A.; Desjardins, R.L.; MacDonald, D.J. Methane emissions from digestate at an agricultural biogas plant. *Bioresour. Technol.* **2016**, *216*, 914–922. [CrossRef] [PubMed]
44. Mateo-Sagasta, J.; Burke, J. *Agriculture and Water Quality Interactions: A Global Overview*; Executive Summary; SOLAW Background Thematic Report-TR08; Food and Agriculture Organization of the United Nations (FAO): Rome, Italy, 2010.

45. World Water Assessment Programme (WWAP). *The United Nations World Water Development Report 2015: Water for a Sustainable World*; United Nations World Water Assessment Programme (WWAP); United Nations Educational, Scientific and Cultural Organization: Paris, France, 2015.
46. World Water Assessment Programme (WWAP). *The United Nations World Water Development Report 2017: Wastewater, the Untapped Resource*; United Nations World Water Assessment Programme (WWAP); United Nations Educational, Scientific and Cultural Organization: Paris, France, 2017.

Article

Experimental Study of Absorbent Hygiene Product Devolatilization in a Bubbling Fluidized Bed

Barbara Malsegna, Andrea Di Giuliano and Katia Gallucci *

Department of Industrial and Information Engineering and Economics (DIIIE), University of L'Aquila, Piazzale E. Pontieri 1-loc. Monteluco di Roio, 67100 L'Aquila, Italy; barbara.malsegna@student.univaq.it (B.M.); diiie.sac@strutture.univaq.it or andrea.digiuliano@univaq.it (A.D.G.)
* Correspondence: katia.gallucci@univaq.it; Tel.: +39-0862-434213

Abstract: This paper aims to investigate the usage of waste from Absorbent Hygienic Products (AHP) as a fuel for gasification or pyrolysis, two attractive routes to obtain valuable products and dispose of this kind of waste. The study experimentally investigated the devolatilization of coarsely shredded materials from diapers, in a laboratory-scale bubbling fluidized bed made of sand, as a representative preparatory step of the above-mentioned thermochemical conversions. Two versions of shredded materials were considered: as-manufactured diapers (AHPam, as a reference), and the cellulosic fraction of sterilized used diapers (AHPus). Results were presented, obtained from physical-chemical characterization of AHPam and AHPus (TGA, CHNS/O, proximate and ultimate analyses, XRF, ICP-AES, SEM-EDS), as well as from their devolatilizations at 500–600–700–800 °C under two different atmospheres (air plus nitrogen, or pure nitrogen as a reference). Generally, temperature influenced syngas composition the most, with better performances under pure nitrogen. At 700–800 °C under pure nitrogen, the highest syngas quality and yield were obtained. For AHPam and AHPus, respectively: (i) H_2 equaled 29.5 vol% and 23.7 vol%, while hydrocarbons equaled 14.8 vol% and 7.4 vol% on dry, dilution-free basis; (ii) 53.7 Nl and 46.0 Nl of syngas were produced, per 100 g of fuel. Overall, AHP emerged as an interesting fuel for thermochemical conversions.

Keywords: absorbent hygiene product; waste; gasification; devolatilization; pyrolysis; fluidized bed; diapers; cellulosic fraction

Citation: Malsegna, B.; Di Giuliano, A.; Gallucci, K. Experimental Study of Absorbent Hygiene Product Devolatilization in a Bubbling Fluidized Bed. *Energies* **2021**, *14*, 2399. https://doi.org/10.3390/en14092399

Academic Editor: Fernando Rubiera González

Received: 7 April 2021
Accepted: 20 April 2021
Published: 23 April 2021

Publisher's Note: MDPI stays neutral with regard to jurisdictional claims in published maps and institutional affiliations.

1. Introduction

Since recent years, research efforts regarding renewable and sustainable energy sources have been more and more intensified because of the complexity of current global energy issues and the urgency of global warming containment [1].

The world population is expected to increase up to more than 11 billion people by the end of the 21st century, geographically concentrated in the currently least developed regions of the world [2]. This will involve net economic growth in those regions, which will in turn cause the growth of world energy demand and the consumption of resources with attendant waste generation, if the current model of linear economy is kept [3].

In September 2018, the World Bank announced that global waste production was predicted to increase by 70% (referring to 2018) by 2050 [4], unless actions were taken to break the "take-make-waste" paradigm of linear economy. Humankind in 2018 (7.6 billion people) produced two billion tonnes of waste per year; therefore, the expected world population growth should be compensated by avoiding the current gross mismanagement of waste [3].

In this framework, the European Union (EU) has promoted thematic strategies regarding waste prevention and recycling, as well as regulations concerning the transition towards a circular and sustainable economy [5]. In this scenario, several governmental programs have sustained actions to contrast climate change, as recently required by the Paris Agreement and the subsequent Conferences of the Parties to the United Nations

Framework Conventions on Climate Change [6–9]. The European Green Deal represents an important response to climate change, setting the final goal of EU climate neutrality by 2050; the EU aims to couple this goal with a new growth strategy and a just transition towards a modern, competitive and resource efficient socio-economic system [10].

The use of waste as a feedstock is fundamental in a circular economy for applications in energy and biofuel production, since it is one of the pillars of the 2030 Agenda for Sustainable Development and contributes to decarbonization and landfill diversion [11]. Thermochemical conversions may serve as an essential part of a sustainable, integrated waste management system, as they are suitable routes to produce energy and fuels from waste [12–14]. In the first place, thermal treatment plants can directly convert the chemical energy content of waste in power and heat [15]. In addition, thermochemical conversions such as pyrolysis and gasification allow obtaining more valuable fuels or chemical products (e.g., by catalytic conversions of syngas [16]), bringing in the advantage of a unified, efficient treatment applicable to different types of solid waste [12–15].

Although biomass utilization in thermochemical processes for energy and chemical production has been extensively studied in the literature [1,17,18], analogue studies concerning solid wastes lead to less unitary conclusions because of the broader heterogeneity of waste materials. This causes the need to specifically study each given class of waste in order to develop the related waste-to-energy/fuel/chemicals chain. This necessity has indeed promoted the birth of various projects, such as: "optiCOM", an industrial project aiming to implement fluidized-bed gasification of several solid wastes at industrial scale; "Lig2Liq", a European research project co-funded by the Research Fund for Coal and Steel of the European Commission, which investigates the co-gasification of lignite with Solid Recovered Fuel (SRF) [13,19]; and more than sixty projects funded in the Biomass, Biofuels and Alternative fuel field, by the EU HORIZON 2020 program [20].

Absorbent Hygiene Products (AHP) constitute one of those class of waste, classified as a sanitary waste according to the European Waste Catalogue code (EWC, Directive 2000/532/EC). AHP consist of a broad set of products, including diapers for babies, sanitary protection pads, tampons, adult incontinence products and personal care wipes. AHP waste currently represents about 2–3% of municipal solid waste and 15–25% of the residual waste stream in some treatment facilities [21].

Nowadays, waste from AHP is quite challenging from the sustainability point of view, since it is usually disposed of via either landfill or incineration, losing its recycling potential [22]. This is a strong motivation of interest in studying possible alternative processes to better exploit this kind of waste. In the literature, some studies reported the separation of AHP waste into its different components, which can then be recycled into secondary raw materials with multiple potential uses (e.g., plastic, cellulose fiber plus super absorbent polymers) [23–25].

In view of the implementation of an integrated and sustainable management of AHP waste, pyrolysis and gasification may represent a viable process option because of their versatility concerning fed feedstock and uniformity of product classes (e.g., syngas or liquid hydrocarbons) [26–29]. In other words, AHP waste is a potential feedstock for thermochemical conversion pathways; this aspect appeared as worthy of a scientific investigation, considering what is stated above about the need of specific studies for the thermochemical conversion of each given kind of waste, and that waste from AHP has not been extensively utilized in this way yet (to our best knowledge).

The present work is the first attempt to fill this gap. The option of the thermochemical conversion of AHP waste materials was studied from the experimental point of view. Two different AHP waste materials were investigated: as-manufactured, coarsely shredded diapers, and the cellulosic fraction of sterilized, used diapers. Devolatilizations of samples from those AHP wastes were performed in a laboratory-scale bubbling fluidized bed made of sand, as a representative preparatory step of pyrolysis or gasification, respectively, intended as thermochemical conversions in the absence of oxidants or in the presence of sub-stoichiometric oxidants (in comparison to combustion) [30]. Physical-chemical characterization of investigated materials was presented and discussed, together with their devolatilization performances, evaluated in terms of syngas yield, carbon conversion, syngas composition and cold gas efficiency. Results were compared, from devolatilizations at different temperatures (from 500 °C to 800 °C), and under fully anoxic or slightly oxidant atmospheres.

These investigations provided novel results concerning the behavior of AHP waste in thermochemical conversions based on fluidized beds, which could be exploited for both scale-up studies and model validations.

2. Materials and Methods

2.1. Waste Materials from AHP for Devolatilization Tests

Two raw materials were selected in this work: (i) as-manufactured, coarsely shredded diapers (AHPam), which represent an industrial post-production waste (Figure 1a); (ii) the cellulosic fraction of sterilized used diapers (AHPus), deriving from a diaper sterilization cycle, followed by the separation of organic and plastic fractions, which are both a post-consumer material (Figure 1b). AHPam and AHPus are classified according to the EWC code as sanitary waste.

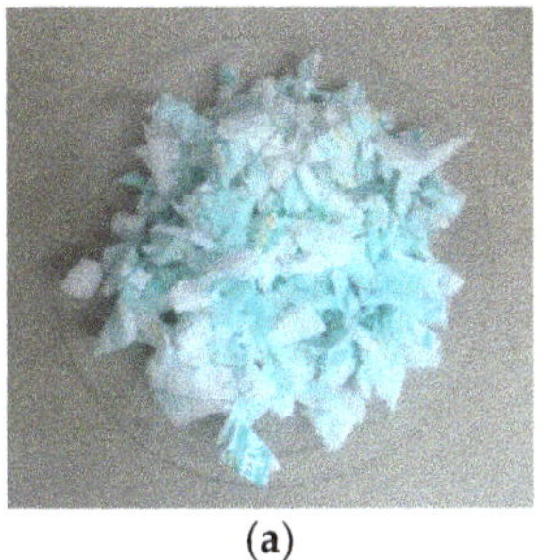

(a)   (b)

Figure 1. Coarsely shredded diapers (AHPam) (**a**); cellulosic fraction of sterilized diapers (AHPus) (**b**).

2.2. Characterization of Waste Materials from AHP

Several physical-chemical characterizations of waste materials from AHP were performed.

Proximate and ultimate analyses were determined by means of Thermo-Gravimetric Analysis (TGA), muffle test, oven test and elemental analysis, which quantified carbon (C), hydrogen (H), nitrogen (N), sulfur (S) and oxygen (O) (from here on named CHNS/O).

TGA tests were carried out by the Linesis STA PT100 equipment, to determine volatile matter and fixed carbon fraction. The tests were performed as specified by the standard ASTM D142/02 [31]. Three samples of AHPam and three of AHPus (masses in the range 13.0–27.0 mg) underwent TGA tests under pure N_2 or air fluxes. The treatment under pure N_2 (16 Nl h^{-1}) consisted of: (i) 10 °C min^{-1} heating rate from room temperature to 105 °C and 15 min dwell at 105 °C; (ii) 10 °C min^{-1} heating rate from 105 °C to 950 °C and a 7 min dwell at 950 °C. The treatment under air (16 Nl h^{-1}) consisted of: (i) 10 °C min^{-1} heating rate from room temperature to 700 °C and 15 min dwell at 700 °C; (ii) 10 °C min^{-1} heating rate from 700 °C to 950 °C and 10 min dwell at 950 °C. For each kind of treatment, blank tests were carried out and used to correct TGA data from actual samples.

Additional TGA tests were carried out, concerning combustion under air flux, in order to evaluate the temperature at which thermal oxidative phenomena could be considered as extinguished. For both AHPam and AHPus, TGA combustion tests were carried out under 10 Nl h^{-1} of air, heating from room temperature up to 950 °C at rates of 5, 10 or 20 °C min^{-1}.

Moisture fractions of AHPam and AHPus were determined by measuring the samples mass variation caused by a drying treatment in an oven (model MPIM factory Oven SL), keeping the temperature at 105 °C for 24 h (BSE EN 15414-3:2011 standard [32]).

Muffle tests (LENTON type ECF 12/10 equipment) were carried out to quantify AHPam and AHPus ash fractions, with reference to the ISO18122:2015 and 1016-104.4:1998 standard [33], according to the thermal program: (i) 8 °C min^{-1} heating rate from room temperature up to 500 °C with 1 h dwell at 500 °C; (ii) 8 °C min^{-1} heating rate from 500 °C up to 815 °C with 2 h dwell at 500 °C.

The elemental analyzer CHNS/O PerkinElmer 2400 Series-II was used to perform the ultimate analysis of AHPam and AHPus, obtaining mass percentages of C, H, N, S and O on dry basis.

The AHPam and AHPus ashes, obtained from the abovementioned muffle treatment at 815 °C, were analyzed by X-Ray Fluorescence (XRF) to determine their semi-quantitative elemental composition. The XRF device was a Spectro Xepos I instrument, with wavelength dispersion method, equipped with a detector that does not allow detecting elements with atomic mass lower than that of magnesium (Mg).

Inductively Coupled Plasma Atomic Emission Spectroscopy (ICP-AES) analyses were carried out with VARIAN 720 ES ICP-AES instrument in order to detect sodium (Na) in the AHPam material, a key element contained in the super-absorbent polymer of diapers (sodium polyacrylate) [34].

XRF and ICP-AES techniques are of interest, since the presence of Na and other mineral elements (e.g., K, Si, Ca, Na, Mg, Al) can lead to the formation of eutectics [35], which can in turn generate low-melting ashes at the temperatures of those processes investigated in this work; these melting species may be responsible for important process issues, as they can develop agglomerates in fluidized beds or obstructions by recondensation in colder areas [36,37].

2.3. Bed Material and Fluid-Dynamic Conditions for Devolatilization Tests

The bed material used for devolatilization tests was sand; its physical properties of fluid-dynamic interest (range of particle diameter d_p and particle density ρ_p) are summarized in Table 1. At experimental conditions of devolatilization tests (see Section 2.4), this sand belongs to Group B of generalized Geldart classification [38], for which minimum fluidization velocity coincides with minimum bubbling velocity, i.e., Group B materials cannot provide homogeneous fluidization. From here on, the term "minimum fluidization velocity (u_{mf})" is used to name the beginning of sand fluidization, which is for sure at bubbling regime. Values of u_{mf} (Table 1), used as references to set devolatilization experimental conditions in this work (see Section 2.4), were determined according to the method described by Di Giuliano et al. [36].

Table 1. Bed material: physical properties of fluid-dynamic interest and minimum fluidization velocity (u_{mf}) as a function of temperature at devolatilization conditions described in Section 2.4.

Bed Material	Sand
d_p (µm)	212–250
ρ_p (kg m^{-3})	2587
T (°C)	u_{mf} (cm s^{-1})
500	2.9
600	2.7
700	2.5
800	2.3

2.4. Devolatilization Tests: Experimental Rig and Procedure

Devolatilization tests were performed for both waste materials from AHP, introduced in Section 2.1 (AHPam and AHPus). Each material was investigated at four temperature levels (500, 600, 700, 800 °C), with two fluidizing agents (pure N_2 or air diluted by N_2 to obtain 1.3–2.3 vol% of O_2 at the bed inlet). Devolatilizations with pure N_2 may be considered as preparatory experiments for a pyrolysis process, while those with diluted air have the same role related to gasification.

The laboratory-scale plant for devolatilization tests is described in detail elsewhere [36]. For the sake of brevity and clarity, only basic information is given here about this experimental rig (Figure 2). It included (Figure 2) one fluidized bed quartz reactor (5 cm internal diameter), heated by a cylindrical electric furnace, in turn controlled by a thermocouple submerged in the bed. The reactor was fed upward by the fluidizing agent, with its flow rate set to always develop 1.4 times u_{mf} (Table 1) at chosen test conditions. The mass of sand in the reactor was selected to obtain 7.5 cm high beds (1.5 times the bed diameter).

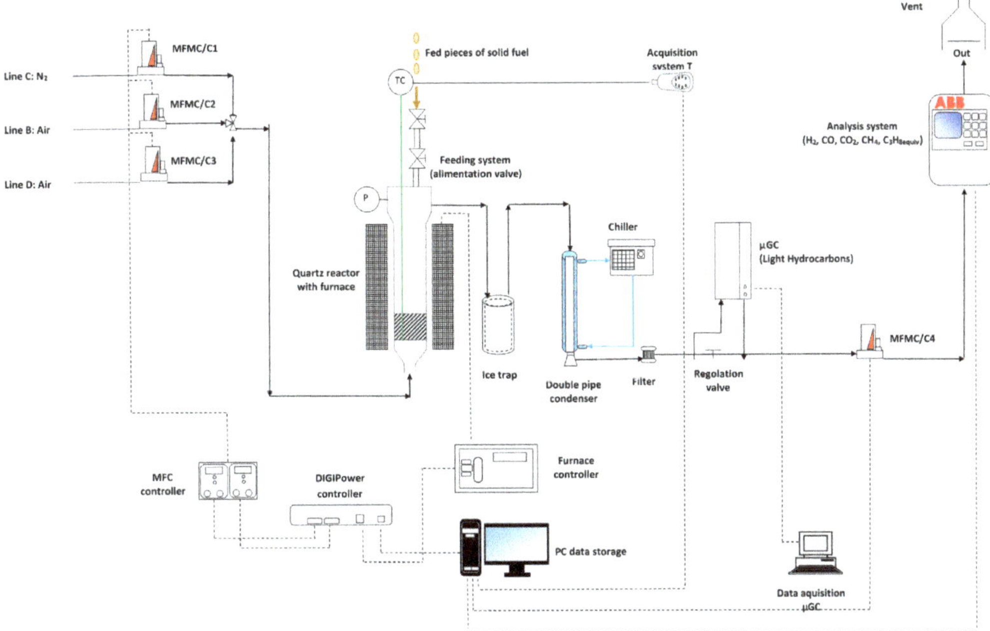

Figure 2. Schematic view of the laboratory-scale experimental apparatus for devolatilization tests.

Devolatilization products left the reactor from its top, together with the fluidizing agent which also served as a carrier; condensable species were separated by an ice trap and a double-pipe glass condenser at 0 °C. The composition of dry product gases was measured online by an ABB analysis system and a micro gas-chromatograph (μCG) AGILENT 490 (see [36] for further details). The ABB system measured the volume percentages of CO, CO_2, CH_4 and H_2, and the overall hydrocarbons content as ppm_v of equivalent C_3H_8. The raw values of equivalent C_3H_8 measured by the ABB system included CH_4 contribution; in this work, the quantity presented as "equivalent C_3H_8" is intended as already reduced by the contribution of CH_4. From here on, "equivalent C_3H_8" means "hydrocarbons other than methane".

Due to the fluffy nature of raw waste materials from AHP (Figure 1), prior to devolatilization AHPam and AHPus were compacted in the form of circular tablets by means of a press; those circular tablets were then cut in half to obtain individual pieces small enough to be fed into the reactor (about 0.2–0.4 g per piece).

AHPam and AHPus pieces were fed individually by hand, thanks to a vertical double valve system at the top of the reactor (Figure 2, see [36] for further details). Each piece experienced an abrupt heating, from room temperature to the one of the fluidized bed, just in the small duration of piece falling. This fast temperature increase could not be performed by TGA equipment.

For each set of parameters "material kind/fluidizing agent/bed temperature", three individual waste pieces were devolatilized, constituting three repetitions of a given test condition. The thermal decomposition of each piece was carried out until completion, before feeding the following; therefore, the process could be considered as a semi-batch, intrinsically at unsteady state.

Outlet molar flow rates ($F_{i,out}$, with i = H_2, CO, CO_2, CH_4 and equivalent C_3H_8) as functions of time (t) were obtained from ABB measurements, thanks to the assumption of the N_2 inlet flow rate as the internal standard. With regard to tests involving diluted air, it is reasonable to hypothesize that fed O_2 was mostly converted and, in any case, it represented a negligible flow rate in comparison to the N_2 one; therefore, fed N_2 was always assumed as the internal standard. Once $F_{i,out}$, masses of fuel pieces (m_p) and their compositions (see Section 3.1) were known, mole balances were performed. On the basis of these balances, for each devolatilization, integral-average values ("av" superscript in Equations (1)–(3)) were calculated for the following quantities:

- Gas yield (η^{av}, Equation (1));
- Percentage of i on dry and dilution free basis (Y_i^{av}, Equation (2));
- Carbon conversion (χ^{av}_C, Equation (3), with n_j as the number of C atoms in C-containing species j).

For those quantities defined in Equations (1)–(3), averages out of the three repetitions were calculated for each set "material kind/fluidizing agent/bed temperature", provided with related standard deviations.

$$\eta^{av} = \frac{\Sigma_i \int F_{i,out}\, dt}{m_p}$$
$$with\ i = H_2,\ CO,\ CO_2,\ CH_4\ and\ equivalent\ C_3H_8 \tag{1}$$

$$Y_i^{av} = \frac{\int F_{i,out}\, dt}{\Sigma_i \int F_{i,out}\, dt} 100$$
$$with\ i = H_2,\ CO,\ CO_2,\ CH_4\ and\ equivalent\ C_3H_8 \tag{2}$$

$$\chi_C^{av} = \frac{\left(12\ g\ mol^{-1}\right) \times \Sigma_j \left(n_j \int F_{j,out}\, dt\right)}{m_p \left(\frac{100 - wt\%ar\ of\ moisture}{100}\right)\left(\frac{wt\%db\ of\ C}{100}\right)} 100$$
$$with\ j = CO,\ CO_2,\ CH_4\ and\ equivalent\ C_3H_8 \tag{3}$$

Cold gas efficiency (ξ, Equation (4), with $LHV_{g,i}$ as the Lower Heating Value of gaseous species i in the product gas, LHV_p as Lower Heating Value of the solid fuel) was calculated as an additional overall evaluation parameter, out of the three repetitions for each set "material kind/fluidizing agent/bed temperature".

$$\xi = \frac{\Sigma_k \left[\left(\Sigma_i \left(LHV_{g,i} \int F_{i,out} dt\right)\right)_k / m_{p,k}\right]}{3 \cdot LHV_p}$$
$$with\ i = H_2,\ CO,\ CO_2,\ CH_4\ and\ equivalent\ C_3H_8;\ k = repetitions\ 1,\ 2,\ 3 \tag{4}$$

Lower heating values in Equation (4) were found in the literature for the syngas components [39,40] and tested waste materials from AHP [34] (Table 2).

Table 2. Lower heating values of gas components ($LHV_{g,i}$) and waste materials from AHP (LHV_p), defined for Equation (4).

Gas	$LHV_{g,i}$ [MJ kmol^{-1}]	Ref
CO	282.99	[39]
CO$_2$	0	.
H$_2$	241.83	[39]
CH$_4$	802.34	[39]
C$_3$H$_8$	46.2	[40]

Material	LHV_p [MJ kg^{-1}]	
AHPam	23.60	[34]
AHPus	10.36	[34]

2.5. Characterization of the Fluidized Bed after Devolatilization Tests

Some morphological and topological aspects were observed by microscopy techniques, namely: (i) Scanning Electron Microscopy (SEM) with Zeiss GeminiSEM 500 microscope equipped with Energy Dispersive X-ray Spectrometry (EDS) for elemental analyses; (ii) observation by means of a stereomicroscope Leica s8 Apo model.

Results of EDS analyses are mentioned in Section 3.3, but only related SEM micrographs are shown. Both microscopy techniques were exploited to study possible agglomeration or sintering phenomena in the fluidized bed, which can interfere with the fluidization quality by increasing the average diameter of bed particles, therefore affecting the performance of thermochemical conversion [39].

This characterization was carried out on the spent fluidized sand at the end of the devolatilization tests (see Section 2.3); the sampling of spent bed particles was performed in different zones: coarse agglomerates of the sand bed (Figure 3a), top of the sand bed (Figure 3b), bottom of the sand bed (Figure 3c), intermediate part of the sand bed (Figure 3d).

Figure 3. Examples of samples from the spent bed after the devolatilization tests: (**a**) coarse agglomerates of the bed, (**b**) top of the bed, (**c**) bottom of the bed, (**d**) intermediate part of the bed.

3. Results

3.1. Physical-Chemical Characterizations of AHPam and AHPus

Table 3 shows the numerical results of proximate and ultimate analyses of AHPam and AHPus, obtained by TGA, oven tests, muffle tests, and the CHNS/O analyzer (see Section 2.2).

Table 3. Proximate and ultimate analyses of AHPam and AHPus (wt% ar = weight percent on as-received basis; wt% db = weight percent on dry basis).

	wt% ar	
	AHPam	**AHPus**
Dry weight	94.2	93.4
Moisture	5.8	6.6
Ashes	7.5	10.7
Volatile matter	82.8	81.5
Fixed carbon	3.9	1.2
	wt% db	
C	65.3	43.5
H	10.3	6.6
N	0.0	0.1
S	1.1	1.8
O	23.3	48.0

Table 4 summarizes the semi-quantitative elemental compositions of AHPam and AHPus, measured by XRF (see Section 2.2).

Table 4. Semi-quantitative elemental composition of AHPam and AHPus, measured by X-ray fluorescence (XRF) (abs. err. = absolute error).

Element	AHPam		AHPus	
	(wt %)	**abs. err. (wt %)**	**(wt %)**	**abs. err. (wt %)**
Mg	1.007	0.031	<0.0020	0.0
Al	<0.0020	0.0	<0.0020	0.0
Si	0.1016	0.0024	1.272	0.008
P	0.0581	0.001	0.5238	0.0034
S	0.03947	0.00049	4.467	0.005
Cl	0.0079	0.00016	1.425	0.002
K	0.04641	0.00069	2.935	0.003
Ca	11.99	0.01	1.157	0.002
Ti	1.402	0.001	0.08108	0.00029
V	0.05397	0.0006	0.00465	0.00013
Cr	<0.00010	0.0	0.00796	0.00006
Mn	0.01207	0.00008	0.00906	0.00007
Fe	0.3068	0.0009	0.2159	0.0006

Table 5 summarizes temperatures at which oxidative decomposition phenomena were no longer detectable during combustions in TGA equipment, for both AHPam and AHPus.

Table 5. End decomposition temperature of AHPam and AHPus, measured during TGA combustions under air flux.

Heating Rates ($^{\circ}$C min^{-1})	5	10	20
Temperature of decomposition end for AHPam ($^{\circ}$C)	690	660	540
Temperature of decomposition end for AHPus ($^{\circ}$C)	530	520	540

3.2. Devolatilization Tests

Figures 4 and 5 show four examples of devolatilizations at 800 °C, one for each combination "fluidizing agent/material kind". As already noted by Di Giuliano et al. [36] about similar experiments with biomass pellets, the gaseous products (H_2, CO, CO_2, CH_4, equivalent C_3H_8 hydrocarbons) were released by an unsteady state process, as evidenced by the characteristic asymmetric shape of $F_{i,out}$ (t) peaks. The numerical integration of $F_{i,out}$ (t) curves with respect to time (t) allowed the calculation of integral-average parameters defined by Equations (1)–(3) (see Section 2.4); these values were then averaged out of the three repetitions of pellet devolatilizations for each set of condition "material kind/fluidizing agent/bed temperature". In addition, cold gas efficiencies (Equation (4)) were calculated on the basis of abovementioned numerical integrations.

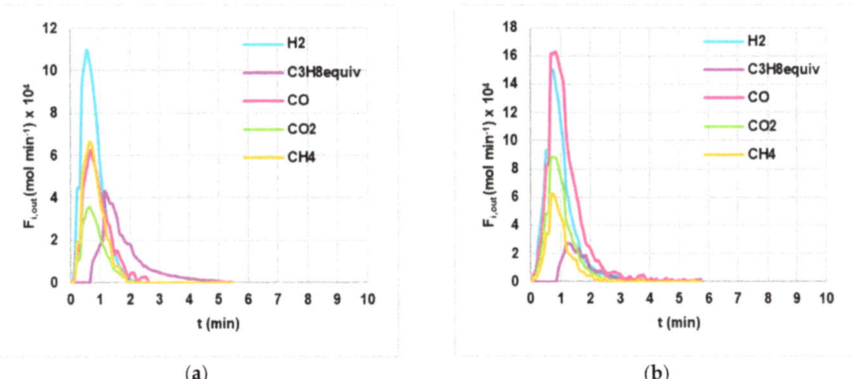

Figure 4. Example of H_2, CO, CO_2, CH_4 and equivalent C_3H_8 outlet molar flow rates ($F_{i,out}$) as functions of time (t), produced by devolatilizations of AHPam (**a**) and AHPus (**b**), in the sand fluidized bed, at 800 °C, under nitrogen atmosphere.

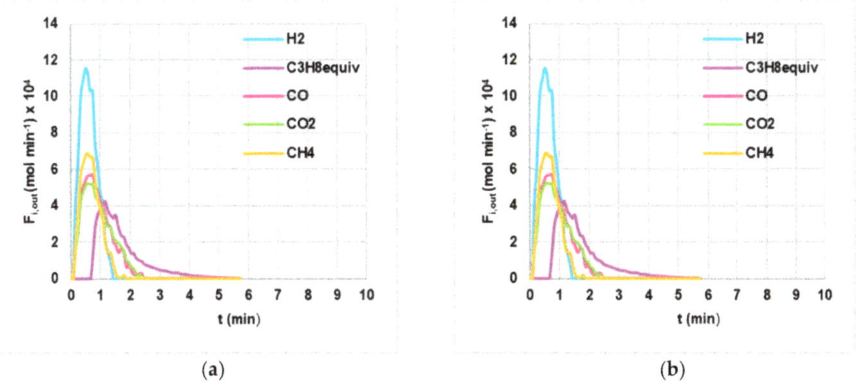

Figure 5. Example of H_2, CO, CO_2, CH_4 and equivalent C_3H_8 outlet molar flow rates ($F_{i,out}$) as functions of time (t) produced by devolatilizations of AHPam (**a**) and AHPus (**b**), in the sand fluidized bed, at 800 °C, under nitrogen plus air atmosphere (diluted air).

Bar charts were used to summarize the devolatilization performances of AHP materials under pure nitrogen (Figure 6) or diluted air (Figure 7), in terms of syngas composition Y_i^{av} (Equation (2), Figures 6a,b and 7a,b), carbon conversion χ^{av}_C (Equation (3), Figures 6c,d and 7c,d), and gas yield η^{av} (Equation (1), Figures 6e,f and 7e,f). In both figures, the left column concerns AHPam, while the right one concerns AHPus.

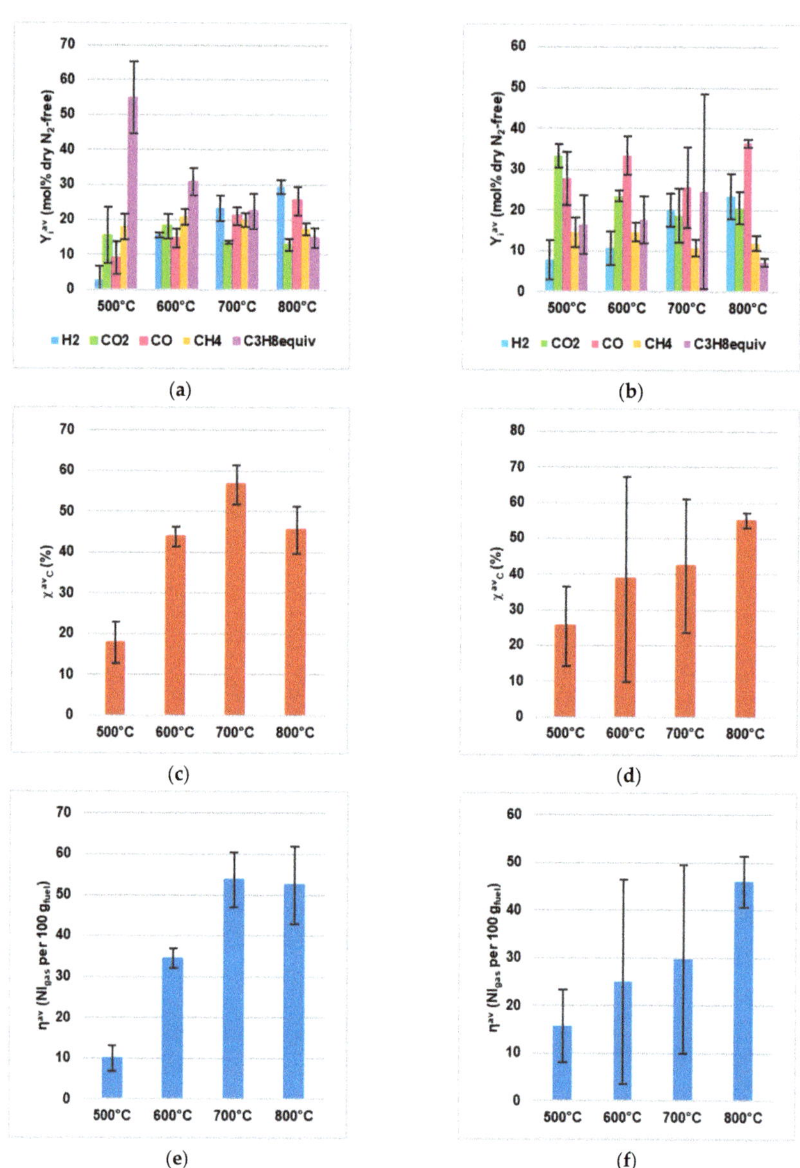

Figure 6. Experimental results of devolatilization tests under nitrogen atmosphere, in the sand fluidized bed, as functions of temperature: integral-average mol% dry, dilution-free (Y_i^{av}, Equation (2)) of AHPam (**a**) AHPus (**b**); integral-average carbon conversion (χ^{av}_C, Equation (3)) of AHPam (**c**) AHPus (**d**); integral-average gas yield (η^{av}, Equation (1)) of AHPam (**e**) AHPus (**f**); colored bars represent average values out of the three repetitions for each set "material kind/fluidizing agent/bed temperature", errors bars quantify the related standard deviations.

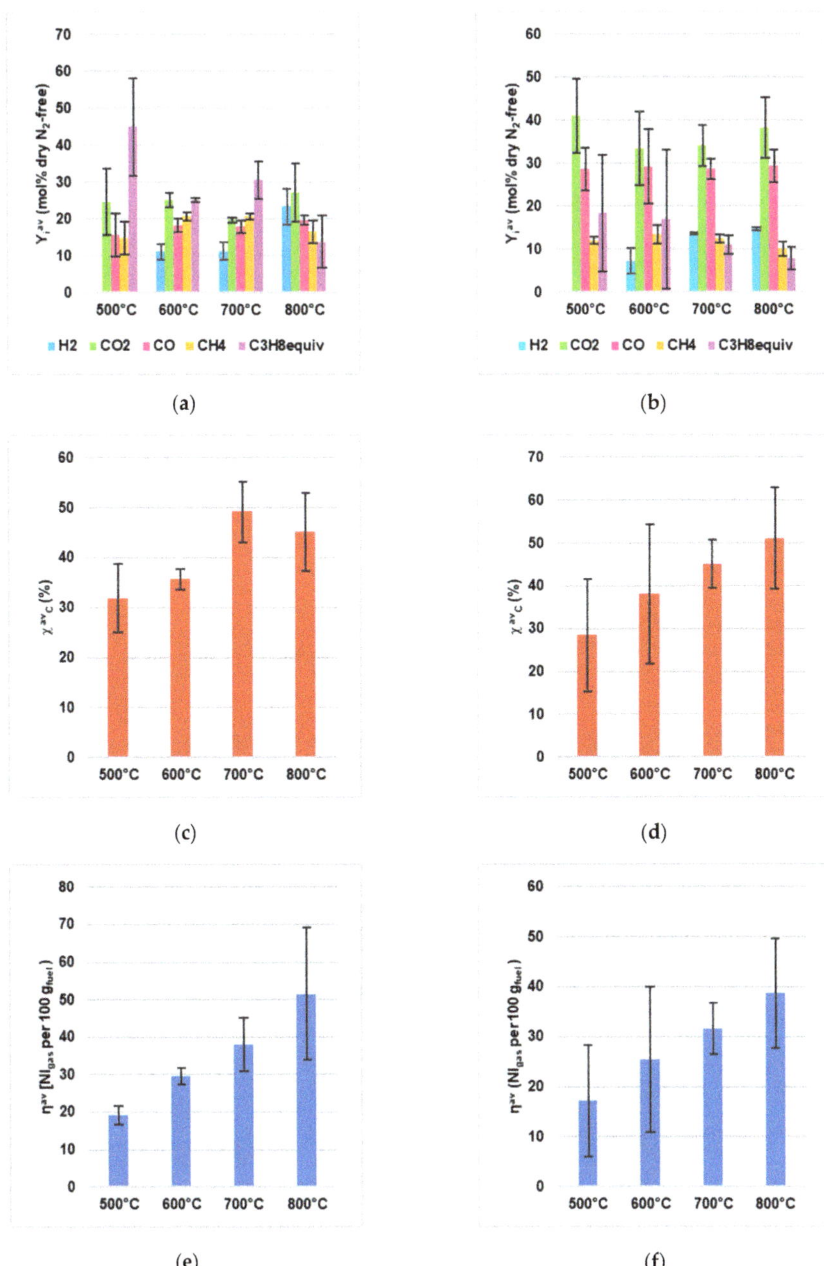

Figure 7. Experimental results of devolatilization tests under nitrogen plus air atmosphere, in the sand fluidized bed, as functions of temperature: integral-average mol% dry, dilution-free (Y_i^{av}, Equation (2)) of AHPam (**a**) AHPus (**b**); integral-average carbon conversion (χ^{av}_C, Equation (3)) of AHPam (**c**) AHPus (**d**); integral-average gas yield (η^{av}, Equation (1)) of AHPam (**e**) AHPus (**f**); colored bars represent average values out of the three repetitions for each set "material kind/fluidizing agent/bed temperature", errors bars quantify the related standard deviations.

Colored bars in Figures 6 and 7 represent average values out of the three repetitions for each set "material kind/fluidizing agent/bed temperature"; errors bars quantify the related standard deviations.

Numerical data showed in Figures 6 and 7 are detailed in Appendix A of this work.

Table 6 summarizes numerical values of cold gas efficiency (ξ, Equation (4)), calculated for each set "material kind/fluidizing agent/bed temperature", as described in Section 2.4.

Table 6. Cold gas efficiency (ξ, Equation (4)) for each set "material kind/fluidizing agent/bed temperature".

T	Tests with N_2		Tests with N_2 Plus Air	
(°C)	AHPam (%)	AHPus (%)	AHPam (%)	AHPus (%)
500	2.6	16.5	6.8	13.8
600	27.6	27.1	14.3	23.9
700	29.1	32.2	18.6	29.7
800	28.8	51.6	23.8	33.1

3.3. Characterizations of Spent Fluidized Bed

After devolatilization tests ended, the spent bed material was recovered and sampled by the procedure described in Section 2.5. Figure 8 shows pictures taken by the stereomicroscope on coarse sand agglomerates (Figure 3a). Figures 9–12 show SEM micrographs of different samples from the spent bed; more in detail, these picture reveal morphological and topological aspects of coarse agglomerates (Figure 9), bed bottom (Figure 10), bed intermediate zone (Figure 11) and bed top (Figure 12).

(a)   (b)
(c)   (d)

Figure 8. Stereo-microscope images of agglomerates from the spent sand bed after the last test, magnified at: 1× (**a**); 2× (**b**); 4× (**c**); 4× (**d**).

Figure 9. SEM micrographs of the coarse agglomerates from the spent sand bed after the last test, magnified at: 80× (**a**); 320× (**b**); 339× (**c**).

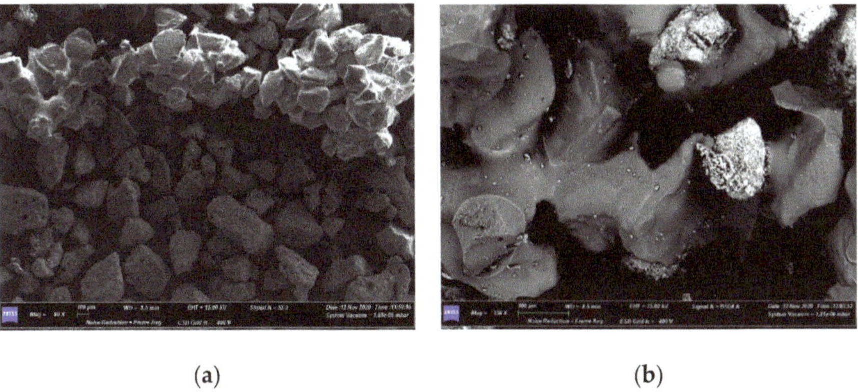

Figure 10. SEM micrographs of bottom of spent sand bed after the last test, magnified at: 80× (**a**); 336× (**b**).

Figure 11. SEM micrographs of intermediate part of spent sand bed after the last test, magnified at: 80× (**a**); 336× (**b**); 500× (**c**).

Figure 12. SEM micrographs of top of spent sand bed after the last test, magnified at: 80× (**a**); 334× (**b**); 500× (**c**).

4. Discussion

4.1. Physical-Chemical Characterizations

Table 3 shows results from proximate and ultimate analyses of AHPam and AHPus. A high volatile content resulted from proximate analyses of both materials (82.8 wt% ar for AHPam and 81.5 wt% ar for AHPus), as well as an important ash fraction (12.5 wt% for AHPam and 11.4 wt% for AHPus), for instance, higher than that of SRF [19]. The high content of volatiles was considered to be a positive factor, since the conversion of these species is favored in the thermochemical processes considered in this work [39]. The ultimate analysis highlighted the absence of nitrogen or its presence in traces (0.0 wt% db in AHPam and 0.1 wt% db in AHPus, Table 3): this could be an interesting feature in an oxyfuel conversion, as it may allow expecting no development of nitrogen-based pollutants. In general, the ultimate analysis (Table 3) for AHPam and AHPus makes them appropriate for thermochemical processes.

The semi-quantitative elemental composition, measured by XRF (Table 4), identified the main elements in the ashes of AHPam and AHPus. Ashes from AHPam mainly contained Mn, Ca and Ti, while in ashes from AHPus there were mainly Si, P, K, Cl, and Ca. With regard to S, a not negligible quantity was found in AHPus, higher by one order of magnitude than S in AHPam.

ICP-AES analyses indicated that Na content in AHPam was in the range of 2–3 wt%; it is reasonable to assume a similar Na content in AHPus.

The Na-Si phase diagram [35] shows the formation of an eutectic in the range of temperature 500–800 °C, corresponding to 45 wt% of Na in Si. This range of temperature is exactly the same considered in devolatilizations of this work. Even though only 2–3 wt% of Na was detected or assumed in waste materials from AHP, this value could have become locally higher, therefore triggering the agglomeration of sand particles. This phenomenon was demonstrated by SEM-EDS analyses of bed samples after devolatilization tests (see Section 3.2, Figures 8–10).

4.2. Experimental Data from Devolatilization Tests

With regard to devolatilization tests, it is worth to stress again that they were carried out with two fluidizing agents: pure N_2 or N_2 plus air (i.e., diluted air). Tests with pure N_2 represented a preliminary experimentation which simulated the pyrolysis of waste materials from AHP, while those with diluted air did the same for the gasification process [30]. In addition, and more interestingly for the purposes of this work, tests with pure N_2 served as a reference to evaluate effects due to the slightly oxidizing atmosphere of experiments with diluted air.

The chosen devolatilization temperatures matched well with results from the combustion tests performed in TGA (Table 5), which highlighted how the combustion process ended in a range of temperatures between 500 and 600 °C. From the analysis of these results at 20 °C min^{-1} (the fastest heating rate performed in TGA, Table 5), decomposition by combustion extinguished at 540 °C for both AHPam and AHPus; considering that devolatilization tests involved a definitively more abrupt temperature increase from room to 500, 600, 700 or 800 °C and that thermal decomposition phenomena tend to overlap as the heating rate is increased, it is reasonable to assume that conditions in the laboratory-scale fluidized bed were sufficient to induce a substantial thermal decomposition of AHP-derived fuels.

The results from devolatilizations were presented in Section 3.2. With both fluidizing agents and both AHPam and AHPus, some general effects emerged due to the temperature increase, which were quantified by means of the parameters defined in Section 2.4.

Integral-average gas yield (η^{av}, Equation (1)) for AHPam had maximum values (net of standard deviations) at 700–800 °C in N_2 atmosphere (Figure 6e) and at 800 °C in N_2 plus air atmosphere (Figure 7e), while η^{av} always increased as the temperature was increased for AHPus with both fluidizing agents, reaching maximum values at 800 °C (Figures 6f and 7f).

Carbon conversion (χ^{av}_C, Equation (3)) for AHPam had its maximum value at 700 °C with both fluidizing agents (Figures 6c and 7c); for AHPus, an increasing trend was observed as the temperature was increased, with the highest values at 800 °C, even though there were important standard deviations at 600 and 700 °C for AHPus in diluted air (Figures 6d and 7d). These observations highlighted a significant differentiation of behaviors due to the solid fuel nature: with AHPam, it is not needed to heat a fluidized bed up to 800 °C to obtain the maximum χ^{av}_C.

As for integral-average gas compositions (Y_i^{av}, Equation (2), Figures 6a,b and 7a,b), generally the temperature increase involved an increase in H_2 percentage and a decrease in hydrocarbons fraction (equivalent C_3H_8), with the variations of their two Y^{av}_i appearing in reciprocal correlation; it is sensible to hypothesize that hydrogen atoms which were converted in gaseous H_2 came from the cellulosic and plastic matrix of the AHP materials [34].

Differences emerged concerning the formation of gaseous hydrocarbons, when comparing the gas developed from the two materials: under both atmospheres, the decomposition of AHPam produced a syngas with higher equivalent C_3H_8 fraction (Figure 6a vs. Figure 6b, Figure 7a vs. Figure 7b). This difference was exalted at 500 °C, the lowest investigated temperature: under N_2, $Y^{av}_{C3H8equiv}$ was 54.8 vol% for AHPam (Figure 6a) and 16.5 vol% for AHPus (Figure 6b) on dry and dilution-free basis; the same trends and orders of magnitude were observed in the N_2 plus air case (Figure 7a vs. Figure 7b). This aspect could be correlated to the composition of the material, since the AHPam contained higher carbon and hydrogen fractions than those of AHPus (elemental analyses, Table 3). This effect, due to the nature of waste materials from AHP, decreased as the temperature was increased.

A deeper qualitative insight about hydrocarbons' nature came from the μCG analyses. Peaks with retention times compatible with those of ethylene/acetylene and propene were often detected at all tested temperatures, while those of benzene were occasionally found in syngas from tests at 800 °C under diluted air. In any case—as evidenced by Di Giuliano et al. for similar tests on the same rig of this work—those μGC analyses are only an optional means to identify some of the hydrocarbons quantified by the ABB system as "equivalent C_3H_8"; it must be taken into account that the μCG sampling was manually started and occurred on a flow produced by an unsteady process [36].

As for oxidized carbon in gases (CO and CO_2), it is reasonable to assume that oxygen could come only from the solid phases in tests under pure N_2, while fed O_2 acted as an additional source in tests with N_2 plus air. This assumption found confirmation in the experimental results: for both waste materials and all temperatures, devolatilizations under diluted air produced syngas with higher fractions of CO and CO_2 (Figure 7a,b), in comparison with those obtained from devolatilizations under pure N_2 (Figure 6a,b). On the other hand, under both atmospheres, AHPus developed higher fractions of CO and CO_2 (Figures 6b and 7b) than those obtained from AHPam (Figures 6a and 7a), in agreement with the oxygen content of the two solid fuels (Table 3). The non-linear behavior of CO_2 concentration with respect to temperature could be ascribed to the network of exothermic and endothermic reactions and the competitive trend of their thermodynamic and kinetic peculiarities.

Numerical data represented in Figures 6 and 7 are reported in Appendix A of this work (Tables A1 and A2): sometimes, important standard deviations were observed. This variability was ascribed to: (i) the macroscopic inhomogeneity of the AHP materials, which made it difficult to obtain solid pressed tablets which were fully representative of the average composition of each whole solid bulk inventory; (ii) the low density of the analyzed materials, together with small allowed dimensions of pressed tablets, involved small absolute masses of fed solid fuel per each devolatilization test, then implying a more significant relative weight of the inherent experimental errors on determined quantities.

The cold gas efficiency (ξ, Equation (4), Table 6) was used as an overall parameter to better evaluate and compare the experimental devolatilization performances. The results in Table 6 confirmed that the temperature increase produced positive effects, since it made the ζ grow for both materials and both atmospheres in the range 500–800 °C. At a given temperature, AHPus devolatilizations under both atmospheres resulted in higher ζ values than those related to AHPam (with the minor exception for pure N_2 at 600 °C, Table 6). The highest ζ occurred for AHPus, at 800 °C: 56.5% in N_2 atmosphere and 33.1% in N_2 plus air atmosphere, respectively (Table 6). This evidence was corroborated by the higher volatile content in AHPus samples (Table 3). These results are in further agreement with the detected improvements of the composition of the produced gas (Figures 6a,b and 7a,b), i.e., an increase in gaseous compounds with a higher calorific value (H_2, CO and CH_4) as the temperature was increased, with the concomitant increase in gas yield η^{av}. This suggested that the endothermic reactions of devolatilization were favored as the temperature was increased, and that the light hydrocarbon compounds, collectively reported as equivalent C_3H_8, were widely decomposed at 800 °C.

Overall, among all tested conditions, the range 700–800 °C appears as the best for the thermal decomposition of both waste materials from AHP, with the best performances in terms of cold gas efficiencies achieved with AHPus at 800 °C.

As a side note, the experimenters of this work would like to underline that important and progressive deposition of allegedly carbonaceous species occurred on the reactor walls and inside downstream pipelines. This made it impossible and unrealistic to close carbon balances; as a first approximation, it can be assumed that the share ($100-\chi^{av}{}_C$) is attributable to these deposits. This is also a significant issue, which must be considered in hypothetical scaled-up processes of the thermochemical conversion of waste materials from AHP.

4.3. Characterization of Spent Fluidized Bed

The SEM-EDS and stereoscopic microscope analyses allowed us to deepen the insight about the agglomeration of ashes and bed particles during the devolatilization tests (see Section 2.5).

Figure 8 shows pictures taken by the stereomicroscope on coarse sand agglomerates, shown at 1:1 scale in Figure 3a. Figure 8a is an overall view of bed particles after the whole devolatilization campaign under N_2 and air atmosphere: many sand particles in the diameter range of 212–250 μm (Table 1) appeared as stuck in a very compact block, i.e., agglomerated into coarse groups in the order of 1–2 cm. This phenomenon could negatively affect the fluidization quality of the bed. Magnifications of stereomicroscope pictures (Figure 8b–d) evidenced the occurrence of Si particles fusion (blue circles) and carbon deposition (red circle).

Figures 9–12 show SEM micrographs of the spent sand bed after the last test; more in detail, these pictures revealed morphological and topological aspects of coarse agglomerates (Figure 9), bed bottom (Figure 10), bed intermediate zone (Figure 11) and bed top (Figure 12).

Figure 9a shows irregularly shaped particles (dark grey) and some agglomerates (light grey, top right). Those agglomerates were detailed by Figure 9b,c, which shows: smooth bridges and jagged areas characterized by crusts and holes, together with melted covers adhering to the original grains. EDS microanalyses highlighted the presence of Si as the main constituent of those bridges, with a lower amount of Na and traces of other elements such as Al, Mg, K and Fe, documented elsewhere as possible causes of sintering phenomena [41].

In its upper part, Figure 10a shows a relatively huge chain of agglomerates, quite evident by contrast with the smaller, well segregated particles below. Figure 10b evidences the presence of the same kind of smooth bridges already found in Figure 9b,c, and areas covered by melted layers (brighter zone in Figure 10b), mainly containing Si, Na and traces of K, Al and Ca, according to EDS analyses. In addition, EDS determined that the lightest zone in Figure 10b consists of a porous structure mainly made up of Ca.

Intermediate (Figure 11) and top (Figure 12) zones of the bed revealed a very similar morphology, also at the single-particle level, as can be seen by magnified micrographs in Figures 11c and 12c. A superficial, spread deposition of fine particles was observed (Figures 11b,c and 12b,c), but no sintering occurred on the whole. SEM-EDS results confirmed that the spread fine particles were made up of Si and traces of Al, Mg, Na, K and Ca.

Overall, the following observations could be inferred from this microscopic characterization:

1. A priori, it was expected to find a greater number of melted particles/agglomerates in the upper part of the bed, because it is the zone of the first contact between its free surface and the falling pieces of waste materials from AHP; however, the exact opposite was found, as the smaller particles had moved to the top of the bed, while the larger ones, allegedly formed on the top, were found on the bottom. This is proof of the good mixing within the fluidized bed;

2. Agglomerations did not diffusely involve the whole bed inventory. This suggested that only where the bed had experienced locally high Na concentrations the eutectic had formed [35], even though the quantity of Na in the fed fuel was low (2–3 wt%) compared to Na eutectic concentration in Si (45 wt%) [35].

5. Conclusions

This work dealt with devolatilizations of two kinds of waste generated from Absorbent Hygiene Products (AHP): shredded scraps of as-manufactured diapers (AHPam) and shredded cellulosic fraction of used and sterilized diapers (AHPus). Tests were carried out in a laboratory-scale bubbling fluidized bed made of sand, at temperatures between 500 °C and 800 °C, under inert and mildly oxidizing atmospheres. This experimental campaign served as a preparatory step for possible scaling-up of thermochemical conversion processes with waste materials from AHP as the solid fuel feedstock.

Devolatilization performances were evaluated in terms of gas yield, carbon conversion and cold gas efficiency.

Temperature in the range 500–800 °C appeared as the most influential process parameter, followed by the oxygen content in the fluidization agent and the nature of solid fuel. Devolatilization tests performed at 700 and 800 °C, using pure nitrogen as the fluidizing agent, allowed us to obtain the highest values of gas yield (53.7 Nl 100 g_{fuel}^{-1} for AHPam at 700 °C; 46.0 Nl 100 g_{fuel}^{-1} for AHPus at 800 °C), carbon conversion (56.5% for AHPam at 700 °C; 55.0% for AHPus at 800 °C), and cold gas efficiency (28.8% for AHPam, 51.6% for AHPus both at 800 °C).

The oxidative atmosphere determined poorer performances in terms of cold gas efficiency (e.g., at 800 °C, 23.8% for AHPam and 33.1% for AHPus), sensibly because of the more pronounced oxidation of gaseous products, leading to a reduction in the lower heating value of syngas.

According to cold gas efficiency, AHPus appeared as the more interesting fuel feedstock when the main aim of thermochemical conversion is the production of power and heat.

SEM-EDS analyses evidenced the occurrence of agglomeration and sintering of the bed particles, ascribed to low-melting elements in the fuel ashes, such as Na, Ca, Si and K. Nevertheless, this phenomenon did not spread within the whole bed inventory and the bed mixing was kept during devolatilizations.

Important depositions of allegedly carbonaceous species were noted inside the reactor and on internal walls of downstream pipelines.

The presented experimental investigation provided results and useful information concerning the behavior of waste materials from AHP in thermochemical conversions occurring in fluidized beds. Technical issues, such as sand particle agglomeration and dirt depositions, were documented in view of the possible scale-up of thermochemical conversion processes of waste materials from AHP. In addition, devolatilization experimental results were made available for possible model validations in further studies, which could also be used to investigate process economics at hypothetical industrial scales.

Overall, AHP waste appeared as an interesting feedstock for thermochemical processes, especially if aimed at power/heat production.

Author Contributions: Conceptualization, K.G. and A.D.G.; methodology, K.G. and A.D.G.; software A.D.G. and B.M.; validation, K.G. and A.D.G.; formal analysis, K.G. and A.D.G.; investigation, K.G., A.D.G. and B.M.; resources, K.G.; data curation, K.G., A.D.G. and B.M.; writing—original draft preparation, B.M.; writing—review and editing, K.G., A.D.G. and B.M.; visualization, K.G., A.D.G. and B.M.; supervision, K.G.; project administration, K.G.; funding acquisition, K.G. All authors have read and agreed to the published version of the manuscript.

Funding: This research was funded by ALMACIS s.r.l.

Institutional Review Board Statement: Not applicable.

Informed Consent Statement: Not applicable.

Data Availability Statement: The data presented in this study are available on request from the corresponding author.

Acknowledgments: The provision of AHP materials by ALMACIS s.r.l. is acknowledged. The authors warmly thank Giampaolo Antonelli for his technical support, and Fabiola Ferrante for XRF and CHNS/O analyses and management.

Conflicts of Interest: The authors declare no conflict of interest. The funders had no role in the design of the study; in the collection, analyses, or interpretation of data; in the writing of the manuscript, or in the decision to publish the results.

Appendix A

This appendix summarizes the numerical results discussed in this manuscript. Tables A1 and A2 show the experimental data of devolatilization tests, discussed in the main text and reported as bar charts in Figures 6 and 7.

Table A1. Experimental results of devolatilization tests under nitrogen atmosphere, as functions of temperature (av3: average of results from the three individual pellets; sd3: standard deviation of results from the three individual pellets).

Sample	η^{av}		χ^{av}_C		Y^{av}_{H2}		$Y^{av}_{C3H8equiv}$		Y^{av}_{CO}		Y^{av}_{CO2}		Y^{av}_{CH4}	
	(Nl Per 100 g_{fuel})		(%)		(mol% Dry N_2-Free)		(mol% Dry N_2-Free)		(mol% Dry N_2-Free)		(mol% Dry N_2-Free)		(mol% Dry N_2-Free)	
	av3	sd3	av3	sd3	av3	sd3	av3	sd3	av3	sd3	av3	sd3	av3	sd3
500 °C														
AHPam	10.0	3.11	17.8	5.1	2.80	3.90	54.8	10.2	9.00	4.62	15.5	7.91	17.8	3.63
AHPus	15.8	7.67	25.4	11.1	7.90	4.80	16.5	7.26	27.8	6.50	33.3	3.00	14.5	9.96
600 °C														
AHPam	34.5	2.43	43.8	2.5	15.6	0.63	30.8	3.96	14.7	2.71	18.1	3.50	20.8	2.24
AHPus	25.0	21.4	38.6	28.7	10.7	4.07	17.7	5.86	33.4	4.66	23.5	1.40	14.7	0.38
700 °C														
AHPam	53.7	6.66	56.5	4.9	23.3	3.72	22.5	5.03	21.0	2.55	13.5	0.38	19.8	2.02
AHPus	29.8	19.8	42.2	18.7	20.1	4.07	24.7	23.8	25.7	9.79	18.7	6.62	10.8	4.00
800 °C														
AHPam	52.5	9.51	45.5	5.8	29.5	2.05	14.8	2.80	25.4	4.04	12.9	1.74	17.5	1.73
AHPus	46.0	5.40	55.0	2.1	23.5	5.53	7.40	0.98	36.4	0.89	20.7	3.95	12.0	1.36

Table A2. Experimental results of devolatilization tests under nitrogen plus air atmosphere, as functions of temperature (av3: average of results from the three individual pellets; sd3: standard deviation of results from the three individual pellets).

Sample	η^{av}		χ^{av}_C		Y^{av}_{H2}		$Y^{av}_{C3H8equiv}$		Y^{av}_{CO}		Y^{av}_{CO2}		Y^{av}_{CH4}	
	(Nl Per 100 g_{fuel})		(%)		(mol% Dry N_2-Free)		(mol% Dry N_2-Free)		(mol% Dry N_2-Free)		(mol% Dry N_2-Free)		(mol% Dry N_2-Free)	
	av3	sd3	av3	sd3	av3	sd3	av3	sd3	av3	sd3	av3	sd3	av3	sd3
							500 °C							
AHPam	19.2	2.53	31.9	6.9	0.00	0.00	45.0	13.3	15.7	5.90	24.6	9.00	14.8	4.40
AHPus	17.2	11.2	28.4	13.2	0.00	0.00	18.4	13.6	28.6	5.00	41.0	8.50	12.0	0.80
							600 °C							
AHPam	29.6	2.14	35.7	2.00	11.1	2.1	25.0	0.60	18.2	1.90	25.0	2.10	20.6	1.20
AHPus	25.5	14.6	38.0	16.3	7.3	3.3	16.9	16.2	29.1	8.70	33.4	8.60	13.3	2.10
							700 °C							
AHPam	38.0	7.11	49.1	6.00	11.3	2.30	30.5	5.03	17.9	1.70	19.7	0.60	30.5	0.80
AHPus	31.6	5.14	45.1	5.50	13.6	0.30	11.1	2.20	28.6	2.50	34.2	4.80	11.1	1.00
							800 °C							
AHPam	51.5	17.6	45.1	7.90	23.3	5.00	13.8	7.10	19.6	1.20	27.0	7.90	16.3	3.10
AHPus	38.7	10.9	51.1	11.8	14.7	0.40	7.80	2.60	29.3	3.80	38.2	7.10	10.0	1.70

References

1. Perea-Moreno, M.A.; Samerón-Manzano, E.; Perea-Moreno, A.J. Biomass as renewable energy: Worldwide research trends. *Sustainability* **2019**, *11*, 863. [CrossRef]
2. Bongaarts, J. Development: Slow down population growth. *Nature* **2016**, *530*, 409–412. [CrossRef] [PubMed]
3. Ellen MacArthur Foundation Circular Economy-UK, USA, Europe, Asia & South America-The Ellen MacArthur Foundation. Available online: https://www.ellenmacarthurfoundation.org/ (accessed on 23 March 2021).
4. The World Bank What a Waste: An Updated Look into the Future of Solid Waste Management. Available online: https://www.worldbank.org/en/news/immersive-story/2018/09/20/what-a-waste-an-updated-look-into-the-future-of-solid-waste-management (accessed on 11 March 2021).
5. EC European Commission–Environment–Circular Economy Strategy. Available online: http://ec.europa.eu/environment/circular-economy/index_en.htm (accessed on 11 March 2021).
6. Obama, B. The irreversible momentum of clean energy: Private-sector efforts help drive decoupling of emissions and economic growth. *Science* **2017**, *355*, 126–129. [CrossRef] [PubMed]
7. Marris, E. Why young climate activists have captured the world's attention. *Nature* **2019**, *573*, 471–473. Available online: https://www.nature.com/articles/d41586-019-02696-0 (accessed on 11 March 2021). [CrossRef] [PubMed]
8. United Nations Climate Change The Paris Agreement I UNFCCC. Available online: https://unfccc.int/process-and-meetings/the-paris-agreement/the-paris-agreement (accessed on 11 March 2021).
9. COP 24 I UNFCCC. Available online: https://unfccc.int/process-and-meetings/conferences/past-conferences/katowice-climate-change-conference-december-2018/sessions-of-negotiating-bodies/cop-24 (accessed on 11 March 2021).
10. A European Green Deal I European Commission. Available online: https://ec.europa.eu/info/strategy/priorities-2019-2024/european-green-deal_en (accessed on 3 January 2021).
11. Johnston, R.B. Arsenic and the 2030 Agenda for sustainable development. *Arsen. Res. and Glob. Sustain. As* **2016**, 12–14. [CrossRef]
12. Arena, U. Process and technological aspects of municipal solid waste gasification. A review. *Waste Manag.* **2012**, *32*, 625–639. [CrossRef] [PubMed]
13. Lombardi, L.; Carnevale, E.; Corti, A. A review of technologies and performances of thermal treatment systems for energy recovery from waste. *Waste Manag.* **2015**, *37*, 26–44. [CrossRef] [PubMed]
14. Materazzi, M.; Foscolo, P.U. The role of waste and renewable gas to decarbonize the energy sector. In *Substitute Natural Gas from Waste: Technical Assessment and Industrial Applications of Biochemical and Thermochemical Processes*; Academic Press: London, UK, 2019; ISBN 9780128155547.
15. Lig2Liq–Cost Effective Conversion of Lignite and Waste to Liquid Fuels. Available online: https://www.lig2liq.eu/ (accessed on 3 January 2021).
16. Weber, G.; Di Giuliano, A.; Rauch, R.; Hofbauer, H. Developing a simulation model for a mixed alcohol synthesis reactor and validation of experimental data in IPSEpro. *Fuel Process. Technol.* **2016**, *141*, 167–176. [CrossRef]

17. Rapagnà, S.; Gallucci, K.; Foscolo, P.U. Olivine, dolomite and ceramic filters in one vessel to produce clean gas from biomass. *Waste Manag.* **2018**, *71*, 792–800. [CrossRef] [PubMed]

18. Molino, A.; Chianese, S.; Musmarra, D. Biomass gasification technology: The state of the art overview. *J. Energy Chem.* **2016**, *25*, 10–25. [CrossRef]

19. Savuto, E.; Di Carlo, A.; Gallucci, K.; Di Giuliano, A.; Rapagnà, S. Steam gasification of lignite and solid recovered fuel (SRF) in a bench scale fluidized bed gasifier. *Waste Manag.* **2020**, *114*, 341–350. [CrossRef] [PubMed]

20. European Comission Innovation and Networks Executive Agency. Available online: https://ec.europa.eu/inea/en/connecting-europe-facility/cef-transport/projects-by-country/multi-country/2014-eu-tm-0196-s (accessed on 23 March 2021).

21. Treatment of Absorbent Hygiene Products for Improved Recycling of Materials | Green Best Practice Community. Available online: https://greenbestpractice.jrc.ec.europa.eu/node/131#bemp_general_title_id (accessed on 23 March 2021).

22. Waldheim, L. Gasification of Waste for Energy Carriers–A Review. 2018. Available online: https://www.etipbioenergy.eu/databases/reports/353-gasification-of-waste-for-energy-carriers-a-review (accessed on 23 March 2021).

23. Deliverable Number 1.7–Blueprint for the Replication of the AHP Pretreatment Technol; Establishing a Multi-Purpose Biorefinery for the Recycling of the Organic Content of AHP Waste in Circular Economy Domain Grant Agreement No.: 745746. Available online: https://www.embraced.eu/repository/EMBRACED_D1.7_Blueprint_DEF.pdf (accessed on 23 March 2021).

24. Mendoza, J.M.F.; D'Aponte, F.; Gualtieri, D.; Azapagic, A. Disposable baby diapers: Life cycle costs, eco-efficiency and circular economy. *J. Clean. Prod.* **2019**, *211*, 455–467. [CrossRef]

25. Kashyap, P.; Ko Win, T.; Visvanathan, C. Absorbent Hygiene Products-An emerging urban waste management issue. In Proceedings of the Asia-Pacific Conference on Biotechnology for Waste Conversion 2016, Hong Kong, China, 5–8 December 2016.

26. Arena, U.; Ardolino, F.; Gregorio, F.D.I. Sustainability of an Integrated Recycling Process of Absorbent Hygiene Products. In Proceedings of the 15th International Waste Management and Landfill Symposium, Cagliari, Italy, 5–9 October 2015.

27. Gerina-Ancane, A.; Eiduka, A. Research and analysis of Absorbent Hygiene Product (AHP) recycling. In Proceedings of the Engineering for Rural Development, Jelgava, Latvia, 25–27 May 2016; Volume 2016, pp. 904–910.

28. Khanyile, A.; Caws, G.C.; Nkomo, S.L.; Mkhize, N.M. Characterisation study of various disposable diaper brands. *Sustainability* **2020**, *12*, 437. [CrossRef]

29. Khoo, S.C.; Phang, X.Y.; Ng, C.M.; Lim, K.L.; Lam, S.S.; Ma, N.L. Recent technologies for treatment and recycling of used disposable baby diapers. *Process Saf. Environ. Prot.* **2019**, *123*, 116–129. [CrossRef]

30. Roddy, D.J.; Manson-Whitton, C. Biomass Gasification and Pyrolysis. In *Comprehensive Renewable Energy-Biomass & Biofuels 2012*; Elsevier: Amsterdam, The Netherlands, 2012; Volume 5, ISBN 9780080878737.

31. Bassano, C.; Deiana, P.; Ricci, G.; Veca, E.; Rse, R. Analisi Termogravimetrica su Campioni di Carbone. 2009. Available online: https://docplayer.it/2987461-Analisi-termogravimetrica-su-campioni-di-carbone.html (accessed on 23 March 2021).

32. BS EN 15414-3:2011Solid Recovered Fuels. Determination of Moisture Content Using the Oven Dry Method. Moisture in General Analysis Sample. Available online: https://www.en-standard.eu/din-en-15414-3-solid-recovered-fuels-determination-of-moisture-content-using-the-oven-dry-method-part-3-moisture-in-general-analysis-sample/ (accessed on 23 March 2021).

33. BS 1016-104.4-1998: Methods for Analysis and Testing of Coal and Coke. Proximate Analysis. Determination of ash. Available online: https://shop.bsigroup.com/ProductDetail/?pid=000000000030013765 (accessed on 23 March 2021).

34. Ambiente Italia s.r.l. Analisi Ambientale della Raccolta e del Riciclo di Prodotti Sanitari Assorbenti Risultati Sintetici. Available online: https://www.senato.it/application/xmanager/projects/leg17/attachments/documento_evento_procedura_commissione/files/000/000/636/Documentazione_FATER.pdf (accessed on 23 March 2021).

35. Okamoto, H. Supplemental Literature Review of Binary Phase Diagrams: Ag-Nd, Ag-Zr, Al-Nb, B-Re, B-Si, In-Pt, Ir-Y, Na-Si, Na-Zn, Nb-P, Nd-Pt, and Th-Zr. *J. Phase Equilibria Diffus.* **2014**, *35*, 636–648. [CrossRef]

36. Di Giuliano, A.; Lucantonio, S.; Gallucci, K. Devolatilization of Residual Biomasses for Chemical Looping Gasification in Fluidized Beds Made Up of Oxygen-Carriers. *Energies* **2021**, *14*, 311. [CrossRef]

37. Scala, F. Particle agglomeration during fluidized bed combustion: Mechanisms, early detection and possible countermeasures. *Fuel Process. Technol.* **2018**, *171*, 31–38. [CrossRef]

38. Gibilaro, L.G. *Fluidization-Dynamic*; Butterworth-Heinemann: Oxford, UK, 2001; ISBN 978-0-7506-5003-8. Available online: https://books.google.co.jp/books?hl=zh-CN&lr=&id=RlZutdBFy4sC&oi=fnd&pg=PP1&dq=Gibilaro,+L.G.+Fluidization-Dynamic%3B+Butterworth-Heinemann:+2001%3B+ISBN+978-0-7506-5003-8,+doi:10.1016/B978-0-7506-5003-8.X5000-9.&ots=Jn_kGOQKRB&sig=IisegBeCakEykL43CA6MirBVotk&redir_esc=y#v=onepage&q&f=false (accessed on 23 March 2021). [CrossRef]

39. Basu, P. *Combustion and Gasification in Fluidized Beds*; CRC Press/Taylor & Francis: Abingdon, UK, 2006; ISBN 9780849333965.

40. Green, D.W.; Robert, H.P. *Perry's Chemical Engineers' Handbook*, 8th ed.; McGraw-Hill: New York, NY, USA, 2017; ISBN 978007142294.

41. Namkung, H.; Xu, L.H.; Kim, C.H.; Yuan, X.; Kang, T.J.; Kim, H.T. Effect of mineral components on sintering of ash particles at low temperature fouling conditions. *Fuel Process. Technol.* **2016**, *141*, 82–92. [CrossRef]

Article

Design of a 1 MW$_{th}$ Pilot Plant for Chemical Looping Gasification of Biogenic Residues

Falko Marx *, Paul Dieringer, Jochen Ströhle and Bernd Epple

Institute for Energy Systems & Technology, Technische Universität Darmstadt, Otto-Berndt-Str. 2, 64287 Darmstadt, Germany; paul.dieringer@est.tu-darmstadt.de (P.D.); jochen.stroehle@est.tu-darmstadt.de (J.S.); bernd.epple@est.tu-darmstadt.de (B.E.)
* Correspondence: falko.marx@est.tu-darmstadt.de; Tel.: +49-6151-16-23002

Abstract: Chemical looping gasification (CLG) is a promising process for the thermochemical solid to liquid conversion route using lattice oxygen, provided by a solid oxygen carrier material, to produce a nitrogen free synthesis gas. Recent advances in lab-scale experiments show that CLG with biomass has the possibility to produce a carbon neutral synthesis gas. However, all experiments have been conducted in externally heated units, not enabling autothermal operation. In this study, the modification of an existing pilot plant for demonstrating autothermal operation of CLG is described. Energy and mass balances are calculated using a validated chemical looping combustion process model extended for biomass gasification. Based on six operational cases, adaptations of the pilot plant are designed and changes discussed. A reactor configuration using two circulating fluidized bed reactors with internal solid circulation in the air reactor is proposed and a suitable operating strategy devised. The resulting experimental unit enables a reasonable range of operational parameters within restrictions imposed from autothermal operation.

Keywords: chemical looping; biomass; gasification; fluidized bed; autothermal; pilot plant

Citation: Marx, F.; Dieringer, P.; Ströhle, J.; Epple, B. Design of a 1 MW$_{th}$ Pilot Plant for Chemical Looping Gasification of Biogenic Residues. *Energies* **2021**, *14*, 2581. https://doi.org/10.3390/en14092581

Academic Editor: Andrea Di Carlo

Received: 31 March 2021
Accepted: 28 April 2021
Published: 30 April 2021

Publisher's Note: MDPI stays neutral with regard to jurisdictional claims in published maps and institutional affiliations.

1. Introduction

The reduction of greenhouse gas emissions is one of the major challenges in the 21th century. The European Commission sets a minimum share of 14% as a goal for renewable transport fuels produced from non food or feed sources in 2030 [1] in order to combat global warming according to the UNFCCC Paris Agreement. This is a major increase from the less than 0.1% share of renewable transport fuels in 2018 in the European Union (including food grade sources) [2] and necessitates the development of second generation biofuels. Moreover, first generation biofuels mostly utilize biochemical conversion from sugar and starch or physicochemical conversion from plant oil or fat for the production of drop in fuels [3]. However, these processes cannot be used efficiently for the production of second generation biofuels from EU approved biogenic sources—as they are low in sugar, starch, oil and fat and high in cellulosis and lignin—so new production processes are needed.

However, efficient technological pathways for the production of second generation exist only partially and not in an entire process chain, in the form of thermochemical conversion through gasification, methanol or Fischer–Tropsch synthesis and subsequent refining. Gasification, the starting point of the process chain for solid to liquid conversion, is presently used for the generation of heat and electricity [4] and very little for the production of liquid biofuels [5]. It is a well known process which converts solid feedstock in to a high caloric syngas and is considered to have a high potential for the decarbonization of hard to electrify aviation and maritime transport sectors. Additionally, the energy required for the conversion is provided by the biomass feedstock giving the potential of a total carbon neutral drop-in fuel.

As the feedstocks considered by the European Union [1] include seasonally varying types of biomass like husk and straw, as well as more continually sourceable foresting residue, sewage sludge, and biogenic household waste, fluidized bed gasification with its good feedstock flexibility seems a suitable process. Moreover, the good heat and mass transfer characteristics of fluidized bed facilitate complete conversion of the feedstock into syngas, thus achieving a high carbon conversion [6,7] and process efficiency [7,8]. Furthermore, as fuel synthesis requires an N_2-free syngas and thus gasification without the presence of N_2 [9], the subsequent syngas cleaning gives rise to easy carbon capture with storage or utilization making the carbon footprint of the product negative. The N_2-free gasification environment is usually created by the provision of pure oxygen provided by an air separation unit (ASU) [9,10], but in fluidized bed gasification another possibility exists to create an N_2-free atmosphere: dual fluidized bed gasification (DFBG) utilizes two reactors to split the gasification process from the oxidation or combustion process used to generate the necessary heat while avoiding the expensive ASU. Nonetheless, as heated solid bed material circulating between the two reactors is used to transfer the energy for the process, the transport of some amounts of carbon from the feedstock to the gasification reactor is necessary for the combustion reactor to generate the required heat, giving a substantial amount of CO_2-emission from the process.

The chemical looping gasification (CLG) process operates in a similar manner using two coupled fluidized bed reactors. However, instead of transporting residual feedstock from the gasification reactor to one operated with air, it employs a metal oxide to transport oxygen from a reactor operated with air towards the gasification, thus giving the benefit of a process with virtually no CO_2 emission. So far all experiments with continuous operation of the process were conducted in lab and pilot scale with external heating [11–15] and a maximum thermal load of 25 kW [16,17]. Furthermore, autothermal operation has not been demonstrated and problems of process scale up have not been identified and alleviated. Therefore the existing 1 MW chemical looping combustion (CLC) pilot plant located at Technische Universität Darmstadt is modified for the operation and investigation of the CLG process with biomass.

In this work, the design and modifications of the 1 MW pilot plant are described. Starting from the underlying, fundamental gasification process, the existing infrastructural restrictions, and the planed operation range, mass and energy balances are calculated and required adjustments identified and implemented.

2. Theory

2.1. Gasification Fundamentals

Fluidized bed CLG of solid feedstocks comprises, after initial drying and devolatilization, the following main reactions:

$$C + CO_2 \longleftrightarrow 2\,CO \qquad \Delta H = 172.4\,\text{kJ mol}^{-1} \qquad \text{Bodouard reaction} \qquad (R1)$$

$$C + H_2O \longleftrightarrow CO + H_2 \qquad \Delta H = 131.3\,\text{kJ mol}^{-1} \qquad \text{char reforming} \qquad (R2)$$

$$CO + H_2O \longleftrightarrow CO_2 + H_2 \qquad \Delta H = -41.1\,\text{kJ mol}^{-1} \qquad \text{water gas shift} \qquad (R3)$$

$$C + 2\,H_2 \longleftrightarrow CH_4 \qquad \Delta H = -74.8\,\text{kJ mol}^{-1} \qquad \text{methanation} \qquad (R4)$$

Further important reactions between the commonly used gasification agent H_2O [9] and the formed methane is the steam methane reforming reaction:

$$CH_4 + H_2O \longleftrightarrow CO + 3\,H_2 \qquad \Delta H = 206.1\,\text{kJ mol}^{-1} \qquad \text{steam methane reforming} \qquad (R5)$$

$$CH_4 + 2\,H_2O \longleftrightarrow CO_2 + 4\,H_2 \qquad \Delta H = 165\,\text{kJ mol}^{-1} \qquad \text{steam methane reforming} \qquad (R6)$$

where reaction (R6) is the combination of reactions (R3) and (R5).

The influence of reactions (R5) and (R6) largely depends on the formed methane from devolatilization and reaction (R4). These reactions require a high amount of heat, as indicated by the reaction enthalpies, thus greatly contributing to the overall endothermic reaction inside the fuel reactor (FR). Moreover, it is clear that higher gasification temperatures lead to lower amounts of CH_4.

Reactions (R1) and (R2) necessitate a high amount of heat which cannot be balanced by the exothermic reactions (R3) and (R4) and has to be supplied for the gasification process. This heat can either be provided in situ through the oxidation of part of the feedstock (syngas species, volatiles and char) or externally e.g., through supply of a bed material heated in a second reactor enabling an autothermal process. The CLG process, schematically shown in Figure 1, employs both routes to supply the gasification energy. The solid oxygen carrier (OC) material supplies sensible heat to the fuel reactor (FR) while also providing lattice oxygen for the oxidation of part of the feedstock.

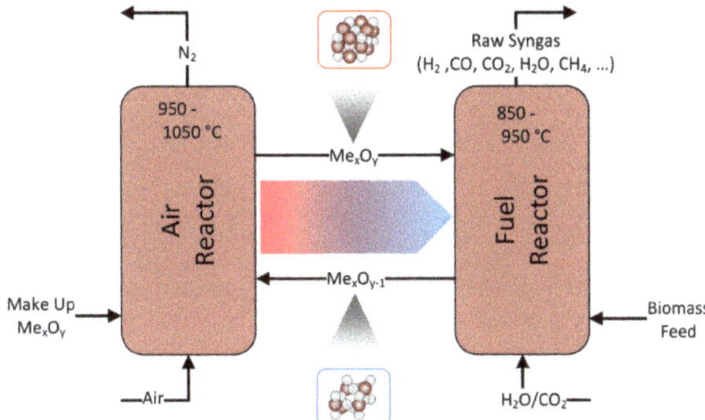

Figure 1. Schematic of the CLG process showing the cyclic reduction and oxidation of an OC material which is oxidized in the air reactor (AR) and reduced in the fuel reactor (FR).

However, additional reactions have to be considered when the bed material is a chemically active part of the feedstock conversion. In the FR where the OC material is reduced the reactions are:

$$Me_x\,O_y + CH_4 \longrightarrow Me_xO_{y-1} + 2\,H_2 + CO \tag{R7}$$

$$4\,Me_x\,O_y + CH_4 \longrightarrow 4\,Me_xO_{y-1} + 2\,H_2O + CO_2 \tag{R8}$$

$$Me_x\,O_y + CO \longrightarrow Me_xO_{y-1} + CO_2 \tag{R9}$$

$$Me_x\,O_y + H_2 \longrightarrow Me_xO_{y-1} + H_2O \tag{R10}$$

$$m\,Me_x\,O_y + C_m\,H_n \longrightarrow m\,Me_x\,O_{y-1} + mCO + n/_2\,H_2O \tag{R11}$$

$$2\,Me_x\,O_y + C \longrightarrow 2\,Me_x\,O_{y-1} + CO_2 \tag{R12}$$

Solid–solid reactions between char and OC (R12) are generally slower than the heterogeneous gas solid reactions (R7) to (R10) and can therefore be neglected [18,19] except for very high reaction temperatures [20]. The reduced OC is then transported to the air reactor (AR), where it is oxidized with air in an exothermic reaction:

$$Me_x\,O_{y-1} + 0.5\,O_2 \longrightarrow Me_xO_y \tag{R13}$$

$$C + O_2 \longrightarrow CO_2 \tag{R14}$$

Inside the AR the combustion of char, reaction (R14) is favored above the oxidation of OC through reaction (R13) [20–22], so residual char transported with the OC from the FR to the AR will be combusted before the oxidation of the OC, adding to the full feedstock conversion and supplying additional heat to the exothermic re-oxidation, (R13). However, (R14) is undesired during CLG, as it reduces the major advantage of a virtually CO_2-free flue gas stream from the AR when compared to DFBG.

CLG has been demonstrated to work as a continuous process in externally heated bench and lab-scale units up to 25 kW [17]. Large-scale experiments at Chalmers

University [23] suffer from the necessity of the AR to supply hot water for building heating and thus the requirement of significant fuel feeding to the AR. Therefore, exhibiting a severe mismatch of reactor dimension of factor 3 to 6 [24] while not depending on (R13) for heat release inside the AR. Hence, these experiments cannot be considered autothermal or even CLG, creating a need for experiments in a bigger scale to confirm the possibility and investigate the performance of autothermal CLG.

Process parameters considered important during the design are the cold gas efficiency:

$$\eta_{CG} = \frac{\dot{n}_{FR,out}(X_{CH_4} \cdot LHV_{CH_4} + X_{CO} \cdot LHV_{CO} + X_{H_2} \cdot LHV_{H_2})}{\dot{m}_{FS} \cdot LHV_{FS}} \tag{1}$$

with X_i being the mole fraction of species i, LHV the lower heating value, and $\dot{n}_{FR,out}$ and $\dot{m}_{FS}$ being the product gas output and the feedstock input, respectively. The oxygen carrier to fuel equivalence ratio is defined by [25]:

$$\phi = \frac{R_{OC} \cdot \dot{m}_{OC} \cdot X_{s,AR}}{\dot{m}_{O,stoich}} \tag{2}$$

$$X_{s,AR} = \frac{m_{OC,AR} - m_{OC,red}}{R_{OC} \cdot m_{OC,ox}} \tag{3}$$

In this definition the oxygen required for full feedstock conversion is $m_{O,stoich}$, R_{OC} is the oxygen transport capacity of the OC material, $X_{s,AR}$ is the oxidation degree of the OC, $m_{OC,red}$ and $m_{OC,ox}$ are the mass of the fully reduced and oxidized state respectively, while the mass of the OC leaving the AR is $m_{OC,AR}$. For gasification ϕ has to be smaller than unity to prevent the full oxidation of the feedstock [25,26]. However, syngas formation is observed even for values of $\phi > 1$ [27]. Values of $\phi < 1$ can be achieved by reducing the mass flow $\dot{m}_{OC}$ or the OC oxidation, i.e., $m_{OC,AR} - m_{OC,red}$. The first option has the disadvantage of also influencing the heat transport $\dot{Q}$ between the reactors:

$$\dot{Q} = \dot{m}_{OC} \cdot c_p \cdot \Delta T \tag{4}$$

with c_p, the heat capacity of the OC and ΔT, the temperature difference of OC particles entering and leaving the FR. The influence of the OC oxidation on c_p is small and can be compensated by adjustments of $\dot{m}_{OC}$ during practical application of option two.

Additionally the fraction of syngas in the dry product gas is defined as:

$$x_{SG} = \frac{X_{CO} + X_{H_2}}{X_{CH_4} + X_{CO} + X_{H_2} + X_{CO_2} + X_{H_2S} + X_{N_2}} \tag{5}$$

2.2. Bed Materials for Chemical Looping Gasification

The selection and testing of bed materials is a crucial task when designing a CLG process. Eight criteria for CLC are given by Adanez et al. [28] and repeated here with notes on how they apply to CLG:

1. Oxygen transport capacity: as gasification processes limit the supply of oxygen below the stoichiometric ratio required for full feedstock conversion, a high oxygen transport capacity is not so important as the process is limited by the sensible heat transported and not the oxygen [16,26]. For CLC the supply of excess oxygen is not critical, for CLG it must be limited without impairing the transport of sensible heat as otherwise, the temperature in the FR would drop, negatively influencing the gasification [26].
2. Thermodynamic suitability: the bed material must be able to oxidise the feedstock at least partially while not releasing molecular oxygen. Thus chemical looping with oxygen uncoupling (CLOU) materials cannot be used for CLG.
3. High reactivity over multiple reduction-oxidation cycles: activation over multiple cycles can increase or decrease reactivity.

4. Stability: the expected lifetime of the bed material should be as long as possible, as losses through attrition need to be compensated by a make-up stream. This make-up stream requires heating to process temperature, thus always leading to an efficiency drop. Measurement and calculation of OC lifetime is not straightforward and can vary by a factor of 3.2 for one experiment depending on the method used [29].

5. Carbon deposition: carbon transport towards the AR with subsequent combustion negatively impacts carbon utilization and capture efficiency. However, Adanez et al. [28] note that no carbon deposition has been found in relevant studies.

6. Fluidization properties: formation of agglomerates or low melting compounds with parts of the feedstock must be avoided. This becomes difficult if a herbaceous feedstock—high in ash and alkali metals—is used and might require mitigation measures like pre-treatment [30,31] or feedstock mixing [32].

7. Cost: the current production cost for synthetic materials make them non competitive when compared to naturally occurring minerals or waste materials.

8. Toxicity: deployment of environmental friendly and non-toxic OC material avoids special and costly requirements during handling and disposal of deactivated OC material.

Moreover, the design for pilot and demonstration plants need to consider an additional point:

9. Availability: the selected material must be available in the required quantity. Synthesized OC materials are not available on a commercial scale yet. So a natural ore or a waste material must be used.

Especially OC materials which are categorized as materials for syngas production [33–35] are problematic as they are either synthetic materials not available in the required quantities, expensive or toxic to humans and the environment. However, even materials with full oxidization capability for combustion can be used for the production of high calorific syngas when suitable control concepts are employed [11,16,26]. While lots of operating experience with bed materials for DFBG in the range above 1 MW exists [36], there is little experience with OC materials in the same power range [23,25]. However, even those experiments do not give a good indication of their process performance, as the AR—or rather combustor, as it is always fed with fuel—used is oversized by a factor of 3 to 6 [24], effectively creating a reservoir of OC and sensible heat more dependent on the required energy for heating supply than the CLG process. Moreover, higher attrition rates of e.g., ilmenite are reported for CLG when compared with CLC [11] but if the effects are the same in a bigger CLG plant is still an open question. Due to the small size of lab-scale reactors, the OC material undergoes more oxidation/reduction cycles per hour, thus giving higher stress from chemical conversion when compared to the mechanical stress from the transport through the reactors and coupling elements.

Depending on the requirements of the targeted application for the syngas, a last point is to be considered when selecting the OC bed material:

10. Catalytic properties: selecting a material (or additive) which catalytically reduces the formation of unwanted components like tars [37,38] and CH_4 [39] or binds elements to the solid fraction (e.g., sulphur in form of gypsum) as a primary method. Secondary gas cleaning methods might therefore not be necessary or can be designed much smaller.

Tar production is of major concern for subsequent syngas treatment especially for biomass gasification where tar production is high [40]. Bed height, bed material, temperatures, velocities, feedstock, and feedstock feeding location [37] have an influence on the production of tars. Existing kinetic models for the prediction of tar production are not applicable as they are developed for a very specific process and reactor size [15], need fitting against the actual reactor performance [41], or are not reliable in the prediction of tars formed [42–44]. Furthermore, no model was developed for CLG yet.

3. Process Design

In the following, the CLG technology fundamentals described in Section 2 are combined with boundary conditions from the existing pilot plant as well as feedstock properties and hydrodynamic characteristics yielding a process design suitable for the demonstration of autothermal CLG in the existing 1 MW$_{th}$ pilot plant.

3.1. Existing Pilot Plant

The heart of the CLG pilot plant consists of two refractory lined circulating fluidized bed (CFB) reactors which are coupled using two loop seals and one J-valve and have properties indicated in Table 1. The CFB400-reactor of the pilot plant has been used as gasifier for High Temperature Winkler (HTW^{TM}) gasification [45,46] and as FR in the CLG-related processes for chemical looping combustion, while the CFB600-reactor has been used as AR [47–51]. Thus major components can be reused for CLG by combining elements from the CLC and the HTW^{TM} process configurations. Nonetheless, major adaptations are made, as the HTW^{TM} configuration is build for lower fluidization velocities and with 0.5 MW$_{th}$ [45] also for lower thermal input.

Table 1. Reactor properties of the 1 MW CLG pilot plant.

Reactor	AR—CFB600	FR—CFB400	Unit
Height	8.66	11.35	m
Inner diameter	0.59	0.28 to 0.4	m
Outer diameter	1.3	1.0	m
Temperature	1050	950	°C
Fuel feeding	in bed (propane lance), return leg of LS 4.5 (solids)	in bed via screw (solids)	

Furthermore, as electrical preheating temperatures of fluidization media are limited to 400 °C, process stream heating has to be done inside the reactors, negatively impacting cold gas efficiency which would be optimized in an industrial plant using heat integration. The cooling system sets a limit of 1 MW$_{th}$ which can be safely handled for CLC. However, as a major part of the energy of the feedstock remains as heating value in the product gas, feedstock input above the 1 MW$_{th}$ is possible for CLG.

Therefore the following case has been set as design specifications for the investigations of CLG for which mass and energy balances were calculated, required changes to the pilot plant identified and modifications designed.

- As the cooling system is designed to handle a thermal load of 1 MW safely, the design power of the pilot plant is is set to 1 MW$_{th}$.
- Ilmenite as OC: For the selected thermal power, a total inventory of about 1000 kg was used during CLC experiments in the pilot plant [50], and the same can be expected for CLG. Thus, of the points listed in Section 2.2, the availability is a major concern for experiments in that scale, and a natural ore or a widely available waste material had to be selected. Recent studies show promising results for ilmenite in continuous units [11], and operating experience with ilmenite in the pilot plant exists [49,50]. Moreover, ilmenite has been shown to catalytically reduce tars [25,52].
- Temperatures for the AR of 1050 °C and 950 °C are considered the maximum viable temperatures. Higher FR temperatures will yield a higher H$_2$/CO ratio at the expense of lower cold gas efficiency. So slightly lower FR temperatures might be desired in industrial application. Moreover, as OC ash interaction may lead to problems at high temperatures [31] and the temperature difference between the reactors is an important parameter for process control [26], the FR temperature is not fixed and considered an important variable in the planned experiments.
- Industrial wood pellets as feedstock as described in Section 3.2.

3.2. Feedstocks

As model feedstock for the calculation of the heat and mass balances and the design of modifications, industrial wood pellets have been selected, as they are widely available and allow for easy comparison with existing gasification technologies in pilot and demonstration scale where wood based materials are gasified [36]. Additionally, wheat straw, as a seasonal varying biomass source, and pine forest residue, as a more constant source, are selected as feedstocks from the EU-approved list [1] for experimental investigations.

Initial investigations of wheat straw by Di Giuliano et al. [31] indicate that it is a difficult feedstock for CLG, due to its low ash softening point and the possibility to cause agglomerates and bed defluidization, so that it requires at least some pre-treatment. However, as fluidization velocities in the CFB reactors are two magnitudes higher than the investigated fluidization velocities, the required pre-treatment cannot be directly inferred, but a higher fluidization velocity seems to lower the required pre-treatment effort [31]. Moreover, reaction kinetics for pelletized wheat straw in various bed materials are similar to pellets of pine forest residue [53] opening up possibilities to switch between these feedstocks during gasifier operation. Nevertheless, additional investigations on the pre-treatment of wheat straw are needed to be able to give accurate information on the fuel properties—which are indicated in Table 2 for the planed feedstocks—as they vary with pre-treatment. It is assumed that pre-treatment of wheat straw will make handling and gasification easier, as it reduces agglomeration tendencies (additivation, torrefaction) and water content (drying, torrefaction). Thus, raw wheat straw is the most difficult to gasify and can be used as a lower end in feedstock quality.

Table 2. Proximate and ultimate analysis of feedstocks.

	Component	Wood Pellets	Pine Forest Residue	Wheat Straw
Proximate Analysis in wt. − %	Moisture	6.5	7	7
	Ash (d.b.)	0.7	1.86	7.5
	Volatiles (d.b.)	85.1	78.86	81.5
	Fixed carbon (d.b.)	14.2	12.28	11
Ultimate Analysis in wt. − %	C (d.a.f.)	50.8	52.7	48.2
	H (d.a.f.)	6	6.4	6.5
	N (d.a.f.)	0.07	0.39	0.43
	O (d.a.f.)	43.2	40.5	44.9
	S (d.a.f.)	0.008	0.05	0.11
	Cl (d.a.f.)	0.006	0.007	0.05
Net calorific value in MJ kg^{-1}		17.96	18.41	17.12

3.3. Heat and Mass Balances

Heat and mass balances for the pilot plant were calculated considering reaction kinetics of ilmenite and reactor hydrodynamics using a validated Aspen Plus^{TM} model for CLC [51] extended to cover biomass gasification via a Langmuir–Hinschelwood mechanism [26]. However, instead of an equilibrium model used by Dieringer et al. [26], the more realistic original reaction kinetics for ilmenite were used for the OC gas reaction. As a starting point, the CLC case was selected in terms of reactor dimensions, temperatures, solid inventories, pressure, and loop seal (LS) fluidization. The feedstock flow $\dot{m}_{FS}$ (industrial wood pellets) was selected as $1\,\text{MW}_{\text{th}}$, and the heat losses were assumed to be $110\,\text{kW}$ which falls in the reported range of $60\,\text{kW}$ to $200\,\text{kW}$ [48,50]. Furthermore, heat losses are considered to be dependent on reactor temperature and independent of feedstock input. LS fluidization with CO_2 is set to $84.2\,\text{kg h}^{-1}$ based on previous operating experience [49,50]. To obtain autothermal operation at these conditions, the oxygen availability inside the AR was varied through the inlet feed rate of air into the AR, while the heat transport between both reactors was controlled through the global metal oxide solid circulation rate ($\dot{m}_{OC,AR,out}$, $\dot{m}_{OC,FR,out}$), until both reactors were in heat balance. The hydrodynamic constraints related to the required solid entrainment (calculated as suggested by Kunii and Levenspiel [54], described in detail elsewhere [51,55]) from each reactor were achieved by varying the steam inlet flow $\dot{m}_{H_2O}$ and flue gas recirculation inlet flow $\dot{m}_{AR,reci}$ for the FR and AR, respectively, while setting internal solid circulation to zero. All boundary conditions are listed in Table 3, and the corresponding results are in Table 4. The listed streams are visualized in the reactor configuration in Figure 2.

Table 3. Boundary conditions for the simulation of autothermal CLG operations of the 1 MW pilot plant.

Property	Value	Unit	Property	Value	Unit
$d_{p,50}$	154	µm	$T_{Gas,in}$	400	°C
Δp_{FR}	61	mbar	Δp_{AR}	90	mbar
p_{FR}	1	bar	p_{AR}	1	bar
d_{FR}	0.28 to 0.4	m	d_{AR}	0.59	m
h_{FR}	11.35	m	h_{AR}	8.66	m

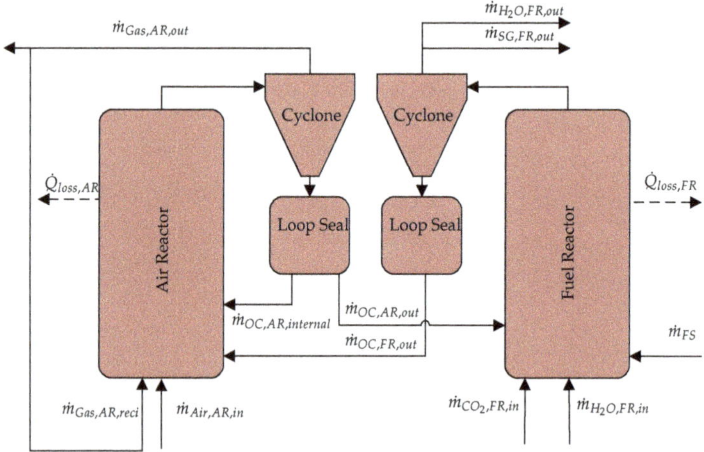

Figure 2. Streams for the calculation of mass and energy balances of the CLG process.

In small scale units where the energy is supplied via furnace heating, the oxygen supply can be controlled via the circulation. However, in the 1 MW pilot plant the heat is supplied only via the circulation of the bed material. From the results of the Reference case, it can be seen that the transport of oxygen must be limited in order to obtain a good gasification process, while the solid circulation must remain high as indicated by the substantial amount of recirculated gas fed to the reactor. Thus, a new control method for the oxygen transport must be realised, decoupling the transport of oxygen from the transport of sensible heat as described by Dieringer et al. [26]. Moreover, the superficial gas velocity u_0 in the AR is below the range of a CFB, as shown in Figure 3, while the calculated solid flux G_S is also below the range commonly observed in commercial CFB units [56]. Indeed, past operation of the AR showed good performance with superficial gas velocities of approximately $3.5\,\mathrm{m\,s^{-1}}$ to $5\,\mathrm{m\,s^{-1}}$. In the pilot plant, the installation of a (partial) flue gas recirculation for the AR is used to increase u_0 while also supplying the inert fluidization medium required for the process control. Increasing the solids discharge from the AR—while keeping the global solids circulation constant—creates the need for an internal solid circulation in the AR—where material not transported through the J-valve is returned via the LS—which is not common in smaller units. In fact, most lab- and bench-scale units have internal solids recirculation for the FR to enhance carbon conversion [57] or no solids recirculation at all [11,17]. Nonetheless, this solution comes with a penalty, as additional fluidization medium is needed and has to be heated to process temperature. However, reducing the diameter of an existing, refractory lined reactor is costly and time consuming, so the efficiency penalty has to be accepted; yet it also opens the possibility to use the start-up burner for fast temperature adjustments in-between experimental set points without severe impact on reactor hydrodynamics. In a commercial unit, the diameter would be designed according to process specification and corresponding hydrodynamics, requiring the flue gas recirculation only for process control. Nonetheless, this initial estimation shows that CLG is possible in the existing 1 MW pilot plant.

Table 4. Simulation results for autothermal CLG operations of the 1 MW pilot plant. Stream names correspod to Figure 2. Boundary conditions deviating from the reference case are underlined. The first block contains the thermodynamic and hydrodynamic constraints and results, the second block contains the process streams (in some cases with composition). The third block gives information on solid composition for the FR, while the last block shows general process performance parameters.

Stream	Reference	HT1	HT2	HF	HP1	HP2	Unit
T_{AR}	1025	1050	1050	1025	1025	1025	°C
T_{FR}	900	900	950	900	900	900	°C
$\dot{m}_{FS}$	200.4	200.4	200.4	200.4	240.48	280.56	kg h^{-1}
$u_{0,AR}$	3.42	3.12	3.97	5.03	5.01	5.01	m s^{-1}
$u_{0,FR}$	6.25	5.46	7.64	5.67	6.23	6.75	m s^{-1}
$\dot{Q}_{loss,AR}$	48.5	49.4	49.9	49.3	50.5	49.6	kW
$\dot{Q}_{loss,FR}$	59.8	61.8	59.8	61.3	59.4	61.2	kW
$\dot{m}_{OC,AR,out}$	7180	5690	9979	6244	7257	8285	kg h^{-1}
$\dot{m}_{OC,FR,out}$	7130	5649	9906	6175	7199	8236	kg h^{-1}
$\dot{m}_{OC,AR,internal}$	0	0	0	12,120	10,994	9971	kg h^{-1}
$\dot{m}_{Air,AR,in}$	640	600	745	730	760	802	kg h^{-1}
$\dot{m}_{Gas,AR,out}$	950.6	854.6	1062.7	1404.3	1413.6	1425.5	kg h^{-1}
— $X_{CO_2,AR}$	0.111	0.117	0.096	0.098	0.111	0.122	
— $X_{O_2,AR}$	0.004	0.003	0.005	0.005	0.004	0.003	
$\dot{m}_{AR,reci}$	360.7	296.3	391.2	744.3	708.2	672.8	kg h^{-1}
$\dot{m}_{H_2O,FR,in}$	301.53	237.04	383.83	247.14	263.4	276.94	kg h^{-1}
$\dot{m}_{CO_2,FR,in}$	84.2	84.2	84.2	84.2	84.2	84.2	kg h^{-1}
$\dot{m}_{H_2O,FR,out}$	362.9	294.5	451.5	318.8	337.9	354.0	kg h^{-1}
$\dot{m}_{Syngas,FR,out}$	271.1	267.0	287.8	280.5	307.3	334.3	kg h^{-1}
— $X_{CO_2,FR}$	0.466	0.439	0.543	0.531	0.440	0.377	
— $X_{CO,FR}$	0.304	0.317	0.240	0.277	0.324	0.354	
— $X_{CH_4,FR}$	0.092	0.099	0.057	0.072	0.095	0.113	
— $X_{H_2,FR}$	0.139	0.145	0.160	0.119	0.141	0.156	
— $X_{H_2S,FR}$	5.26×10^{-5}	5.22×10^{-5}	5.13×10^{-5}	5.38×10^{-5}	5.48×10^{-5}	5.56×10^{-5}	
$\dot{n}_{Solid,FR,out}$	14.84	11.54	20.60	13.38	15.09	16.90	mol h^{-1}
— $X_{C,out}$	0.04	0.05	0.03	0.05	0.05	0.05	
— $X_{Fe_2O_3,out}$	0.09	0.06	0.10	0.13	0.09	0.07	
— $X_{FeTiO_3,out}$	0.68	0.76	0.66	0.57	0.68	0.74	
— $X_{TiO_2,out}$	0.18	0.12	0.21	0.25	0.18	0.14	
$\dot{n}_{Solid,FR,in}$	15.55	12.11	21.70	14.43	15.93	17.54	mol h^{-1}
— $X_{Fe_2O_3,in}$	0.17	0.16	0.18	0.23	0.19	0.15	
— $X_{FeTiO_3,in}$	0.48	0.53	0.47	0.30	0.44	0.55	
— $X_{TiO_2,in}$	0.35	0.31	0.35	0.46	0.37	0.30	
— $X_{Fe_3O_4,in}$	0.00	0.00	0.00	0.00	0.00	0.00	
η_{CG}	0.474	0.505	0.384	0.396	0.475	0.531	
x_{SG}	0.443	0.462	0.400	0.396	0.465	0.511	
ϕ	0.585	0.412	0.836	0.729	0.537	0.412	

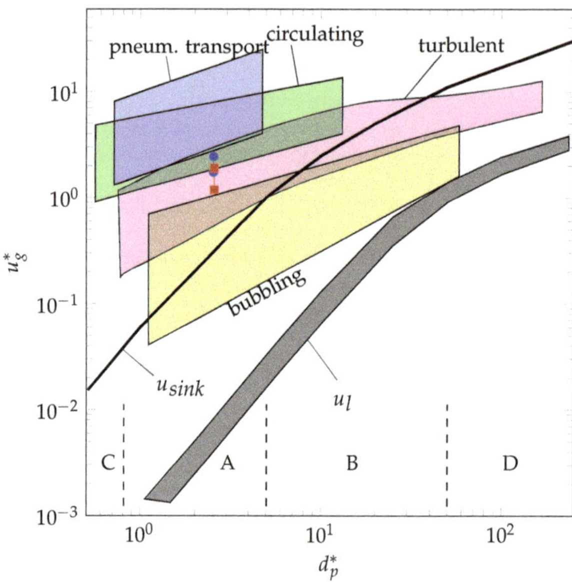

Figure 3. Grace diagram indicating the operation regimes of the FR —●— and AR —■—.

In order to asses the exact limits and to find the corresponding bottle necks where adaptations are needed, some variations on the boundary conditions have been made to be able to decide on equipment alteration and to generate data for the subsequent detailed design. While the simulation of the reference case yields a cold gas efficiency η_{CG} of 0.474, values above 0.8 are reported for externally heated continuous units with a slight increase of η_{CG} with increasing FR temperature [11]. Thus, two additional points with increased AR temperature (HT1) and increase of both reactor temperatures (HT2) were considered to test the feasibility of higher temperatures in the 1 MW pilot plant. The low superficial gas velocity for the AR was raised to $u_{0,AR} = 5\,\mathrm{m\,s^{-1}}$ by increasing the flow of fluidization medium (HF) to see the effect and possibility at higher inlet and outlet streams. This case was also used as a basis for an increase in fuel input to 1.2 MW (HP1) and 1.4 MW (HP2) to reduce the relative impact of heat loss and test the limits of the syngas handling equipment. During experimental operation, the AR superficial velocity would be targeted at slightly above the minimum discharge needed for either stable operation or required by the process—whichever is higher—in order to keep the negative impact of heat demand by fluidization medium low. However, for design purposes, the upper end of the range has to be considered.

From the variation of the reactor temperatures, it is clear that increasing the AR temperature is beneficial to process efficiency, while also increasing the FR, negatively impacts the process performance. For HT1 the increased heating demand in the AR is counteracted by the reduced solids circulation ($\dot{m}_{OC,FR,out}$ and $\dot{m}_{OC,AR,out}$) needed to supply the heat for the gasification process and thus reducing the overall amount of required fluidization medium ($\dot{m}_{Air,AR,in}$, $\dot{m}_{AR,reci}$ and $\dot{m}_{H_2O,FR,in}$) to achieve this lowered solids circulation. The higher FR temperature in HT2 leads to a syngas composition higher in H_2 and lower in CH_4 which is desired, but also requiring significantly higher solids circulation. The corresponding heating requirement of fluidization medium negatively impacts process efficiency. The influence on the syngas quality is caused not only by the raise in gasification temperature, but also in the added steam content from fluidization, influencing reactions (R3), (R5) and (R6). The biggest effect has the increase of the oxygen carrier to fuel equivalence ratio ϕ which raises the relative contribution of oxidation reactions (R8) to (R12).

The increase in AR solids entrainment through higher fluidization velocity (HF) essentially decouples the reactor hydrodynamics of both reactors. Here the model constraint of no internal solids circulation $\dot{m}_{OC,AR,internal}$ is omitted. Instead, the superficial gas velocity $u_{0,AR}$ is targeted at $5\,\mathrm{m\,s}^{-1}$. The higher heat demand for the fluidization medium here has to be supplied by exothermic reaction (R13) leading to a higher degree of OC oxidation as indicated by the increase in ϕ. Consequently, the solids circulation between the reactors is lowered as more oxygen is supplied per OC mass. This leads to lower fluidization requirements and heat demand in the FR reducing the negative impact of the higher fluidization velocity in the AR. Increasing the feedstock input $\dot{m}_{FS}$ while keeping the AR hydrodynamic constant (HP1 and HP2) positively influences process efficiency, as the relative increase in FR fluidization medium required for solids discharge is only about half of the relative increase in feedstock. Thus, only a relatively small part of the additional feedstock is used to cover the energy requirement of the additional fluidization medium, while most of the additional feedstock energy is available for the conversion into syngas making a positive impact on syngas content and cold gas efficiency. This positive influence is mostly caused by more beneficial reactor hydrodynamics and lower relative heat losses of the reactors.

The simulated cases shed light on the process range the reactors can be operated without major modifications, and also highlights the huge impact of heat loss and heat demand in this scale of experiments. It shows that higher FR temperatures in case HT2 require higher fluidization and bigger size of downstream syngas equipment than significant increases in feedstock input (HP2) making this the more critical case to be considered during design. Although the syngas quality increases with higher FR temperature, the cold gas efficiency is drastically reduced, which is in contrast to the observations from Condori et al. [11]. This discrepancy can be explained by the external heating in the lab-scale plant, which can thus compensate the higher heating demands of the process streams. The positive effect of high temperatures for process streams entering the reactors has been shown [26], highlighting the need of good heat recovery and integration for the process. Moreover, the simulated process conditions make clear that individual variations of process parameters like steam to biomass ratio, or oxygen carrier to fuel equivalence ratio ϕ as done by Condori et al. [11] are not possible if no external heating is available. Instead, the CFB mode and the defined solid discharge required for the heat transport also lower the steam to biomass ratio and oxygen carrier to fuel equivalence ratio, as can be seen by the feedstock increase (HP1, HP2). Furthermore, the predicted influence of these combined changes is not necessarily the same as the one observed in small-scale experiments. This can bee seen by the increase of X_{CH_4} with increasing feedstock input, where the accompanying changes in steam to biomass ratio and oxygen carrier to fuel equivalence ratio lead to lower CH_4 in the experiments described by Condori et al. [11].

Confirmation or refutation of either the trends experimentally observed in small scale units or simulated for the existing pilot plant necessitates experiments in the $1\,\mathrm{MW}_{th}$ range where autothermal operation—instead of external electrical heating—becomes necessary. Here, the requirements imposed by autothermal operation of the process limit the range of applicable parameter variation as they are interdependent. Therefore, the existing pilot plant is modified to provide the experimental data needed.

4. Plant Design

The flow sheet in Figure 4 shows a simplified configuration of the designed pilot plant, including major components and important subsystems. Some of the components already available from CLC and HTW^{TM} can be reused, while other subsystems are new or altered. For all subsystems affected by the new CLG process and the alterations a HAZOP analysis has been performed to ensure safe operation.

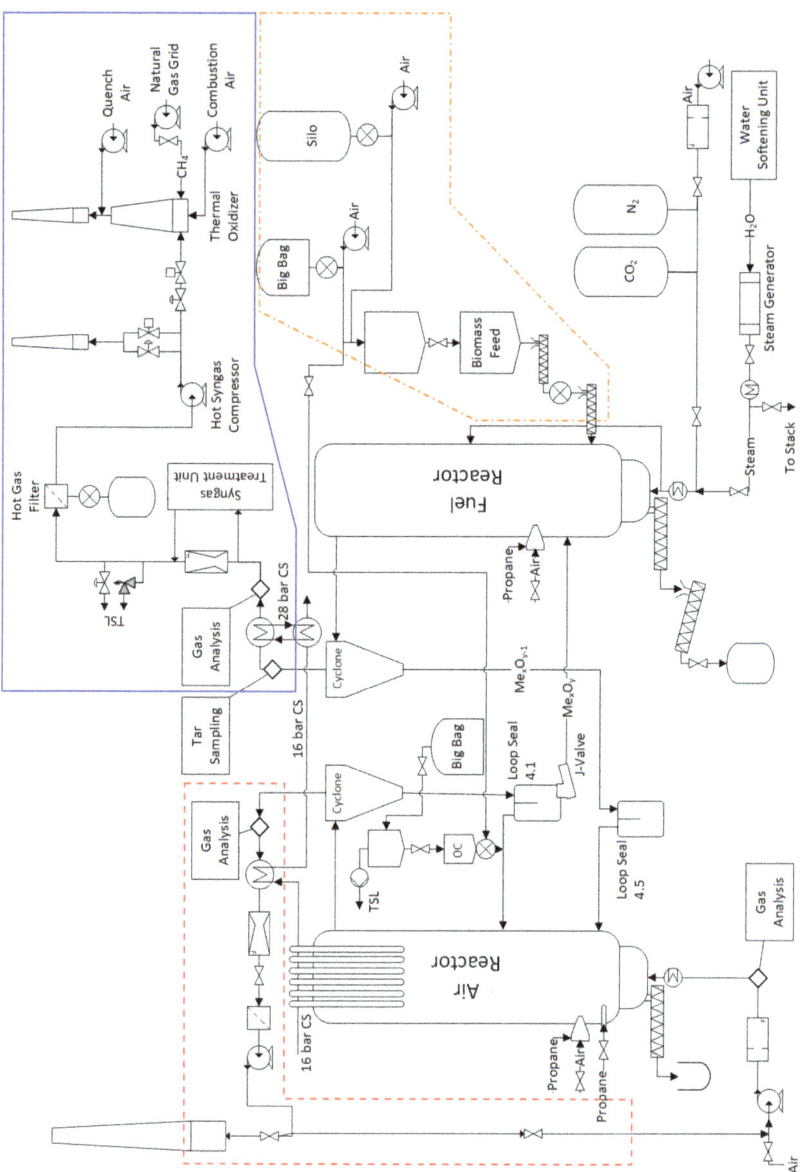

Figure 4. Schematic of the CLG pilot plant showing the main subsystems. CS: cooling system, OC: oxygen carrier, TSL: to safe location. syngas handling: ___, flue gas handling: _ _, biomass feeding: _ . _

4.1. Reactor System

The reactor system (Figure 5) comprises of the two CFB reactors, two LS and a J-Valve as coupling elements. The total inventory of bed material during CLG operation with ilmenite is about 1000 kg with approximately 250 kg in the AR, 80 kg in the FR, and the rest in the coupling elements. Transport of sensible heat to the FR is not facilitated by internal solid circulation and additional fluidization medium would be required, cooling down the reactor and negatively impacting on process efficiency. Thus, no internal circulation is implemented for the FR. Moreover, process simulations show only reduced OC leaving the reactor [26] giving no benefit of returning it from the cyclone to the reactor. However, the Gibbs reactor model employed in this study leads to full conversion, while in reality a mixture of different phases will always be present. Nonetheless, the prevalence of highly reduced phases in both FR and AR has been confirmed in continuous experiments [11].

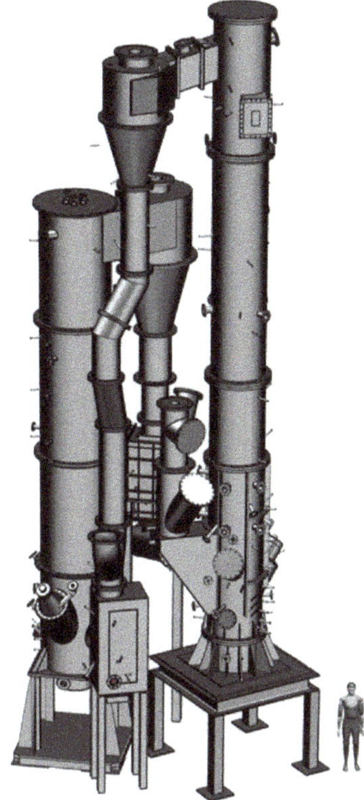

Figure 5. CLG reactor configuration including the main coupling elements.

Although many lab-scale reactor designs feature a FR operating in bubbling mode (e.g., [11,14,17,58]), the used CFB mode of the FR has the advantage of improved gas-solid mixing and thus featuring higher rates of carbon conversion [57], while the requirement ranges for the size and shape of the feedstock is wider [10] opening up possibilities for more feedstocks. Furthermore, the higher solids concentration in the freeboard may enhance tar cracking and methane reforming by supply of additional oxygen and catalytic sites in this region. However, increasing superficial velocities too much will lead to pneumatic transport in the FR (Figure 3) and unstable reactor hydrodynamics.

The disadvantage of having no internal solid recirculation for the FR is the transport of all discharged feedstock particles towards the AR. Furthermore, for the pilot plant, the minimization of heat losses is considered more important than the minimization of carbon slip towards the AR as relative heat losses for the pilot plant are in the range of 0.1 to 0.2. So minimization of coupling elements is used instead of carbon recovery via a carbon stripper. However, carbon slip is assumed to be a minor problem as the biomasses considered for the experiments contain low amounts of fixed carbon [9]. The feeding location is lowered into the dense region of the bed when compared to previous CLC experiments [49] where high carbon slip for hard coal was experienced, which should reduce the carbon slip as char gasification in the densest region is enhanced. Moreover, carbon slip is more pronounced in small reactors and the sometimes utilized carbon strippers might not be required in bigger units [59]. Nonetheless, to maximize residence time of char particles inside the dense region, a variable amount of fluidization medium can be rerouted directly before the wind box of the FR to a second stage fluidization located at approx. one fifth of the reactor height. This increases bed density in the lower region and residence time of OC particles while keeping a high solids discharge in the CFB operation is possible by increasing the reactor inventory. The exact influence must be determined via experimental operation.

Investigations in the FR are the most crucial, as the formation of tars make the process and reactor design more critical to subsequent equipment than the re-oxidation in the AR. Therefore, it is advantageous for experimental operation to handle imbalances of solids discharge between the reactors inside the AR instead of the FR where it would negatively impact temperature and possibly lead to poorer syngas quality. The feedstock input directly in the dense zone of the bed should also reduce the amount of tars formed during initial devolatilization [37].

4.2. Flue Gas Handling

The flue gas composition from the AR is measured by an on-line gas analysis before the flue gas is cooled down in a heat exchanger to approx. 230 °C (Figure 4, red box). The flow rate is measured using a venturi before the fines passing the cyclone are separated by a filter giving a dust-free flue gas. The following induced draft fan is used to control the pressure in the reactor and vents the flue gas via a stack. Part of the flue gas can be recirculated via a controlled butterfly valve to adjust the inlet of the AR fluidization. The variation of flue gas recirculation allows to adjust the superficial gas velocity $u_{0,AR}$ and thereby the entrainment of particles from the AR while keeping the OC to fuel equivalence ratio ϕ constant. This is a small but significant adjustment in converting from a CLC plant to a CLG plant as it allows to control the overall process as described in [26].

4.3. Syngas Handling

Major modifications are needed for the FR off-gases (Figure 4, blue box) when converting a CLC unit into a CLG unit, as all parts need to be designed with the consideration of explosive atmospheres . Moreover, commonly used heat exchangers are either prone to clogging with tars on cold surfaces or the syngas cooling rate is to low, allowing for recombination of syngas species. The process simulation from Section 3.3 show high syngas streams that need to be safely handled and greatly exceed the capacity of the syngas removal deployed for HTW^{TM} gasification [45,46]. The only component reusable is the cooler, a patented tube-in-tube gas liquid heat exchanger from SCHMIDT'SCHE SCHACK consisting of four tubes cooling the gas to approx. 380 °C very fast and without recirculation zones [60] avoiding the recombination of syngas to longer hydro-carbons. The cooling water is pressurised to 28 bar to be able to raise temperature levels to 200 °C in order to avoid excessive condensation of tars inside the tubes of the raw gas cooler.

After the cooler the syngas is available for cleaning. Here part of the syngas can be routed to a syngas treatment unit for cleaning and separation of CO_2, so that it is subsequently available for synthesis. Moreover, test rigs for the fine cleaning of the syngas and the synthesis of higher hydro-carbons are added, creating the unique possibility to investigate the whole solid to liquid value chain.

The return line from the syngas treatment unit consisting of all streams not used for synthesis is merged back, and the gas is routed to a hot gas filter for the removal of solids, resulting in a dust free syngas stream to the hot syngas compressor used to control the pressure in the FR. From here the syngas is transported to a thermal oxidizer for safe venting. The option of a second stack where the FR off gas can be vented is included for start up, shut down and to allow for a restart of the thermal oxidizer in case of failures without the full shut down of the pilot plant. The additional valves before the hot gas filter are installed for safety pressure relief in case the switching between the thermal oxidizer and the second stack fails.

The described syngas line differs substantially from the ones deployed in either industrial scale or lab-scale. While in industrial plants all produced syngas would be cleaned, only the amount of syngas needed for research in gas cleaning is processed in the pilot plant to reduce the cost of the deployed gas cleaning equipment. In lab-scale the small quantities of formed syngas allow for untreated release to a safe location in the environment, which is not possible for streams in the size of the pilot plant, entailing the need for the thermal oxidizer.

All properties of the syngas stream leaving the FR are of major importance for further process development. Thus, sample and measurement sites consisting of an isokinetic dust and tar sampling port, a psychrometric water content measurement, and an on-line gas analysis are integrated into the syngas line. The isokinetic sampling of dust and tars is done before the raw syngas is cooled while ports for the measurement of the water content and gas composition are located before and after the cooler and can be connected as required.

4.4. Solid Feeding

4.4.1. Feedstock

The pilot plant is equipped with various entry points for solid feedstocks (Figure 4, orange box) like a big bag station, a container station (not shown on Figure 4) and a silo capable of introducing pulverized and pelletized feedstocks which are transported pneumatically to a fuel container purged with CO_2. This container discontinuously feeds fuel to a second, weighted container from which the fuel is fed continuously, controlled via screws and a hopper directly in to the bed of the FR. Both containers are pressurised to the bed pressure of the FR at the location of the feed screw to avoid the back flow of syngas into the fuel feeding system. The screw feeder is cooled with thermal oil to ensure that gasification temperatures are only reached in the bed and no gasification occurs inside the screw.

4.4.2. Oxygen Carrier

Initial filling of loop seals with OC is done via a weighted dosing container, a hopper, a screw conveyor, and a series of tubs connected to the stand-pipes. OC materiel is fed into the return leg of LS 4.1 for reactor filling and make-up dosing to compensate losses caused by agglomeration and attrition.

4.5. Cooling and Preheating

The cooling system is designed to handle the full 1 MW of heat released during CLC and therefore has enough capacity for further increase of feedstock as discussed previously. However, for bigger units, where process heat would be used to generate steam and preheat the input streams, changes might be required when compared to CLC to optimize the heat integration. Nonetheless, this is no concern for the pilot plant, where steam generation and preheating is done via independently powered systems. Yet, it limits also the operation range of the pilot plant—seen on simulated case HT2—where higher outlet stream temperatures always lead to a severe process penalty. For the pilot plant, this penalty cannot be alleviated by heat recovery for the preheating of inlet streams. Here the option of higher preheating temperatures would require a substantial increase of heat exchanger surface, for which no space is available at the existing site. Furthermore, the existing electrical infrastructure is already at its limit, so increasing the electrical preheating power is not feasible.

Increasing the fuel input necessitates deeper investigation of the limitation of safe operation in terms of the cooling system, especially when considering that most of the 1000 kg OC material is in a highly reduced state during operation. Here the safety relevant quantity is not the total amount of feedstock input or the reduced OC, but the possible amount of oxygen input to the AR. The oxygen input will first fully oxidise the OC inside the AR, possibly with much higher power than the nominal feedstock input which will set a limit only after full oxidation inside the AR has been reached. Here mitigation measures are an over design of the cooling system and a limitation of oxygen input to safely handable amounts.

4.6. On Line Measurements

4.6.1. Gas Analysis

The main product of the gasification process, the synthesis gas from the FR, is extracted and analyzed continuously as sown in Figure 6 via a heated probe which includes a filter (1), that can be back flushed with CO_2 to prevent blockage.

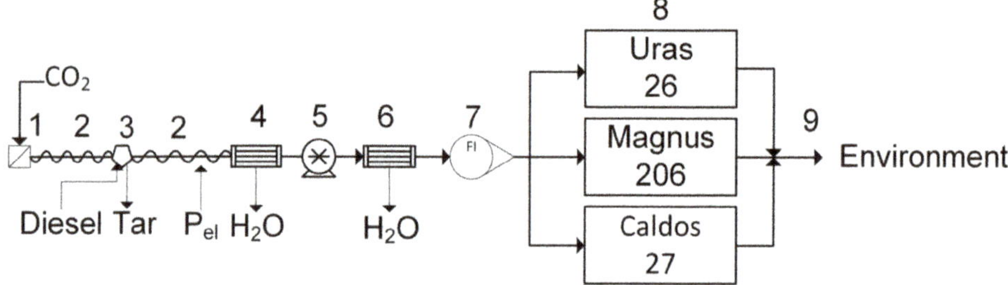

Figure 6. Schematic of the gas analysis equipment: (1) heated probe with filter, (2) heated tube, (3) tar removal (only for FR), (4) condenser for water removal, (5) pump, (6) condeser for water removal, (7) rotameter, (8) measurement equipment, (9) safe location in the environment.

The gas then passes in a heated tube (2)—to prevent the condensation of remaining tars—to a tar removal unit with diesel as solvent (3) and a first condenser unit (4) where the majority of the water and higher hydro-carbons are removed. The measurement gas pump (5) transports the gas through a second condenser unit to remove the rest of the water (6) which is followed by a rotameter (7) measuring the sampling gas flow. The sampling gas is distributed to the commercially available gas analysing equipment from ABB (8) given in Table 5 before being released to a safe location in the environment (9).

O_2 is measured via its paramagnetic quality in an Magnos 206 analyzer while H_2 is determined via thermal conductivity in a Caldos 27 unit. The components CO_2, CO, CH_4, SO_2 and NO are measured by an spectroscopic non-dispersive infra red (NDIR) sensor in an Uras 26 analyzer.

For the AR both gas analysis lines differ in the heated probe which does not include a tar removal unit. The measurement ranges of the equipment is different, as can be seen in Table 5 and H_2 and CH_4 is not measured. At the inlet of the AR, the composition is of interest to control the oxygen feed to the process and the amount of recirculated flue gas.

The water content is measured in both reactor outlets via a psychrometric Hygrophil H4320 unit from Bartec with the sampling gas extraction as shown in Figure 7. The gas is extracted via a heated probe (1) and transported in an electrically heated tube (2) to the analyzer (3) which includes a CO_2-driven ejector pump to facilitate the gas transport. The gas is released to the environment afterwards (4).

Table 5. Listing of gas analysis equipment for all reactors.

Reactor	Equipment	Measurement Principle	Component	Range	Error	Unit
FR	Magnos 206	paramagnetic	O_2	0 to 25	0.9	vol. − %
	Caldos 27	thermal conductivity	H_2	0 to 40	1.8	vol. − %
	Uras 26	NDIR	CO_2	0 to 100	3.0	vol. − %
	Uras 26	NDIR	CO	0 to 40	1.2	vol. − %
	Uras 26	NDIR	CH_4	0 to 20	0.6	vol. − %
	Uras 26	NDIR	SO_2	0 to 5	0.15	vol. − %
	Uras 26	NDIR	NO	0 to 1000	30	ppm
	Hygrophil H4320	psychrometric	H_2O	2 to 100	0.3	vol. − %
AR outlet	Magnos 206	paramagnetic	O_2	0 to 25	0.9	vol. − %
	Uras 26	NDIR	CO_2	0 to 30	0.9	vol. − %
	Uras 26	NDIR	CO	0 to 5	0.15	vol. − %
	Uras 26	NDIR	SO_2	0 to 4000	120	ppm
	Uras 26	NDIR	NO	0 to 1000	30	ppm
	Hygrophil H4320	psychrometric	H_2O	2 to 100	0.3	vol. − %
AR inlet	Magnos 206	paramagnetic	O_2	0 to 25	0.9	vol. − %
	Uras 26	NDIR	CO_2	0 to 100	3.0	vol. − %
	Uras 26	NDIR	CO	0 to 5	0.15	vol. − %
	Uras 26	NDIR	SO_2	0 to 5	0.15	vol. − %
	Uras 26	NDIR	NO	0 to 1000	30	ppm

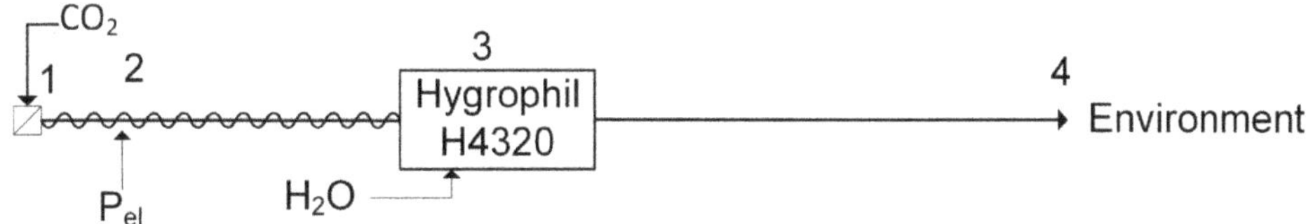

Figure 7. Schematic of the water content analysis equipment: (1) heated probe with filter, (2) heated tube, (3) psychrometric analyzer, (4) safe location in the environment.

Both water content measurements and the three gas analysis are integrated in the process control system of the pilot plant with all measurements available in real time and as trend lines.

4.6.2. Temperature and Pressure

The pilot is equipped with temperature and pressure in all inlet and outlet streams of the reactors including the LS fluidization. Multiple additional measurement sites for pressure and temperature are installed along the reactor height to acquire more insight in the reactor state during operation. The pressure sensors for the AR are differential pressure transducers with the other side open to atmosphere while at the FR all pressure measurements are purged with CO_2 and are mostly differential pressure transducers measuring between different reactor heights. This allows us to control the bed hight and density and to control the influence of the second stage fluidization.

4.6.3. Flow Measurements

The flow rates of all streams entering the reactors and coupling elements are measured either with an orifice plate, a rotameter or are controlled via a mass flow controller. The main streams leaving the reactors are measured via two venturi with side streams for off-line analysis, process control or the syngas treatment unit measured inside the respective analysis or control equipment. The mass flow of solids entering the system is measured via load cells and the corresponding trend line gradients.

4.7. Off-Line Sampling

4.7.1. Solid Sampling

The bottom product removal of the AR transfers the material to an open barrel and is immediately accessible for inspection and sampling, while for the FR it is transferred to a sealed and CO_2-purged container which can be replaced periodically during operation to allow for the collection of samples. The same is implemented for filter dust sampling. The AR filter is equipped with a hopper and an open barrel, while the FR filter has an additional CO_2 purge and the container is sealed.

Both loop seals allow for the collection of solid samples for off-line analysis. The OC samples enable the determination of the exact phase composition of the circulating OC and to balance the reactors individually. Moreover, knowledge of the oxidation level before and after the reactors allows for an additional method for the quantification of solid circulation.

4.7.2. Gas and Tar Sampling

More gas species like COS and higher hydrocarbons can be measured using Fourier transform infrared (FTIR) spectroscopy which can be connected at different locations. These measurements are not considered important during pilot plant operation but are important for the evaluation of the process. At the FTIR a port for gas sample bags and gas mice exists to enable off-line gas analysis.

Additionally, isokinetic sampling is possible in the synthesis gas line allowing for dust and tar sample collection according to tar protocol/CEN TS 15439. Velocity is measured by an S-Pitot tube 550 mm downstream of the sampling lance, both located in the center line of a refractory lined tube. The dust loaded syngas sample stream is transfered via a heated lance towards a heated filter and through six impinger bottles where five are filled with isopropanol as solvent and the last is empty. The impinger bottles are tempered to 40 °C (impinger 1, 2 and 4) and −20 °C (impinger 2, 5 and 6). The sample volume is measured inside a comercially available ST5 isokinetic sampler from Dado lab, which also adjusts the sample volume flow based on the pitot measurement.

5. Plant Operation

The simulations from Section 3.3 show that autothermal CLG experiments are needed to obtain further insights into the process, which are of high relevance for industrial deployment. The modified pilot plant (Section 4) renders these experiments feasible allowing for the generation of the following, required information:

5.1.   Literature, describing the demonstration of autothermal operation of the CLG process, is not yet available. While autothermal CLC has been successfully demonstrated [49,51] the higher prevalence of endothermic reactions impose the need for higher heat transfer to the FR and different control strategies [26].

5.2.   Continuous CLG of residual biomass has been successfully demonstrated in lab-scale [11,15,16]. Nonetheless, upscaling to higher thermal loads is necessary to obtain data for reliable simulations and design of industrial scale units.

5.3.   Due to their interdependence, the key performance indicators achievable in autothermal operation are unknown. This affects the cold gas efficiency η_{CG}, the carbon conversion η_{CC}, the syngas yield x_{SG} and the syngas quality (tars, CH_4, etc.). For example, in electrically heated systems the cold gas efficiency η_{CG} can be theoretically driven to 100% by supplying enough heat through the furnace. However, the exact amount of external heat supplied is seldom reported. The carbon slip depends amongst other on reactor size [59] and data for bigger scale units is not existent.

5.4.   Tar production can presently not be accurately predicted as no model was developed for CLG yet. Especially bed height and feeding location are also dependent on reactor size and their influence cannot be quantified [37]. The pilot plant experiments will give important insight on this matter in industrial like conditions, allowing for inferences for future upscaling endeavours.

5.5. OC life time is difficult to asses with currently available data, as the time of circulation and thus of re-oxidation cycles increases with increasing reactor size while the mechanical erosion is dependent on the transport velocity only. The exact contribution of the two effects is unknown and thus it is likely that the size of the reactor will have an influence on the OC life time.

5.6. Assessment of economic feasibility of the CLG process requires data from bigger scale units to make accurate predictions for e.g., sizing of components and process performance.

For the demonstration of autothermal CLG (item 5.1.) a suitable control concept for the oxygen carrier to fuel equivalence ratio ϕ based on a sub-stoichiometric AR operation (reduced OC oxidation, see Equations (2) and (3)) according to Dieringer et al. [26] is implemented. The corresponding operating strategy considering pilot plant limitations is described in the following.

Chemical Looping Gasification Operation

The start-up sequence of the pilot plant is preheating with electrically heated air, preheating with propane burner, OC filling plus propane burner, CFB combustion, CLC as described in [47]. Afterwards the switch to CLG is achieved by a reduction of air input to the AR while increasing flue gas recirculation, thus reducing ϕ to values smaller than unity. After stable CLG operation is attained, optimization of individual key performance indicators is targeted during experiments. The devised experimental operation of the pilot plant (described hereafter) allows to directly obtain data for items 5.2. and 5.3. while information for items 5.4. and 5.5. can be inferred from additional off-line analysis. Item 5.6. builds on this data but needs additional information, e.g., component and material pricing, which cannot be generated in the pilot plant. The main operation variables through which the process can be controlled are:

- Thermal load: Increasing the thermal load above $1\,\mathrm{MW_{th}}$ decreases the relative heat loss as it depends on reactor temperature and not on thermal load. Therefore, a higher fraction of the feedstock input, $\dot{m}_{FS}$, can be converted into syngas increasing process efficiency. The feedstock input rate $\dot{m}_{FS}$ is directly proportional to the thermal load, but an adjustment requires corresponding changes in fluidization imposed by reactor hydrodynamics and heat balance influencing the steam to biomass ratio. Nonetheless, the simulations in Section 3.3 show also an increase of CH_4 production with increasing thermal load, indicating a tendency to form hydrocarbons including tars.
  The limit for the thermal input is set by the maximum possible feedstock input and the syngas handling and cooling, as higher loads result in a higher amount of product gas which has to be handled safely. During operation a high thermal load is targeted at all operation points to obtain high η_{CG}.

- The OC to fuel equivalence ratio ϕ determines the net heat release from the process. A higher value of ϕ (while keeping everything else constant) results in a higher temperature inside AR and FR. However, the cold gas efficiency η_{CG} will decrease with higher ϕ as does the production of CH_4 and tars. The control of ϕ is straightforward through the control of the oxygen availability inside the AR.
  For experimental investigation the variation of temperatures is important. However, higher temperatures increase the load on the cooling system. Here the limits have to be considered during operation, and a reduction in thermal load (leading to smaller process streams and further decreasing η_{CG}) may be required in order to be able to reach higher gasification temperatures. Moreover, the refractory lining of the AR and/or the ash melting behaviour of the feedstock inside the FR limit the maximum admissible reactor temperatures.
  Actual control of ϕ is achieved via the variable amounts of air and recirculated AR flue gas fed to the AR to obtain a sub-stoichiometric environment inside the AR as it is the most suitable method for large scale operation described in detail by Dieringer et al. [26].

- The global solids circulation $\dot{m}_{OC}$ can be controlled via adjustment of J-valve and FR fluidization and transports sensible heat required in the FR. Depending on the operating state of the AR internal solids recirculation, fluidization of the AR needs adjustment as well to obtain hydrodynamic equilibrium between the reactors. Yet, $\dot{m}_{OC}$ is not directly accessible during pilot plant operation but can be inferred qualitatively from the temperature difference between the reactors. Higher solids circulation reduces the temperature difference ΔT between AR and FR. An accurate determination of $\dot{m}_{OC}$ is possible only indirectly via the oxygen content in the solid samples taken from the loop seals.

  Increasing global solids circulation reduces not only ΔT but also η_{CG} as more fluidization medium and corresponding heating is required. Furthermore, OC residence time inside the reactors is reduced when the solids circulation increases and as higher superficial gas velocities are employed, carbon slip towards the AR might increase. The variable to be controlled is the gasification temperature inside the FR while the limit of the AR temperature might require adjustment via ϕ.

The OC to fuel equivalence ratio ϕ and the global solids circulation $\dot{m}_{OC}$ are used to investigate the inevitable trade-off between cold gas efficiency and syngas quality in the form of produced CH_4, higher hydrocarbons, and tar. In contrast, the maximization of the thermal load is used to boost the process performance η_{CG} for all operation points by allowing for a smaller value of ϕ while at the same time guaranteeing autothermal operation.

While the variables above are used to adjust and stabilize the process and to investigate general trends, two more adjustable parameters exist which can be used to influence the syngas quality:

- Bed pressure drop Δp: The simulations in Section 3.3 are done with a fixed pressure drop Δp for both reactors. However, during operation of the pilot plant, Δp can be varied and is dependent on the exact distribution of bed material between the reactors (controlled by the governing hydrodynamic boundary conditions) as well as the total amount of bed material inside the reactor system. Increasing the pressure drop inside the FR will increase OC particle residence time inside the reactor (and the amount of OC per feedstock input). This will also increase the entrainment from the FR and thereby the solid circulation. However, increasing Δp allows for the reduction of fluidization medium, while keeping the entrainment constant, thus improving process efficiency. Reduction of tar and CH_4 content in syngas is facilitated by the increased availability of catalytic sites for conversion.

  The OC make up stream is used to control the overall amount of OC inside the reactor system, while its distribution is influenced by small adjustments to fluidization medium. The required changes in fluidization are small compared to the changes needed for the operation variables discussed above. The range of Δp is limited by the reactor hydrodynamics and the characteristics of corresponding pripheral equipment (e.g., maximum load of AR primary air fan).

- Second stage fluidization can be varied to enhance the residence time of the feedstock inside the dense zone of the FR as describes in Section 4.1. Rerouting part of the fluidization medium to the second stage fluidization will reduce entrainment and solid circulation, if the total amount of steam is kept constant and can be counteracted by additional bed material. Qualitative effects on synthesis gas are the same as for the bed pressure drop Δp, however, the quantitative influence may vary.

The feedstock types given in Table 2 are an additional parameter for experimental variation. However, the feedstock is not usable as process control variable and is therefore not included in the list above. Furthermore, the other variables must be used to adjust for feedstock variation to keep the process stable.

6. Conclusions

In this article, the design pathway of a $1\,MW_{th}$ chemical looping gasification (CLG) pilot plant, allowing for autothermal, semi-industrial process investigation, has been described in detail. Starting from a process model, considering fundamental CLG characteristics, a suitable operational mode and associated necessary adaptions for an existing $1\,MW_{th}$ chemical looping combustion (CLC) pilot plant have been established. Subsequently, it has been illustrated which inherent interconnections and trade-offs associated to CLG can be further analyzed in such an experimental setup and which strategies towards an optimized process setup, replicable in industry scale, can be pursued with it. These are:

- Calculation of heat and mass balances for autothermal CLG show a significantly reduced range of freely selectable operation parameters (operation temperatures, steam to biomass feed ratio, and oxygen carrier to fuel equivalence ratio), when compared to externally heated lab-scale units, due to the requirements of autothermal operation.
- Process control under autothermal condition can be achieved via three parameters: thermal load, oxygen carrier to fuel equivalence ratio, and global solid circulation. However, due to restrictions imposed by reactor hydrodynamics and autothermal operation, changes in one parameter must be balanced by changes in at least one of the other two. Moreover, the global solids circulation is adjusted indirectly via fluidization velocities and can only be inferred qualitatively from the reactor temperature difference during operation.
- Attempting to attain high cold gas efficiency and good syngas quality through higher gasification temperature inevitably results in high relative heat losses, as heat integration is not reasonably achievable in the $1\,MW_{th}$ scale and the existing unit. This leads to an unavoidable trade-off between cold gas efficiency and syngas quality, e.g., CH_4 and tar content which has to be accepted during experiments.
- Data which are not reliably obtainable from simulation, like tar formation or oxygen carrier (OC) life time, yet are fundamental for scale-up and economic considerations becomes available by conducting experiments in an industry relevant scale in the designed pilot plant.

In summary, future endeavours aiming towards industrial application of CLG are facilitated, through the described design of a $1\,MW_{th}$ CLG pilot plant. Here, the experimental facility lays the foundation to generate a unique robust dataset containing essential information required for up-scaling of CLG to industry size, thus propelling the technology towards market maturity.

Author Contributions: conceptualization, F.M., J.S. and P.D.; simulation, P.D. and F.M.; writing—original draft preparation, F.M.; writing—review and editing, P.D., J.S. and F.M.; visualization, F.M. and P.D.; supervision, B.E. All authors have read and agreed to the published version of the manuscript.

Funding: This work has received funding of the European Union's Horizon 2020—Research and Innovation Framework Programme under grant agreement No. 817841 (Chemical Looping gasification foR sustainAble production of biofuels—CLARA).

Acknowledgments: The authors gratefully acknowledge the support given by Harald Tremmel and Karl Voigtländer from AICHERNIG Engineering GmbH during HAZOP analysis.

Conflicts of Interest: The authors declare no conflict of interest.

Abbreviations

The following abbreviations are used in this manuscript:

AR	air reactor
ASU	air separation unit
CFB	circulating fluidized bed
CLC	chemical looping combustion
CLG	chemical looping gasification
CLOU	chemical looping with oxygen uncoupling
DFBG	dual fluidized bed gasification
FR	fuel reactor
FTIR	Fourier transform infrared
HTW^{TM}	High Temperature Winkler
LS	loop seal
NDIR	non-dispersive infra red
OC	oxygen carrier

Symbols

LHV	$\mathrm{MJ\,kg^{-1}, MJ\,mol^{-1}}$	lower heating value
R_{OC}		oxygen transport capacity
T	K, °C	temperature
X		mole fraction
ΔH	$\mathrm{J\,mol^{-1}}$	reaction enthalpy
Δp	Pa, bar	differential pressure
ϕ		oxygen carrier to fuel equivalence ratio
$\dot{Q}$	W	heat flow
$\dot{m}$	$\mathrm{kg\,s^{-1}}$	mass flow
$\dot{n}$	$\mathrm{mol\,s^{-1}}$	molar flow
η_{CG}		cold gas efficiency
c_p	$\mathrm{J\,kg^{-1}\,K^{-1}}$	specific heat
$d_{p,50}$	m	mean particle diameter
d	m	diameter
h	m	height
m	kg	mass
p	Pa, bar	pressure
u	$\mathrm{m\,s^{-1}}$	velocity
x_{SG}		syngas content

Subscripts

AR	Air Reactor
FR	Fuel Reactor
FS	Feed Stock
OC	Oxygen Carrier
O	Oxygen
internal	internal recirculation
in	stream entering reactor
loss	loss
out	stream leaving reactor
ox	oxidized
reci	recirculation
red	reduced
stoich	stoichiometric

References

1. Directive (EU) 2018/2001 of the European Parliament and of the Council of 11 December 2018 on the Promotion of the Use of Energy from Renewable Sources. p. 128. Available online: https://eurovent.eu/?q=articles/review-directive-eu-20182001-promotion-use-energy-renewable-sources-gen-115400 (accessed on 19 December 2020).

2. International Energy Agency. Data & Statistics. 2020. Available online: https://www.iea.org/data-and-statistics?country=EU2 8&fuel=Energy20transition20indicators&indicator=Biotrans (accessed on 19 December 2020).

3. Kaltschmitt, M. (Ed.) *Energy from Organic Materials (Biomass): A Volume in the Encyclopedia of Sustainability Science and Technology*, 2nd ed.; Springer: New York, NY, USA, 2019. [CrossRef]

4. Carrasco, J.E.; Monti, A.; Tayeb, J.; Kiel, J.; Girio, F.; Matas, B.; Santos Jorge, R. Strategic Research and Innovation Agenda 2020. EERA Technical Report. 2020. Available online: http://www.eera-bioenergy.eu/wp-content/uploads/pdf/ EERABioenergySRIA2020.pdf=AOvVaw012VUhnaqiUbL-yP76cz6s (accessed on 19 December 2020).

5. Molino, A.; Larocca, V.; Chianese, S.; Musmarra, D. Biofuels Production by Biomass Gasification: A Review. *Energies* **2018**, *11*, 811. [CrossRef]

6. Gómez-Barea, A.; Leckner, B. Estimation of Gas Composition and Char Conversion in a Fluidized Bed Biomass Gasifier. *Fuel* **2013**, *107*, 419–431. [CrossRef]

7. Thomsen, T.P.; Sárossy, Z.; Gøbel, B.; Stoholm, P.; Ahrenfeldt, J.; Frandsen, F.J.; Henriksen, U.B. Low Temperature Circulating Fluidized Bed Gasification and Co-Gasification of Municipal Sewage Sludge. Part 1: Process Performance and Gas Product Characterization. *Waste Manag.* **2017**, *66*, 123–133. [CrossRef]

8. Arena, U.; Zaccariello, L.; Mastellone, M.L. Fluidized Bed Gasification of Waste-Derived Fuels. *Waste Manag.* **2010**, *30*, 1212–1219. [CrossRef]

9. De, S.; Agarwal, A.K.; Moholkar, V.S.; Thallada, B. (Eds.) *Coal and Biomass Gasification: Recent Advances and Future Challenges*; Energy, Environment, and Sustainability; Springer: Singapore, 2018. [CrossRef]

10. Higman, C.; van der Burgt, M. *Gasification*, 2nd ed.; Gulf Professional Pub.: Boston, MA, USA; Elsevier: Amsterdam, The Netherlands, 2008.

11. Condori, O.; García-Labiano, F.; de Diego, L.F.; Izquierdo, M.T.; Abad, A.; Adánez, J. Biomass Chemical Looping Gasification for Syngas Production Using Ilmenite as Oxygen Carrier in a 1.5 kWth Unit. *Chem. Eng. J.* **2021**, *405*, 126679. [CrossRef]

12. Huseyin, S.; Wei, G.Q.; Li, H.B.; He, F.; Huang, Z. Chemical-Looping Gasification of Biomass in a 10 kWth Interconnected Fluidized Bed Reactor Using Fe_2O_3/Al_2O_3 Oxygen Carrier. *J. Fuel Chem. Technol.* **2014**, *42*, 922–931. [CrossRef]

13. Guo, Q.; Cheng, Y.; Liu, Y.; Jia, W.; Ryu, H.J. Coal Chemical Looping Gasification for Syngas Generation Using an Iron-Based Oxygen Carrier. *Ind. Eng. Chem. Res.* **2014**, *53*, 78–86. [CrossRef]

14. Wei, G.; He, F.; Huang, Z.; Zheng, A.; Zhao, K.; Li, H. Continuous Operation of a 10 kW_{th} Chemical Looping Integrated Fluidized Bed Reactor for Gasifying Biomass Using an Iron-Based Oxygen Carrier. *Energy Fuels* **2015**, *29*, 233–241. [CrossRef]

15. Samprón, I.; de Diego, L.F.; García-Labiano, F.; Izquierdo, M.T.; Abad, A.; Adánez, J. Biomass Chemical Looping Gasification of Pine Wood Using a Synthetic Fe_2O_3/Al_2O_3 Oxygen Carrier in a Continuous Unit. *Bioresour. Technol.* **2020**, *316*, 123908. [CrossRef] [PubMed]

16. Ge, H.; Guo, W.; Shen, L.; Song, T.; Xiao, J. Experimental Investigation on Biomass Gasification Using Chemical Looping in a Batch Reactor and a Continuous Dual Reactor. *Chem. Eng. J.* **2016**, *286*, 689–700. [CrossRef]

17. Ge, H.; Guo, W.; Shen, L.; Song, T.; Xiao, J. Biomass Gasification Using Chemical Looping in a 25 kW Th Reactor with Natural Hematite as Oxygen Carrier. *Chem. Eng. J.* **2016**, *286*, 174–183. [CrossRef]

18. Brown, T.A.; Dennis, J.S.; Scott, S.A.; Davidson, J.F.; Hayhurst, A.N. Gasification and Chemical-Looping Combustion of a Lignite Char in a Fluidized Bed of Iron Oxide. *Energy Fuels* **2010**, *24*, 3034–3048. [CrossRef]

19. Leion, H.; Mattisson, T.; Lyngfelt, A. The Use of Petroleum Coke as Fuel in Chemical-Looping Combustion. *Fuel* **2007**, *86*, 1947–1958. [CrossRef]

20. Chen, L. The Direct Solid-Solid Reaction between Coal Char and Iron-Based Oxygen Carrier and Its Contribution to Solid-Fueled Chemical Looping Combustion. *Appl. Energy* **2016**, *184*, 9–18. [CrossRef]

21. Leion, H.; Mattisson, T.; Lyngfelt, A. Solid Fuels in Chemical-Looping Combustion. *Int. J. Greenh. Gas Control* **2008**, *2*, 180–193. [CrossRef]

22. Song, Q.; Xiao, R.; Deng, Z.; Zhang, H.; Shen, L.; Xiao, J.; Zhang, M. Chemical-Looping Combustion of Methane with $CaSO_4$ Oxygen Carrier in a Fixed Bed Reactor. *Energy Convers. Manag.* **2008**, *49*, 3178–3187. [CrossRef]

23. Pissot, S.; Vilches, T.B.; Maric, J.; Seemann, M. Chemical Looping Gasification in a 2–4 MWth Dual Fluidized Bed Gasifier. In Proceedings of the 23rd International Conference on Fluidized Bed Conversion, Seoul, Korea, 13–17 May 2018; p. 10.

24. Larsson, A.; Seemann, M.; Neves, D.; Thunman, H. Evaluation of Performance of Industrial-Scale Dual Fluidized Bed Gasifiers Using the Chalmers 2–4-MW_{th} Gasifier. *Energy Fuels* **2013**, *27*, 6665–6680. [CrossRef]

25. Larsson, A.; Israelsson, M.; Lind, F.; Seemann, M.; Thunman, H. Using Ilmenite to Reduce the Tar Yield in a Dual Fluidized Bed Gasification System. *Energy Fuels* **2014**, *28*, 2632–2644. [CrossRef]

26. Dieringer, P.; Marx, F.; Alobaid, F.; Ströhle, J.; Epple, B. Process Control Strategies in Chemical Looping Gasification—A Novel Process for the Production of Biofuels Allowing for Net Negative CO_2 Emissions. *Appl. Sci.* **2020**, *10*, 4271. [CrossRef]

27. Yin, S.; Shen, L.; Dosta, M.; Hartge, E.U.; Heinrich, S.; Lu, P.; Werther, J.; Song, T. Chemical Looping Gasification of a Biomass Pellet with a Manganese Ore as an Oxygen Carrier in the Fluidized Bed. *Energy Fuels* **2018**, *32*, 11.

28. Adanez, J.; Abad, A.; Garcia-Labiano, F.; Gayan, P.; de Diego, L.F. Progress in Chemical-Looping Combustion and Reforming Technologies. *Prog. Energy Combust. Sci.* **2012**, *38*, 215–282. [CrossRef]

29. Linderholm, C.; Knutsson, P.; Schmitz, M.; Markström, P.; Lyngfelt, A. Material Balances of Carbon, Sulfur, Nitrogen and Ilmenite in a 100 kW CLC Reactor System. *Int. J. Greenh. Gas Control* **2014**, *27*, 188–202. [CrossRef]

30. Alabdrabalameer, H.A.; Taylor, M.J.; Kauppinen, J.; Soini, T.; Pikkarainen, T.; Skoulou, V. Big Problem, Little Answer: Overcoming Bed Agglomeration and Reactor Slagging during the Gasification of Barley Straw under Continuous Operation. *Sustain. Energy Fuels* **2020**, *4*, 3764–3772. [CrossRef]

31. Di Giuliano, A.; Funcia, I.; Pérez-Vega, R.; Gil, J.; Gallucci, K. Novel Application of Pretreatment and Diagnostic Method Using Dynamic Pressure Fluctuations to Resolve and Detect Issues Related to Biogenic Residue Ash in Chemical Looping Gasification. *Processes* **2020**, *8*, 1137. [CrossRef]

32. Fernández, M.J. Sintering reduction of herbaceous biomass when blended with woody biomass: Predictive and combustion tests. *Fuel* **2019**, *239*, 1115–1124. [CrossRef]

33. Luo, S.; Zeng, L.; Fan, L.S. Chemical Looping Technology: Oxygen Carrier Characteristics. *Annu. Rev. Chem. Biomol. Eng.* **2015**, *6*, 53–75. [CrossRef] [PubMed]

34. Fan, L.S.; Zeng, L.; Luo, S. Chemical-Looping Technology Platform. *AIChE J.* **2015**, *61*, 2–22. [CrossRef]

35. Zhao, X.; Zhou, H.; Sikarwar, V.S.; Zhao, M.; Park, A.H.A.; Fennell, P.S.; Shen, L.; Fan, L.S. Biomass-Based Chemical Looping Technologies: The Good, the Bad and the Future. *Energy Environ. Sci.* **2017**, *10*, 1885–1910. [CrossRef]

36. Larsson, A.; Kuba, M.; Berdugo Vilches, T.; Seemann, M.; Hofbauer, H.; Thunman, H. Steam Gasification of Biomass—Typical Gas Quality and Operational Strategies Derived from Industrial-Scale Plants. *Fuel Process. Technol.* **2021**, *212*, 106609. [CrossRef]

37. Gómez-Barea, A.; Ollero, P.; Leckner, B. Optimization of Char and Tar Conversion in Fluidized Bed Biomass Gasifiers. *Fuel* **2013**, *103*, 42–52. [CrossRef]

38. Devi, L.; Ptasinski, K.J.; Janssen, F.J. Pretreated Olivine as Tar Removal Catalyst for Biomass Gasifiers: Investigation Using Naphthalene as Model Biomass Tar. *Fuel Process. Technol.* **2005**, *86*, 707–730. [CrossRef]

39. Amin, A.M.; Croiset, E.; Epling, W. Review of Methane Catalytic Cracking for Hydrogen Production. *Int. J. Hydrogen Energy* **2011**, *36*, 2904–2935. [CrossRef]

40. Milne, T.A.; Evans, R.J.; Abatzaglou, N. *Biomass Gasifier "Tars": Their Nature, Formation, and Conversion*; Technical Report NREL/TP-570-25357; United States Department of Energy: Washington, DC, USA, 1998. [CrossRef]

41. Benedikt, F.; Kuba, M.; Schmid, J.C.; Müller, S.; Hofbauer, H. Assessment of Correlations between Tar and Product Gas Composition in Dual Fluidized Bed Steam Gasification for Online Tar Prediction. *Appl. Energy* **2019**, *238*, 1138–1149. [CrossRef]

42. Palma, C.F. Model for Biomass Gasification Including Tar Formation and Evolution. *Energy Fuels* **2013**, *27*, 5. [CrossRef]

43. Wojnicka, B.; Ściążko, M.; Schmid, J.C. Modelling of Biomass Gasification with Steam. *Biomass Conv. Bioref.* **2019**. [CrossRef]

44. Stark, A.K.; Bates, R.B.; Zhao, Z.; Ghoniem, A.F. Prediction and Validation of Major Gas and Tar Species from a Reactor Network Model of Air-Blown Fluidized Bed Biomass Gasification. *Energy Fuels* **2015**, *29*, 2437–2452. [CrossRef]

45. Herdel, P.; Krause, D.; Peters, J.; Kolmorgen, B.; Ströhle, J.; Epple, B. Experimental Investigations in a Demonstration Plant for Fluidized Bed Gasification of Multiple Feedstock's in 0.5 MW Th Scale. *Fuel* **2017**, *205*, 286–296. [CrossRef]

46. Krause, D.; Herdel, P.; Ströhle, J.; Epple, B. HTW™-Gasification of High Volatile Bituminous Coal in a 500 kWth Pilot Plant. *Fuel* **2019**, *250*, 306–314. [CrossRef]

47. Ströhle, J.; Orth, M.; Epple, B. Design and Operation of a 1 MWth Chemical Looping Plant. *Appl. Energy* **2014**, *113*, 1490–1495. [CrossRef]

48. Ströhle, J.; Orth, M.; Epple, B. Chemical Looping Combustion of Hard Coal in a 1 MWth Pilot Plant Using Ilmenite as Oxygen Carrier. *Appl. Energy* **2015**, *157*, 288–294. [CrossRef]

49. Ohlemüller, P.; Busch, J.P.; Reitz, M.; Ströhle, J.; Epple, B. Chemical-Looping Combustion of Hard Coal: Autothermal Operation of a 1 MWth Pilot Plant. *J. Energy Resour. Technol.* **2016**, *138*, 042203. [CrossRef]

50. Ohlemüller, P.; Ströhle, J.; Epple, B. Chemical Looping Combustion of Hard Coal and Torrefied Biomass in a 1 MW Th Pilot Plant. *Int. J. Greenh. Gas Control* **2017**, *65*, 149–159. [CrossRef]

51. Ohlemüller, P.; Alobaid, F.; Abad, A.; Adanez, J.; Ströhle, J.; Epple, B. Development and Validation of a 1D Process Model with Autothermal Operation of a 1 MW Th Chemical Looping Pilot Plant. *Int. J. Greenh. Gas Control* **2018**, *73*, 29–41. [CrossRef]

52. Min, Z.; Asadullah, M.; Yimsiri, P.; Zhang, S.; Wu, H.; Li, C.Z. Catalytic Reforming of Tar during Gasification. Part I. Steam Reforming of Biomass Tar Using Ilmenite as a Catalyst. *Fuel* **2011**, *90*, 1847–1854. [CrossRef]

53. Di Giuliano, A.; Lucantonio, S.; Gallucci, K. Devolatilization of Residual Biomasses for Chemical Looping Gasification in Fluidized Beds Made up of Oxygen-Carriers. *Energies* **2021**, *14*, 311. [CrossRef]

54. Kunii, D.; Levenspiel, O. *Fluidization Engineering*, 2nd ed.; Butterworth-Heinemann Series in Chemical Engineering; Butterworth-Heinemann: Boston, MA, USA, 1991.

55. Ohlemüller, P.; Alobaid, F.; Gunnarsson, A.; Ströhle, J.; Epple, B. Development of a Process Model for Coal Chemical Looping Combustion and Validation against 100 kWth Tests. *Appl. Energy* **2015**, *157*, 433–448. [CrossRef]

56. Grace, J.R.; Avidan, A.A.; Knowlton, T.M. (Eds.) *Circulating Fluidized Beds*, 1st ed.; Blackie Academic & Professional: London, UK; New York, NY, USA, 1997.

57. Schmid, J.C.; Pfeifer, C.; Kitzler, H.; Pröll, T.; Hofbauer, H. A New Dual Fluidized Bed Gasifier Design for Improved in Situ Conversion of Hydrocarbons. In Proceedings of the International Conference on Polygeneration Strategies (ICPS), Vienna, Austria, 30 August–1 September 2011; p. 10.

58. Kronberger, B.; Johansson, E.; Löffler, G.; Mattisson, T.; Lyngfelt, A.; Hofbauer, H. A Two-Compartment Fluidized Bed Reactor for CO_2 Capture by Chemical-Looping Combustion. *Chem. Eng. Technol.* **2004**, *27*, 1318–1326. [CrossRef]

59. Lyngfelt, A.; Leckner, B. A 1000 MWth Boiler for Chemical-Looping Combustion of Solid Fuels—Discussion of Design and Costs. *Appl. Energy* **2015**, *157*, 475–487. [CrossRef]
60. Hetzer, J.; Kulik, R.; Rothenpieler, K.; Stückrath, K.; Weidenfeller, D.J. Design, Simulation and Practical Experience of the Largest Syngas Cooler in Operation for Coal Gasification. In Proceedings of the 8th International Freiberg Conference, Cologne, Germany, 12–16 June 2016.

energies

Article

Production of Fuel-Like Fractions by Fractional Distillation of Bio-Oil from Açaí (*Euterpe oleracea* Mart.) Seeds Pyrolysis

Douglas Alberto Rocha de Castro [1], Haroldo Jorge da Silva Ribeiro [1], Lauro Henrique Hamoy Guerreiro [2], Lucas Pinto Bernar [1], Sami Jonatan Bremer [3], Marcelo Costa Santo [1], Hélio da Silva Almeida [4,5], Sergio Duvoisin, Jr. [6], Luiz Eduardo Pizarro Borges [7] and Nélio Teixeira Machado [1,4,*]

[1] Graduate Program of Natural Resources Engineering of Amazon, Rua Corrêa N° 1, Campus Profissional-UFPA, Belém 66075-110, Brazil; douglascastro87@hotmail.com (D.A.R.d.C.); harold_lr@hotmail.com (H.J.d.S.R.); lucas.bernar7@gmail.com (L.P.B.); marcelo.santos@ufra.edu.br (M.C.S.)
[2] Graduate Program of Chemical Engineering, Rua Corrêa N° 1, Campus Profissional-UFPA, Belém 66075-110, Brazil; guerreirolauro@hotmail.com
[3] Hochschule für Technik und Wirtschaft Berlin, Wilhelminenhofstrasse 75A, 12459 Berlin, Germany; jonatan-bremer-berlin@web.de
[4] Faculty of Sanitary and Environmental Engineering, Rua Corrêa N° 1, Campus Profissional-UFPA, Belém 66075-900, Brazil; helioalmeida@ufpa.br
[5] Graduate Program of Civil Engineering, Rua Corrêa N° 1, Campus Profissional-UFPA, Belém 66075-110, Brazil
[6] Faculty of Chemical Engineering-UEA, Avenida Darcy Vargas N°. 1200, Manaus 69050-020, Brazil; sjunior@uea.edu.br
[7] Laboratory of Catalyst Preparation and Catalytic Cracking, Section of Chemical Engineering-IME, Praça General Tibúrcio No. 80, Rio de Janeiro 22290-270, Brazil; luiz@ime.eb.br
* Correspondence: machado@ufpa.br; Tel.: +55-91-984-620-325

Citation: Rocha de Castro, D.A.; da Silva Ribeiro, H.J.; Hamoy Guerreiro, L.H.; Pinto Bernar, L.; Jonatan Bremer, S.; Costa Santo, M.; da Silva Almeida, H.; Duvoisin, S., Jr.; Pizarro Borges, L.E.; Teixeira Machado, N. Production of Fuel-Like Fractions by Fractional Distillation of Bio-Oil from Açaí (*Euterpe oleracea* Mart.) Seeds Pyrolysis. *Energies* **2021**, *14*, 3713. https://doi.org/10.3390/en14133713

Academic Editors: Andrea Di Carlo and Elisa Savuto

Received: 8 February 2021
Accepted: 26 April 2021
Published: 22 June 2021

Publisher's Note: MDPI stays neutral with regard to jurisdictional claims in published maps and institutional affiliations.

Abstract: This work investigates the effect of production scales (laboratory, bench, and pilot) by pyrolysis of Açaí (*Euterpe oleracea* Mart.) seeds at 450 °C and 1.0 atmosphere, on the yields of reaction products and acid value of bio-oils. The experiments were carried out in batch mode using a laboratory scale reactor of 143 mL, a bench scale reactor of 1.5 L, and a pilot scale reactor of 143 L (≈1:10:1000). The bio-oil was obtained in pilot scale, fractionated by distillation to produce biofuel-like fractions. The distillation of bio-oil was carried out in a laboratory column. The physical-chemistry properties (density, kinematic viscosity, acid value, and refractive index) of bio-oils and distillation fractions were determined. The qualitative analysis was determined by FT-IR and the chemical composition by GC-MS. The pyrolysis showed bio-oil yields from 4.37 to 13.09 (wt.%), decreasing with reactor volume. The acid value of bio-oils varied from 68.31 to 70.26 mg KOH/g. The distillation of bio-oil produced gasoline, light kerosene, and kerosene-like fuel fractions, and the yields were 16.16, 19.56, and 41.89 (wt.%), respectively. The physical-chemistry properties of distillation fractions increase with temperature. The FT-IR analysis of bio-oils and distillation fractions identified the presence of functional groups characteristic of hydrocarbons (alkenes, alkanes, aromatics, and aromatics rings) and oxygenates (carboxylic acids, ketones, esters, ethers, alcohols, phenols). The GC-MS identified 48.24 (area.%) hydrocarbons and 51.76 (area.%) oxygenates in the bio-oil produced in bench scale and 21.52 (area.%) hydrocarbons and 78.48 (area.%) oxygenates in the bio-oil produced in pilot scale. The gasoline-like fraction was composed by 64.0 (area.%) hydrocarbons and 36.0 (area.%) oxygenates, light kerosene-like fraction by 66.67 (area.%) hydrocarbons and 33.33 (area.%) oxygenates, and kerosene-like fraction by 19.87 (area.%) hydrocarbons and 81.13 (area.%) oxygenates.

Keywords: Açaí; residual seeds; pyrolysis; bio-oil; distillation; gasoline; light kerosene; kerosene-like fuel

1. Introduction

Açaí (*Euterpe oleracea* Mart.) is a native palm of natural occurrence in tropical Central and South America [1]. The palm gives a dark-purple, berry-like fruit, clustered into

---

bunches [2]. The fresh fruits are traditionally processed by crushing and/or extracting the pulp and skin with warm water to produce a thick, purple-colored beverage/juice or a paste [3,4]. The fruit is a staple food in rural and urban areas of the Amazon River estuary, particularly in the State Pará (Pará-Brazil), with great economic importance for both rural areas and at regional levels [5]. It has become one of the most important export products of the Amazon River estuary to other parts of Brazil [5], as well as overseas [6].

Of the total 1.228.811 tons/year of fruits produced by the State Pará, between 85% [7] and 83% (wt.) [8] is a residue (Açaí seeds), thus producing between 1.019.913 and 1.044.489 tons/year of a residue. The mechanical processing of Açaí fruits in nature produces around 175.7 tons residue/day in off-season crop and 448.0 tons residue/day in in-season crop in the metropolitan region of Belém (Pará-Brazil), posing a complex environmental problem of solid waste management [9,10]. The Açaí fruit is a small, dark-purple, berry-like fruit, almost spherical, weighing between 2.6 to 3.0 g [11], with a diameter around 10.0 and 20.0 mm [11], containing a large core seed that occupies almost 85% (vol.%) of its volume [3]. Açaí fruit has an oily-fiber seed, rich in lignin-cellulose material [12–15].

Pyrolysis makes possible the use of low quality lignin-cellulosic material to produce not only bio-oils, but also gaseous fuels, and a carbonaceous rich solid phase, as reported in the literature [16–73]. Studies include biomass pyrolysis [23,24,26,45,56,57,62,67,68], bio-oil chemical upgrading techniques [26,45,50], bio-oil physical-chemical properties [21,25,26,28,34,35,43,57,62,63], as well as separation and/or purification processes to improve bio-oil quality [17–22,30–33,36–41,46–48,51,53,54,59–61,65,66,70–73].

The bio-oil produced by pyrolysis is a multicomponent liquid mixture presenting water, carboxylic acids, aldehydes, ketones, alcohols, esters, ethers, aliphatic hydrocarbons, aromatic hydrocarbons, anhydrous-sugars, furans, phenols derivatives, among others chemical functions [16,17,20,38,44,47,48,53,60,61,73]. In addition, its organic fraction has a wide distribution of polarity, molecular weight [47], as well as differences in thermo-physical and transport properties of chemical compounds, as reported by the simulation of organic liquid compounds [74], posing challenges to the efficient separation and/or purification processes [47,74].

In the last years, several thermal and physical separation processes were applied to remove oxygenates from biomass-derived bio-oils including molecular distillation [30,33,36–39,71], fractional distillation [17–21,40,41,46–48,53,59,60,66,70,72,73], liquid-liquid extraction [22,31,61], and fractional condensation [51,54,65]. In addition, chemical methods such as catalytic upgrading of bio-oil vapors have been applied to improve bio-oil quality [19,29,64].

The fractional distillation studies were carried out in micro/bench scale [17,46,47], laboratory scale [41,53,66,70,72,73], and pilot scale [21], under atmospheric [17,18,46–48,53,66,70,72,73], or under vacuum conditions [18,19,41,48,53]. Açaí seeds are the only fruit specie whose centesimal and elemental composition is completely different from wood biomass (aspen poplar wood, eucalyptus, maple wood, and softwood bark) [17–19,21,53], residues of cereal grains (corn Stover, rice Rusk) [41,46,47,66,70,72], jatropha curcas [46], and horse manure and switch-grass [53]. However, until the moment, no systematic study investigated the physicochemical properties (density, kinematic viscosity, refractive index, and acid value) and chemical composition of Açaí seeds bio-oil distillation fractions [73]. The fractional distillation studies are summarized as follows [21–23,25,45,50–52,57,70,72,73].

Adjaye et al. [21], investigated the distillation of aspen poplar wood high-pressure liquefaction bio-oil. High-pressure liquefaction was carried out at 5.0 MPa and 360 °C, using 100 g of feed material ($\varnothing_{Particle}{\sim}1.0$ mm), 10 g of Na_2CO_3, 500 g distilled H_2O, under CO atmosphere, and 2 h reaction time. The reaction liquid products consist of an organic (bio-oil) and an aqueous phase, showing a bio-oil yield of 30 (wt.%), containing 1.5 (wt.%) H_2O. The bio-oil distilled within the boiling temperature ranges 85 °C $<$ T^{Boiling} $<$ 250 °C, using 4.0 g of bio-oil. The distillations carried out at 85, 115, 140, 165, 175, 190, 200, 220, and 250 °C. The yields of distillation fractions varied between 21.0 and 62.3 (wt.%), while those of bottom products were between 37.7 and 79 (wt.%). The content of hydrocarbons in the distillation fractions varied within the range 38.5 and 47.4 (area.%), increasing between 85 and 175 °C, reaching a maximum of 47.4 (area.%) at 175 °C, decreasing between 175 and 250 °C. The concentration of oxygenates lies in the range 43 and 56.6 (area.%), showing a tendency to increase between 85 and 250 °C, presenting a maximum of 56.6 (area.%) at 250 °C. In addition, the concentration of phenols in the distillation fractions increases with temperature. The GC-MS analysis of bio-oil identified 85 compounds including carboxylic acids (formic, acetic, and propionic acids), cyclic alcohols, aliphatic alcohols, aldehydes, ketones, aromatic hydrocarbons, aliphatic hydrocarbons, polycyclic hydrocarbons, unsaturated hydrocarbons, substituted furans, substituted phenols, and methoxy phenols (phenol, guaiacol, *p*-cresol, *p*-guaiacol, *o*-guaiacol, iso-eugenol, and catechol).

Carazza et al. [22] investigated the distillation of Eucalyptus tar, recovered from the carbonization retort process. The physical-chemical properties of Eucalyptus tar show a density of 1.180 g/cm^3 and viscosity of 87.17 mm^2/s, with an acidity of 8.70 (% HAc). The distillation consists of a 3000 L boiler, coupled to a fractionation column of 04 stages, a reflux system, 2 condensers, a homogenization system (centrifuge pump), sample units, a collecting unit, and a vacuum system. The Eucalyptus tar (bio-oil) distilled within the boiling temperature ranges from 110 °C $<$ T^{Boiling} $<$ 300 °C, using a 2000 L feed at 70 mmHg, without reflux. The reaction products consist of an aqueous phase (T^{Boiling} $<$ 110 °C), an organic phase (bio-oil) (110 °C $<$ T^{Boiling} $<$ 300 °C), and a pitch, showing an average bio-oil yield of 26.9 (wt.%), an aqueous phase yield of 17.3 (wt.%), a pitch yield of 50.9 (wt.%), and 4.9 (wt.%) losses. The GC-MS identified 0.63 (area.%) acetic acid, 4.03 (area.%) ketones, and 85.48 (area.%) phenol derivatives, as well as 9.86 (area.%) non-identified chemical compounds. In a secondary fractionation step, the bio-oil re-distilled at 340 °C and 10 mmHg, with reflux ration of 3:1 and 7:1, being the distillates fractionated into 95 samples. The GC-MS identified in all the samples only phenol derivatives (the distribution of phenol derivatives determined for all the samples and/or fractions was phenol, ethyl phenol, cresol, *o*-cresol, *p*-cresol, *m*-cresol, guaiacol, ethyl guaiacol, phenyl guaiacol, syringol, ethyl syringol, methyl syringol, and propyl syringol).

Adjaye and Bakhshi [23], investigated the distillation of maple wood bio-oil. Rapid thermal process (RTP) was carried out at 525 °C and residence times were between 0.45 and 0.50 s. The yield of the bio-oil was 74 (wt.%), containing 21 (wt.%) H$_2$O. The bio-oil density and viscosity at 25 °C were 1.12 g/cm^3 and 9.0 × 10^{-2} Pa s (80.36 mm^2/s), respectively. The bio-oil distilled under a vacuum of 172 Pa and at 150, 200, and 250 °C, heating rate of 40 °C/min, using a Buchi GKR-56 distillation unit. Vacuum distillation of bio-oils then yielded two fractions: a volatile (distillates) and a non-volatile fraction (bottoms), which were 54.6 (wt.%) of distillate and 45.4% (wt.) residue at 150 °C, 63.4 (wt.%) distillate and 45.4% (wt.%) residue at 200 °C, and 54.6 (wt.%) distillate and 45.4 (wt.%) residue at 250 °C. The GC-MS analysis of volatile (distillates) fraction identified carboxylic acids, esters, alcohols, aliphatic and aromatic hydrocarbons, aldehydes, ketones, amines, ethers, furans, phenols, with 6.0 (area.%) hydrocarbons.

Boucher et al. [25], studied the distillation of softwood bark bio-oil produced by vacuum pyrolysis. The pyrolysis was carried out under vacuum in a pilot-scale reactor at 500 °C and 14 kPa, using 92 kg softwood bark residues with a particle diameter $\varnothing_{Particle}$ $<$ 14.0 mm, in batch mode. The reaction products yielded 20 (wt.%) bio-oil, 24 (wt.%) aqueous phase, 35 (wt.%) coke (solid phase), and 20 (wt.%) gaseous phase. The aqueous phase composed by 84 (wt.%) H$_2$O and 16 (wt.%) soluble/dissolved organic compounds. The density and kinematic viscosity of bio-oil were 1.066 g/cm^3 (20 °C) and 38.0 mm^2/s (40 °C), respectively, while the flash point was $>$90 °C. The distillation was carried out in a 250 mL glass flask, coupled to a fractionation column, a condenser, and an electric heater, using 150 g of bio-oil at 140 °C, 1.0 atmosphere. The yield of distillation (aqueous phase + organic phase) was 17.0 (wt.%), while that of organic fraction, with initial boiling point (T^{IBP} = 140 °C), was 11.7 (wt.%). At boiling temperature T^{Boiling} $<$ 100 °C, the distillation curve exhibits the evaporation of H$_2$O and low boiling point compounds, reaching a distillation yield of approximately 10.0 (wt.%). After removal of H_2O, the slope of the distillation curve increases as molecules of high molecular weight evaporate.

Zheng and Wei [45] studied the distillation, under vacuum, of fast pyrolysis rice husk bio-oil at 80 °C, 15 mmHg, using a distillation round bottom glass apparatus of 1000 cm^3. The pyrolysis products include an aqueous phase (H_2O, volatile organic acids, and oxygenate compounds), a bio-oil, and a residue. The yields of bio-oil, aqueous phase, and residue were 61 (wt.%), 29 (wt.%), and 10 (wt.%), respectively. The density of bio-oil was 1.270 g/cm^3. The content of linear and aromatic carboxylic acids in fast bio-oil (HCOOH, CH_3COOH, $C_6H_4(COOH)_2$) decrease from 14.35 (wt.%) to 1.01 (wt.%) after distillation. The pH increased from 2.8 to 6.8, proving that distillation was effective to de-acidify the bio-oil. In addition, distillation causes a deoxygenation of bio-oil, as the concentration of oxygenates decreased from 50.3 to 9.2 (wt.%). In order to investigate the stability of bio-oil and distilled bio-oil, experiments were carried out to analyze variations on the kinematic viscosity (20 °C) within a period of 30 days. The results show that viscosity (20 °C) of bio-oil is almost constant ($\sim$210 mm^{2}s^{-1}) after distillation, and that of fast pyrolysis bio-oil increases from 130 to 240 mm^{2}s^{-1}, showing that distillation produces a chemical stable distilled bio-oil.

Majhi et al. [50] investigated the distillation of jatropha curcas cake pyrolysis bio-oil. The pyrolysis was carried out in a fixed bed stainless steel tubular reactor (ϕ_{id} = 40 mm, H = 240 mm, $V_{Reactor}$ = 301.6 mL) in laboratory scale at 550 °C, 5 °C/min heating rate, N_2 flow rate of 50 cm^3/min, using 250 g biomass with a particle diameter $\varnothing_{Particle}$ 0.5 and 0.8 mm. The distillation carried out in a 500 mL glass flask, coupled to a Hempel fractionating packed column and a condenser, using 300 mL of bio-oil. The temperature of gaseous phase was measured at the column outlet using a glass thermometer. The bio-oil was distilled within the boiling temperature ranges $T^{Boiling} < 140$ °C, 140 °C $< T^{Boiling} < 250$ °C, and $T^{Boiling} > 250$ °C. The bio-oil density, kinematic viscosity, and flash point were 1.100 g/cm^3, 3.96 mm^2/s, and 180 °C, respectively, containing 15.4 (wt.%) H_2O. The bio-oil distilled fraction within the boiling temperature range $T_{Boiling} < 140$ °C had a density of 0.8735 g/cm^3, kinematic viscosity of 2.35 mm^2/s, and flash point of 41 °C, containing less than 0.05 (wt.%) H_2O, and acidity less than 0.05 mg KOH/g. FT-IR analysis identified the presence of alkenes and mono and polycyclic substituted aromatic groups, confirming that $T^{Boiling} < 140$ °C distillation fraction contains no polar compounds. In addition, GC analysis identified the presence of non-polar compounds including hexane, methyl cyclopentane, 3,3-dimethyl cyclopentane, 3-methylhexane, methyl cyclohexane, and toluene.

Zhang et al. [51] studied the distillation of fast co-pyrolysis of rice Rusk and ADR (Atmospheric Distillation Residue). The fast pyrolysis was carried out in a downdraft fixed-bed reactor, under N_2 atmosphere at flow rate of 0.16 L/min, reaction time between 1–2 s, at 450 °C, using 5.0 g of dried rice Rusk with a particle diameter $\varnothing_{Particle} < 88.0$ µm. The co-pyrolysis was carried out by mixing dried rice Rusk and ADR powder ($\varnothing_{Particle} < 300.0$ mm), and the ADR powder content was set at 15, 20, 25, 30 and 35 (wt.%) of feed. The distillation was carried out in a round-bottom flask placed, using an oil bath heating system, under vigorous magnetic stirring, at 240 °C and 1.0 atmosphere. The distillation products include the distilled bio-oil, a residual solid phase (ADR), non-condensable gases, and residual bio-oil. The yields of distillation fractions (FI, FII, FIII, FIV, FV, and FVI) increased with temperature, varying between 33 and 51.86 (wt.%), within the interval 145 °C $< T^{Boiling} < 240$ °C. The content of carboxylic acid (acetic acid, propionic acid, and *n*-butyric acid) decreased in distilled fractions FI [2.74 (wt.%)], FII [1.026 (wt.%)], and FIII [0.529 (wt.%)], increased in distilled fractions FIV [0.627 (wt.%)] and FV [1.092 (wt.%)], and decreased again in fraction FVI [0.475 (wt.%)], the same behavior observed for the content of furfural and phenols derivatives (phenol, o-cresol, p-cresol, and guaiacol). In addition, the H_2O content present in the distilled fraction FVI was 37.13 (wt.%), higher than that of raw bio-oil, equal to 30.3 (wt.%), showing that dehydration reactions occurred. For the distillation of fast co-pyrolysis bio-oils obtained by mixing rice Rusk and ADR, one can observe that yields of bio-oil remain almost constant ($\sim$46–47) (wt.%), while those of char increase from 37 to 41 (wt.%) and those of non-condensable gases decrease from 18 to 14 (wt.%) with increasing ADR content in feedstock. Finally, experimental results based on the elemental analysis and chemical identification by GC-MS show that distillation of bio-oils can be considered a reactive distillation process, as the solid phase (ADR) contains substance of high molecular weight and carbon chain length, not detected and/or identified in the raw bio-oil by GC-MS.

Capunitan and Capareda [52] investigated the distillation of corn Stover bio-oil under atmospheric and vacuum conditions (0.5 bar). The bio-oil produced by pyrolysis of corn Stover using a high-pressure reactor at 400 °C, in batch mode, contains 20.3 (wt.%) H_2O. The aqueous phase acidity was 24.6 mg KOH/g oil. The distillation set-up consisted of distillation flasks, a fractionating column, a condenser, and collecting flasks, using 10 g bio-oil. For the distillation in atmospheric conditions, bio-oil was distilled within the boiling temperature ranges $T^{Boiling} < 100$ °C, 100 °C $< T^{Boiling} < 180$ °C, and 180 °C $< T^{Boiling} < 250$ °C, while that under vacuum was within the boiling temperature ranges $T^{Boiling} < 80$ °C, 80 °C $< T^{Boiling} < 160$ °C, and 160 °C $< T^{Boiling} < 230$ °C. Atmospheric distillation of bio-oil yields 15.0 (wt.%) of an organic phase and 18.7 (wt.%) of an H_2O phase at 100 °C, 4.7 (wt.%) of an organic phase between 100 °C $< T^{Boiling} < 180$ °C, and 45.3 (wt.%) of an organic phase between 180 °C $< T^{Boiling} < 250$ °C. The vacuum distillation yields 10.3 (wt.%) of an organic phase and 16.2 (wt.%) of a H_2O phase at 80 °C, 5.9 (wt.%) of an organic phase between 80 °C $< T^{Boiling} < 160$ °C, and 40.9 (wt.%) of an organic phase between 160 °C $< T^{Boiling} < 230$ °C. The acid values of distillation fractions under atmospheric conditions were 4.1, 15.1, and 7.4 mg KOH/g oil, while those under vacuum were 3.0, 13.9, and 5.0 mg KOH/g oil. Chemical functions characteristic of hydrocarbons (alkanes, aromatics) and oxygenates (aldehydes, alcohols, and phenols) were identified by FT-IR. The GC analysis identified in the distillation fraction obtained under atmospheric and vacuum condition, the presence of paraffin's, naphthenics, aromatics, oxygenates, furans, and phenols, being the content of hydrocarbons 23.1, 17.1, and 7.1 (area.%) under atmospheric conditions, and 22.2, 18.4, and 7.4 (area.%) under vacuum conditions.

Elkasabi et al. [57] studied the fractional distillation of tail-gas reactive pyrolysis bio-oil of horse manure (TGPHM), switch grass (TGPSG), and eucalyptus (TGPE), and the fast pyrolysis bio-oil of horse manure (BHM), switch grass (BSG), and eucalyptus (BE) to fractionate and enrich chemical compounds. TGPHM, TGPSG, and TGPE were produced by pyrolysis of tail-gas with recycle rates of 70, 70, and 50 (vol.%), respectively. A vacuum-jacketed distillation apparatus of 100 mL (Vigreux), operating in batch mode, coupled to a Liebig condenser, was used to carry out the distillation experiments under vacuum and 1.0 atm. The distillation apparatus was connected to a 100 mL round-bottomed glass flask, inserted inside a heating mantle with digital temperature control. The vapor temperature was measured at the column top outer joint. The reduced pressure (vacuum) distillation was applied when the bottom flask temperature reached 350 °C. The distillation yields of tail-gas reactive pyrolysis bio-oils TGPHM, TGPSG, and TGPE were 56.3, 55.8, and 54.1 (wt.%), respectively. The distillation yields of fast pyrolysis bio-oils BHM, BSG, and BE were 24.2, 33.1, and 38.9 (wt.%), respectively. The acid values of all distillation fractions of TGPHM, TGPSG, and TGPE decrease with increasing vapor temperature. The acid values of lighter distillation fractions are very high, compared to the initial acid value of TGPSG and TGPE, showing that fractional distillation was not effective in decreasing the acidity of tail-gas reactive pyrolysis bio-oil with initial high acid values.

Huang et al. [70], investigated the fluidized bed pyrolysis with in-line distillation of rice Rusk under atmospheric conditions. The bio-oil was produced by pyrolysis of rice Rusk using a fluidized bed pyrolyzer (ϕ_{id} = 50 mm, H = 500 mm, $V_{Reactor}$ = 981.7 mL) at 490 and 590 °C, N_2 flow rate of 10.0 mL/min, gas fluidization velocity of 0.36, 0.46, and 0.56m/s, and feed flow rate of 10.0 g/min, with a particle diameter 1.68 mm $< \varnothing_{Particle} < 3.36$ mm. The distillation column contains 04 stages, 4.75 cm internal diameter, and 43.0 height, connected to the fluidized bed pyrolyzer. The distillation fractions at the bottom stage were collected at 136 °C. In addition, 02 condenser at 20 °C and 0 °C was used to complete the fractionation of bio-oil distillation. For the pyrolysis experiments carried out at 590 °C and 0.36 m/s, the composition of distillation fraction collected in the first stage (bottom) at 136 °C contained 63.32 (area.%) phenols, 1.47 (area.%) furans, and 4.87% (area.%) hydrocarbons. Inside the condenser at 20 °C, the composition of condensates contained 63.34 (area.%) acetic acid, 14.85 (area.%) phenols, 8.04 (area.%) ketones, and 4.33 (area.%) furfural, and inside the condenser at 0 °C, the composition of condensates contained 28.66 (area.%) acetic acid, 34.48 (area.%) phenols, 3.31 (area.%) ketones, 4.84 (area.%) furans, and 2.73 (area.%) hydrocarbons.

Huang et al. [70,72], investigated the fluidized bed pyrolysis with in-line distillation of rice Rusk under atmospheric conditions. The bio-oil was produced by pyrolysis of rice Rusk using a fluidized bed pyrolyzer (ϕ_{id} = 50 mm, H = 500 mm, $V_{Reactor}$ = 981.7 mL) at 500 and 600 °C, N_2 flow rate of 10.0 mL/min, gas fluidization velocity of 0.255, 0.340, and 0.425 m/s, and feed flow rates of 10.0 and 20.0 g/min, particle diameter 1.68 mm < $\varnothing_{Particle}$ < 3.36 mm. The distillation column contains 04 stages, 4.75 cm internal diameter, and 43.0 height, and was connected to the fluidized bed pyrolyzer. The distillation fractions were collected in the first (bottom) stage at 120 °C and the third stage at 80 °C. In addition, 02 condensers at 20 °C and 0 °C were used to complete the fractionation of bio-oil distillation. For the pyrolysis experiment carried out at 500 °C, 10.0 g/min, and 0.340 m/s, the composition of distillation fraction collected in the first stage (bottom) at 120 °C contained 34.12 (area.%) phenols, 7.59 (area.%) ketones, 0.56 (area.%) anhydride acetic, and 3.41 (area.%) hydrocarbons. The distillation fraction collected in the third stage at 80 °C contained 40.96% (area.%) phenols, 9.78% (area.) ketones, 0.78% (area.%) anhydride acetic, and 3.87% (area.%) hydrocarbons. In addition, 02 condensers at 20 °C and 0 °C were used to complete the fractionation of bio-oil distillation, which was the composition of condensates inside the condenser at 20 °C, composed by 35.59 (area.%) acetic acid, 10.81 (area.%) phenols, 15.91 (area.%) ketones, 6.35 (area.%) furfural, 2.19 (area.%) anhydride acetic, and 4.69 (area.%) hydrocarbons. Inside the condenser at 0 °C, the condensates were composed by 10.17 (area.%) acetic acid, 27.17 (area.%) phenols, 8.03 (area.%) ketones, 8.28 (area.%) furfural, 1.57 (area.%) anhydride acetic, and 4.12 (area.%) hydrocarbons.

The work investigates systematically the influence of production scales (laboratory, bench, and pilot) by pyrolysis of pretreated (drying, milling, and sieving) Açaí seeds at 450 °C and 1.0 atmosphere, on the yields of reaction products (bio-oil, aqueous phase, coke, and gas) and the acid value of bio-oils. The bio-oil obtained in the pilot scale was submitted/subjected to fractional distillation using a laboratory-scale column (Vigreux) of 30 cm to study the feasibility of producing fuel-like fractions (gasoline, light-kerosene, and kerosene), determine the chemical composition of bio-oils and distillation fractions, as well as to perform the physical-chemistry (density, kinematic viscosity, acid value, and refractive index) characterization of distillation fractions.

2. Materials and Methods

2.1. Methodology

The process flow sheet shown in Figure 1 summarizes the applied methodology, described in a logical sequence of ideas, chemical methods, and procedure to produce fuel-like fractions (gasoline, light kerosene, and kerosene) by fractional distillation of bio-oil produced by thermal degradation (pyrolysis) of Açaí seeds at 450 °C and 1.0 atm, in laboratory, bench, and pilot scales. Initially, the Açaí seeds are collected. Afterwards, they are subjected to pretreatments of drying to diminish moisture content, followed by milling and sieving. The pyrolysis carry out in different production scales (laboratory, bench, and pilot) to investigate the influence on reaction products (coke, bio-oil, H_2O, and gas), particularly the bio-oil, as well as on the acid value of bio-oil. The bio-oil was produced in pilot scale, submitted to fractional distillation to produce gasoline, light kerosene, and kerosene-like fractions. The physical-chemistry properties and chemical composition of distillation fractions were determined.

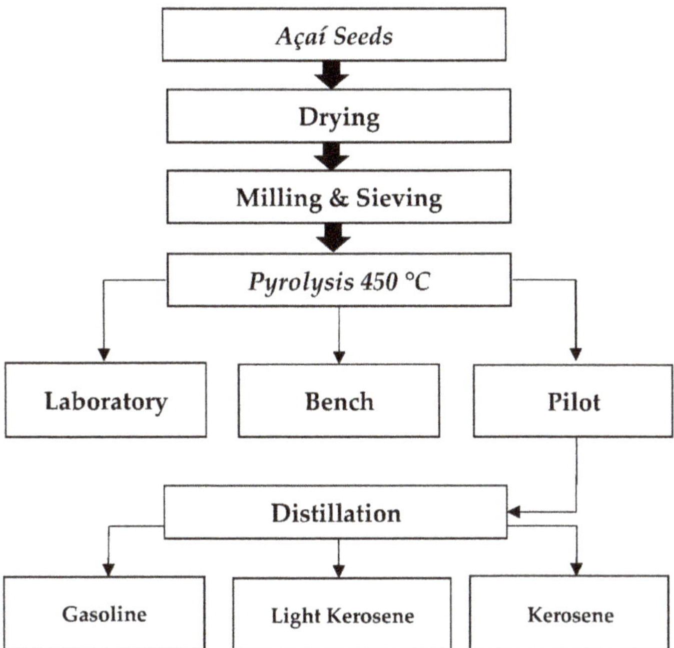

Figure 1. Process flow sheet by the production of fuel-like fractions (gasoline, light kerosene, and kerosene) obtained by fractional distillation of bio-oil obtained in laboratory, bench, and pilot scales.

2.2. Materials

By the processing of Açaí fruits *in nature* with warm water to produce a purple-colored juice or a paste [3,4], a byproduct, the *Açaí seeds*, is generated [3,7,8]. The seeds of Açaí were obtained in a small store of Açaí commercialization, located in the City of Belém-Pará-Brazil [73].

2.3. Pre-Treatment of Açaí Seeds in Nature

The seeds had a high moisture content, not only because the pulping process uses water to extract the Açaí juice, but also due to its incorrect disposal, exposed to the environment. The high moisture content of raw materials favors the generation of undesirable products. In this sense, it was necessary to dry the seeds. Next, 750 kg of Açaí seeds weighed and separated into five loads/charges of 150 kg. Afterwards, the charges were subjected to the drying process in a pilot thermal oven with air recirculation and analog temperature control (FABBE. Ltda, São Paulo-Brazil, Model: 170), at 110 °C for 24 h. The dried seeds were grounded using a pilot knife mill (TRAPP, Model: TRF 600). Bottom sieves with 0.8 mm and 5.0 mm opening diameters were fixed at the cutting mill exit. Two charges with 50.0 kg of dried seeds were sieved, the first using a bottom sieve of 0.80 mm, while the second used a bottom sieve of 5.0 mm. Afterwards, the dried and grounded seeds were sieved using a 0.6 mm sieve to remove the excess fibers from Açaí seeds. Figure 2 shows the material after the drying, milling and sieving process of Açaí seeds to carry out the pyrolysis experiments [Seeds + Fibers (a); Comminuted Seeds + Fibers (b); Fibers (c); and Comminuted Seeds (d)]. The pretreatments of drying, milling, and sieving increases the biomass contact-surface for carbonization and the uniformity (particle size distribution) of the raw material. The pyrolysis experiments were carried out with comminuted seeds, as shown in Figure 2d.

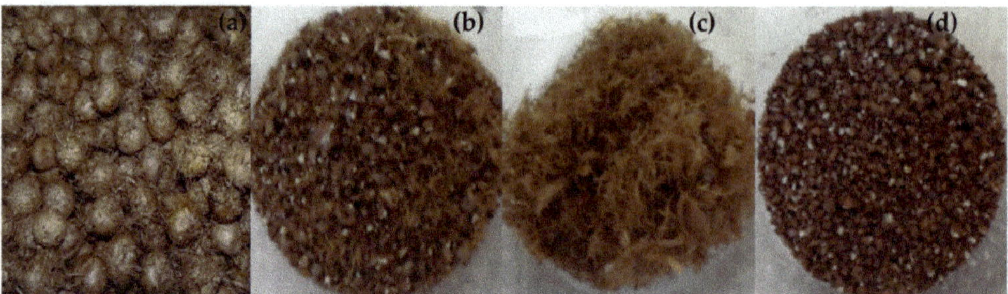

Figure 2. Material after drying, milling, and sieving process of Açaí seeds [Seeds + Fibers (**a**); Comminuted Seeds + Fibers (**b**); Fibers (**c**); and Comminuted Seeds (**d**)].

2.4. Centesimal Characterization of Açaí Seeds

The dried seeds in nature were physico-chemically characterized for moisture, volatile matter, ash, and fixed carbon according to official methods ASTM D 3173-87, ASTM D 3175-04, ASTM D 3174-04, and ASTM D3172-89 [73]. In a previous study [74], the dried Açaí seeds were physico-chemically characterized for lipids, proteins, fibers, cellulose, hemicellulose, and insoluble lignin according to official methods [73–76].

2.5. Experimental Apparatus and Procedures
2.5.1. Pyrolysis Units

The experiments were carried out in laboratory, bench, and pilot scales at 450 °C and 1.0 atmosphere, and all the apparatuses in laboratory, bench, and pilot scales are described in detail elsewhere [77].

The pyrolysis unit in laboratory scale consists of a cylindrical borosilicate glass reactor of 200 cm^3. The reactor was inserted in a cylindrical oven with a ceramic resistance of 800 W, with a digital temperature and heating rate control system (THERMA, São Paulo-Brazil, Model: TH90DP202-000), and the temperature was measured with the aid of a K-type thermocouple sensor (Ecil, São Paulo-Brazil, Model: QK. 2). A borosilicate glass Liebig condenser was connected to the reactor exit using a Y shaped connection and used to liquefy the gaseous phase, with cooling H_2O supplied by a thermostatic recirculation bath (VWR) with digital temperature control. The liquid phase products were collected in a 50 mL borosilicate glass flask. The non-condensable gaseous products (CH_4, CH_3-CH_3, CH_3-CH_2-CH_3, O_2, CO, CO_2, H_2, etc.) flow through an opening in the 90° curve, coupled between the Liebig condenser and the collection flask, to the flare system. Figure 3 illustrates the pyrolysis unit in laboratory scale.

The bench pyrolysis unit, mounted on a movable metallic structure, consists of a cylindrical Reactor (R-1) constructed by AISI 304 Stainless steel of 2.0 L. The experimental apparatus has a control unit with a PLC (Programmable Logic Controller), making it possible to control the reactor temperature, heating rate, and the mechanical stirrer angular velocity. The reactor operating temperature is programmed using a temperature controller (NOVUS, Model: N1100), coupled to a K-type thermocouple inside the reactor (R-1). The reactor connected to a stainless steel double pipe condenser (DN $\frac{1}{2}$″) with a heat exchange area of 0.05 m^2, coupled to a thermostatic recirculation bath with digital temperature control, using H_2O as cooling fluid. The liquefied products were withdrawn using a 2.0 L stainless steel collection vessel (C-01). The non-condensable gaseous products flow through an escape valve located between the condenser and the collection vessel, and burned in the flare system. Figure 4 illustrates the pyrolysis unit in bench scale.

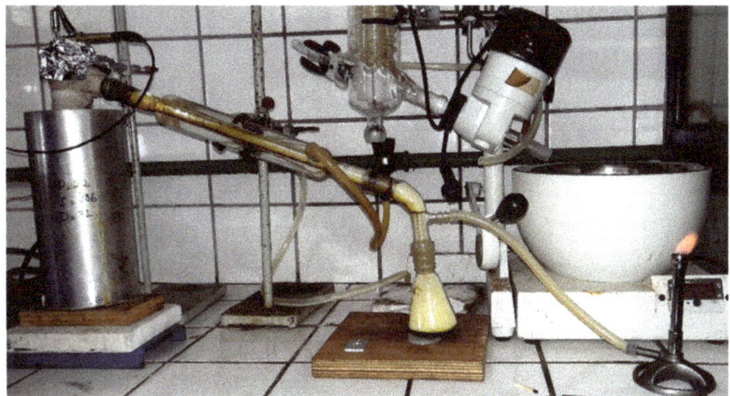

Figure 3. View of the pyrolysis unit in laboratory scale.

Figure 4. View of the pyrolysis unit in bench scale.

The pilot unit is described/presented in detail elsewhere [77]. The unit is composed of seven operational sections (feed/pumping section, pre-heating section, reacting section, cooling section, condensation section, separation & collection, and burning of non-condensable gases, instrumentation and control section). The reacting section consists of the jacketed stainless steel reactor with mechanical stirring system (R-01) of 145 L, operating pressure 1.0 atmosphere, and operating temperature 500 °C. LPG supplies the thermal energy to the reactor (R-01), with the aid of a digital controlled burning system (HOLLAMAQ-MB-20/AB-R, SERIES 05-10), inlet pressure of 0.0–45.0 mbar and energy capacity of 5.000–30,000 Kcal/h, with the mixture LPG-air burning within the annular space concentric to R-01. Figure 5 illustrates the pyrolysis unit in pilot scale. The cooling section consists of 03 polyethylene tanks of 500 L, with H_2O at 25 °C, connected to the condenser in a closed loop. H_2O flows continuously using 01 (one) centrifugal pump (DANCOR, CHS-17) of 1.0 Hp. The condensation section has a multi-tube stainless steel heat exchanger (C-01) of 1.30 m^2, maximum mass flow of condensing gases of 25 kg/h, H_2O flow rate at 25 °C of 4.5 m^3/h, maximum temperature and operating temperatures of 100 and 50 °C, pipe side design pressure and test pressure 1.0 atmosphere, operating in countercurrent mode. The separation & collection, and burning of non-condensable gases contains a stainless steel collection vessel (VC-02) of 30.0 L, operating pressure 1.0 atmosphere. The non-condensable gases exhausted using piping lines of 25 mm and gate valves at the top of VC-02 collection vessel. The non-condensable gases burned in the flare system.

Figure 5. View of the pyrolysis unit in pilot scale.

2.5.2. Experimental Procedures

The pyrolysis of dried Açaí seeds carried out in laboratory, bench, and pilot scale. Pyrolysis in laboratory scale performed using approximately 50.0 g seeds, weighed on a semi-analytical balance (QUIMIS, São Paulo-Brazil, Model: Q-500L210C), placed inside the reactor. Afterwards, the reactor inserted in the jacketed cylindrical furnace. The control system programmed the reaction time, the heating rate, and the process temperature. The heating rate was set equal to 10 °C/min. The reactor was maintained for 10 min at 450 °C after reaching the set point temperature. The liquid product subjected to filtration to separate the aqueous/bio-oil phases and compute the yields of bio-oil and aqueous phases. The coke weighed and the mass of gaseous phase computed by difference. This separation carried out with aid of a glass funnel and a porous filter medium (Qualitative Filter Paper 24.0 Ø; J. PROLAB). Pyrolysis in bench scale performed using approximately 900.0 g seeds, weighed on a semi-analytical balance (QUIMIS, São Paulo-Brazil, Model: Q-500L210C), placed inside the reactor. The control unit programmed the reaction time, the heating rate, and the process temperature. The heating rate was set equal to 10 °C/min. This separation carried out with aid of a glass funnel and a porous filter medium (Qualitative Filter Paper 24.0 Ø; J. PROLAB). The liquid product subjected to filtration to separate the aqueous/bio-oil phases and compute the yields of bio-oil and aqueous phases. The solid phase weighed and the mass of gaseous phase computed by difference. Pyrolysis in pilot scale is performed using approximately 30.0 kg seeds, weighed with the aid of a digital scale (DIGI-TRON, Curitiba-Brazil, Class III). Then, the reactor fed manually by the upper inlet. Afterwards, LPG gas cylinder weighed in order to compute the fuel consumption. The control unit programmed the process temperature. The rector temperature monitored every 10 min to register the formation of liquid and gaseous products. As the temperature rises, the gaseous reaction products condensed inside the collection vessel (VC-02). The non-condensable gases burned in the flare system. The reaction time computed from the moment the reactor reached the operating temperature (set point). The liquid (aqueous fraction + bio-oil) and the solid products weighed to compute the process yields. The liquid (aqueous fraction + bio-oil subjected to filtration using a glass funnel and a porous filter medium (Qualitative Filter Paper 24.0 Ø; J. PROLAB).

2.5.3. Distillation Unit

The fractional distillation of bio-oil was performed using an experimental apparatus similar to those described in the literature [73,77–79]. The distillation apparatus, illustrated in Figure 6, is described in detail by Ferreira et al. [79]. The distillation fractions were subjected to the pretreatment of decantation to separate the aqueous and organic phases, and the organic phase submitted to filtration to remove small solid particles.

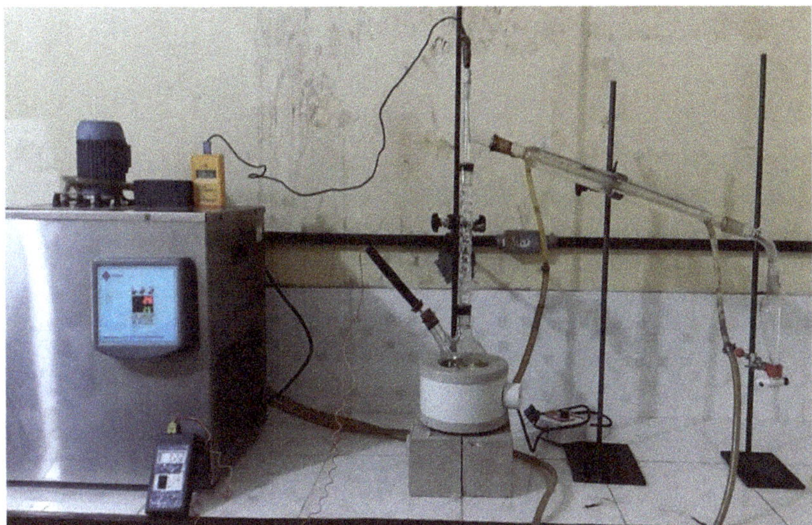

Figure 6. View of distillation column (Vigreux).

2.6. Physical-Chemistry Analysis and Chemical Composition of Bio-Oils and Distillation Fractions

2.6.1. Physical-Chemistry Analysis of Distillation Fractions

The bio-oils and the distillation fractions were (gasoline, light kerosene, and kerosene) physical-chemistry analyzed for acid value, density at 25 °C, kinematic viscosity at 40 °C, and refractive index according to official methods, as described elsewhere [73,77–79]. The analysis of chemical functions (carboxylic acids, aliphatic and aromatic hydrocarbons, ketones, phenols, aldehydes, furans, esters, ethers, etc.) present in distillation fractions determined by FT-IR [73,77–80].

2.6.2. Chemical Composition of Bio-Oils and Distillation Fractions

The composition bio-oils and distillation fractions determined by CG-MS and the equipment and operational procedures described in details elsewhere [73]. The identification of compounds in bio-oils and distillation fractions performed by CG-MS, and the equipment details described in the literature [73]. The GC-MS operating conditions illustrated in Table 1. The intensity, retention time, and compound identification were analyzed according to the NIST mass spectra library. The concentration of all oxygenates and hydrocarbons in each sample was expressed in each area, as no internal standard was injected to compare the peak areas.

Table 1. The GC-MS operating conditions.

Injetor	Injection	T (°C)	Flow Rate (mL/min)	Split
			6.0	1:50
		250	Heating Rate (°C/min)	Volume (µL)
			10	1.0
Oven	Heating Rate (°C/min)	T (°C)		Residence Time (min)
	-	60		1
	5	200		2
	20	230		10
	10	280		39
Detector (MS)	T (°C)	Carrier Gas	Flow Rate (mL/min)	T (°C) Quadrupole
	230	He	30.0	150

3. Results

3.1. Pre-Treatments and Centesimal Characterization of Açaí Seeds

The yields of drying, comminution and sieving pre-treatments were 58.92, 94.34, and 84.36 (wt.%), showing a water content of 41.08 (wt.%). The global yield of drying, comminution and sieving pre-treatments was 46.84 (wt.%). The moisture, volatile matter, ash, and fixed carbon content of Açaí seeds, determined according to official methods [73], were 12.45, 85.98, 0.42, and 13.60 (wt.%).

3.2. Pyrolysis of Açaí Seeds

3.2.1. Material Balances, Operating Conditions, and Yields of Reaction Products

Table 2 summarizes the material balance, process conditions, and yields of reaction products by pyrolysis of Açaí seeds at 450 °C and 1.0 atmosphere, in laboratory, bench, and pilot scales. Figure 7 shows the yields of reaction products by pyrolysis of Açaí seeds carried out in different production scales (laboratory, bench, and pilot), using reactors of 140, 1500, and 143,000 mL, that is, volume scales of approximately ≈ 1:10:1000 [77]. One observes that the yield of bio-oil lies between 4.37 and 13.09 (wt.%), decreasing with increasing reactor volume, while the yields of coke and gas increase with increasing reactor volume. Since the pyrolysis experiments in different production scales used no mechanical stirring system, the higher the reactor volume, the worse the energy transport by conduction in the solid phase (seeds) and convection in the fluid phase (air), and hence the lower the carbonization grad of biomass. The lower the carbonization grade of biomass, the lower the bio-oil yield. This explains a decrease of bio-oil yield with increasing reactor volume. The yield of gas ranged from 24.14 to 27.0 (wt.%), showing no differences between laboratory and bench scales and an increment of ~10% compared to the pilot scale.

3.2.2. Physical-Chemistry Characterization of Bio-Oil

The physico-chemical properties of bio-oil obtained by pyrolysis of dried Açaí (*Euterpe oleracea* Mart.) seeds at 450 °C and 1.0 atmosphere, in different production scales (laboratory, bench, and pilot), are illustrated in Table 3. The measured bio-oil density in pilot scale was 1.043 g/mL, close to the density of (1.066 g/mL, 20 °C) softwood bark residues bio-oil [21], and the density of (1.030 g/mL, 20 °C) palm empty fruit bunches bio-oil [81]. The bio-oil density was lower than the densities of corn Stover bio-oil [25], rice husk bio-oils [28,41,69], and loblolly pine wood chips bio-oil [82]. This is probably due to the high hydrocarbon content in bio-oil [35], but also due to the absence of dissolved acidified aqueous phase in bio-oil after the separation and purification steps of decantation and filtration [35].

Table 2. Mass balances, process conditions, and yields of reaction products (coke, bio-oil, H_2O, and gas) in laboratory, bench, and pilot scales.

	Temperature		
Process Parameters	**450 ($^\circ$C)**	**450 ($^\circ$C)**	**450 ($^\circ$C)**
	Pilot	**Bench**	**Laboratory**
Mass of Açaí Seeds (kg)	30	900×10^{-3}	50.49×10^{-3}
Mass of LPG (g)	13.33	-	-
Cracking Time (min)	160	90	53
Initial Cracking Temperature ($^\circ$C)	117	115	310
Mass of Liquid (Bio-Oil + H_2O) (kg)	11.40	394.07×10^{-3}	24.95×10^{-3}
Mass of Coke (kg)	10.50	283.97×10^{-3}	13.35×10^{-3}
Mass of Bio-Oil (kg)	1.31	50.40×10^{-3}	6.61×10^{-3}
Mass of H_2O (kg)	10.09	343.67×10^{-3}	18.34×10^{-3}
Mass of Gas (kg)	8.10	221.96×10^{-3}	12.19×10^{-3}
Yield of Bio-Oil (wt.%)	4.37	6.60	13.09
Yield of Coke (wt.%)	35.00	31.55	26.44
Yield of H_2O (wt.%)	33.63	38.19	36.32
Yield of Gas (wt.%)	27.00	24.66	24.14

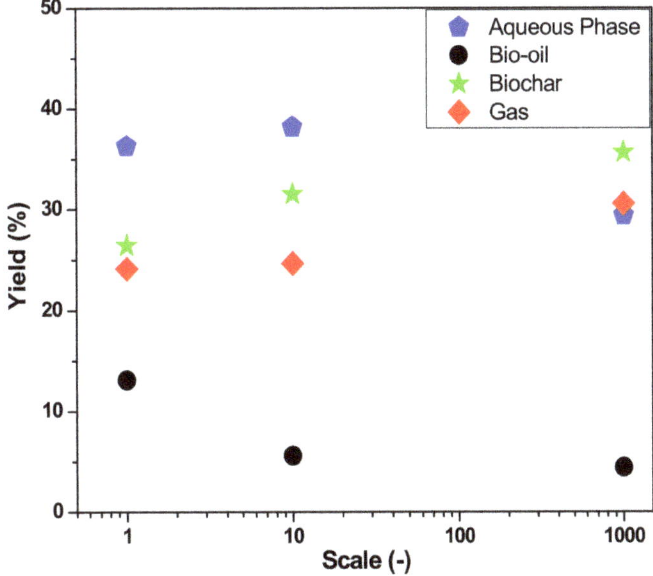

Figure 7. Yields of reaction products in laboratory, bench, and pilot scales.

Table 3. Physico-chemical properties of bio-oil obtained by pyrolysis of dried Açaí seeds at 450 °C and 1.0 atmosphere, compared to similar data reported in the literature [21,25,28,41,69,81,82].

Physicochemical Properties	450 °C Pilot	450 °C Laboratory	450 °C Bench	[21]	[25]	[28]	[41]	[69]	[82]	[81]	ANP No 65
	Bio-Oil	Bio-Oil	Bio-Oil	Bio-Oil	Bio-Oil	Bio-Oil	Bio-Oil	Bio-Oil	Bio-Oil	Bio-Oil	
ρ (g/cm³), 30 °C	1.043	ND	ND	1.066	1.250	1.140	1.190	1.1581	1.200	1.030	0.82–0.85
I. A (mg KOH/g)	70.26	70.25 ± 1.00	68.31 ± 0.90	-	-	-	-	-	-	-	-
ν (mm²/s), 40 °C	68.34	ND	ND	38.0	148.0	13.2	40.0 *	5.0–13.0	12.0	-	2.0–4.5

I.A = Acid Value; ANP: Brazilian National Petroleum Agency, Resolution No 65 (Specification of Diesel S10); ND = Not Determined.
* (Measured at 60 °C).

The Açaí seeds' bio-oil viscosity (68.34 mm^2/s) is lower than that of (148 mm^2/s, 60 °C) corn Stover [25]. The bio-oil viscosity (68.34 mm^2/s) is higher than those of (38.0 mm^2/s) softwood bark residues [21], (13.2 mm^2/s) rice husk [28], (40.0 mm^2/s, 60 °C) rice husk [41], (5.0–13.0 mm^2/s, 40 °C) rice husk [69], and (12.0 mm^2/s, 40 °C) loblolly pine wood chips [82]. The results for density and kinematic viscosity are according to similar data reported elsewhere [27], where average density of wood bio-oil is 1.2 g/cm^3 and the kinematic viscosity of wood bio-oils at 40 and 50 °C lies between 40 and 150 mm^2/s [21,25,28,41,69,81,82].

The effect of reactor scale on the acid value of bio-oil by pyrolysis of Açaí seeds pyrolysis at 450 °C and 1.0 atmosphere, in laboratory, bench, and pilot scales illustrated in Figure 8. The acid value of bio-oils ranged from 68.31 to 70.26 (wt.%), showing no significant differences between laboratory, bench, and pilot scales.

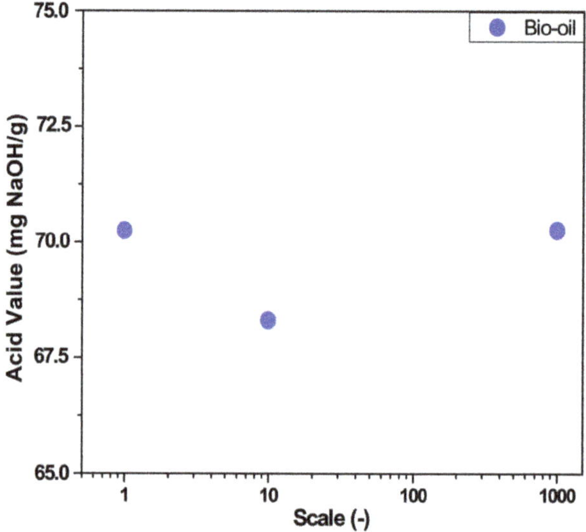

Figure 8. Acid value of bio-oil by pyrolysis of Açaí seeds pyrolysis at 450 °C and 1.0 atmosphere, in laboratory, bench, and pilot scales.

The acid value of bio-oils obtained by pyrolysis of dried Açaí seeds at 450 °C and 1.0 atmosphere, in different production scales (laboratory, bench, and pilot), summarized in Table 4, showed no significate deviation, varying between 67.41 and 71.25 mg KOH/g, close to the acid value of (70.50 mg KOH/g) corn Stover bio-oil [43]. The acid value of bio-oils were lower than that of (95.0 mg KOH/g) corncobs [43], (82.0 mg NaOH/g) sugarcane bagasse [83], Douglas fir (124.0 mg KOH/g) [84], Hardwood (91.7 mg KOH/g) [84], oak (133.0 mg KOH/g) [84], poplar (129.0 mg KOH/g) [84], pine (91.6 mg KOH/g) [84], softwood (115.0 mg KOH/g) [84], switch-grass (125.0 mg KOH/g) [84], and wheat straw (94.9 mg KOH/g) [84], and higher than the acid value of (47.7 mg NaOH/g) softwood bark bio-oil [85], and the acid value of (24.0 mg KOH/g) corn Stover bio-oil [48]. The high acidity of bio-oils is due to the presence of oxygenates compounds, such as carboxylic acids, phenols, cresols, ketones, and aldehydes [35]. This is according to the results reported by Oasma et al. [34], who stated that acidity of fast pyrolysis bio-oils, is due to not only volatile carboxylic acids but also other functional groups such as phenols, resin acids, and hydroxyl acids [34].

Table 4. Acid value of bio-oil obtained by pyrolysis of dried Açaí seeds at 450 °C and 1.0 atmosphere, compared to similar data reported in the literature [43,48,83–85].

| Physicochemical Properties | 450 °C Pilot | 450 °C Laboratory | 450 °C Bench | [43] | [48] | [83] | [84] | [84] | [84] | [84] | [84] | [84] | [84] | [84] | [85] |
	Bio-Oil	Bio-Oil	Bio-Oil	Bio-Oil	Bio-Oil	Bio-Oil	Douglas Fir	Hardwood	Oak	Poplar	Pine	Softwood	Switch-Grass	Wheat Straw	Bio-Oil
I. A [mg KOH/g]	70.26	70.25 ± 1.00	68.31 ± 0.90	95.00	24.00	82.00	124.00	91.70	133.00	129.00	91.60	115.00	125.00	94.90	47.70

I.A = Acid Value.

3.3. Distillation of Bio-Oil from Pyrolysis of Açaí Seeds

Table 5 summarizes the material balance and yields by fractional distillation of bio-oil, and the distillation fractions and bottoms are illustrated in Figure 9. The yields of fuel-like fractions (gasoline, light kerosene, and kerosene) were 16.16, 19.56, and 41.89 (wt.%), respectively, giving a total distillation yield of 77.61 (wt.%). The results are according to similar studies for distillation of biomass derived bio-oil in the literature [17–19,21,41,46–48,53,66,70,72]. The yield of distillation fractions is higher than those reported in the literature for both atmospheric and vacuum conditions [17–19,21,41,46–48,53,66,70,72].

Table 5. Material balance and yields by fractional distillation of bio-oil.

| Distillation: Vigreux Column | Bio-Oil (g) | Gas | Raffinate | Distillates (g) | | | | | Yield (wt.%) | | | | |
	(g)	(g)	(g)	H₂O	G	LK	K	LD	H₂O	G	LK	K	LD
(30–215 °C)	307.53	0	69.87	0	49.48	59.91	128.27	0	0	16.16	19.56	41.89	0

G = Gasoline, LK = Light Kerosene, K = Kerosene, LD = Light Diesel.

Figure 9. Distillation fractions [gasoline (yellow), light kerosene (red), and kerosene (red dark)-like boiling range temperature fossil fuels] and bottoms [Raffinate (black solid)] obtained by fractional distillation of bio-oil produced by pyrolysis of Açaí (*Euterpe oleracea*, Mart.) seeds at 450 °C and 1.0 atmosphere, in pilot scale.

Zheng and Wei [41] reported distillation of fast pyrolysis bio-oil at 80 °C under vacuum (15 mmHg) and a distilled bio-oil yield of 61 (wt.%). Zhang et al. [47] reported atmospheric distillation of fast pyrolysis bio-oil, an accumulated distillate of 51.86 (wt.%). Zhang et al. [47] observed that as the distillation temperature reached 240 °C, condensation reactions take place, generating water, a behavior not observed during the course of distillation as illustrated in Table 5. Capunitan and Capareda [48] reported, for the distillation in atmospheric conditions, an organic phase (Distillates) yield of 15.0 (wt.%) at 100 °C, 4.7 (wt.%) between 100 °C < $T^{Boiling}$ < 180 °C, and 45.3 (wt.%) between 180 °C < $T^{Boiling}$ < 250 °C, while vacuum distillation yielded 10.3 (wt.%) of an organic phase at 80 °C, 5.9 (wt.%) between 80 °C < $T^{Boiling}$ < 160 °C, and 40.9 (wt.%) between 160 °C < $T^{Boiling}$ < 230 °C. Elkasabi et al. [53] reported by distillation of tail-gas reactive pyrolysis (TGRP) bio-oil, yields ranging from 55 to 65 (wt.%).

Physico-Chemical Characterization of Distillation Fractions

The physical-chemical properties of distillation fractions (gasoline, 80–175 °C; light kerosene, 175–200 °C; and kerosene-like fraction, 200–215 °C) of bio-oil are illustrated in Table 6.

Table 6. Physico-chemical properties of distillation fractions of bio-oil produced by pyrolysis of Açaí seeds at 450 °C and 1.0 atmosphere, in pilot scale.

Physico-Chemical Properties	450 °C			ANP N° 65
	G	**LK**	**K**	
ρ (g/cm^3), 30 °C	0.9146	0.9191	0.9816	0.82–0.85
I. A (mg KOH/g)	14.94	61.08	64.78	
I. R (-)	1.455	1.479	1.497	
ν (mm^2/s), 40 °C	1.457	3.106	4.040	2.0–4.5

I.A = Acid Value, I.R = Refractive Index.

It can be observed that acidity of distillation fractions increases with boiling temperature. However, the acidity of gasoline-like fraction is much lower than that of raw bio-oil (70.26 mg KOH/g), as described in Table 4. The high acid value of bio-oil is due to the presence of 78.48 (area.%) oxygenates [35]. The same behavior was observed for the densities, kinematic viscosities, and refractive indexes of gasoline, light kerosene, and kerosene-like like fractions with increasing boiling temperature. This is probably due to the high concentration of higher-boiling-point compounds in the distillate fractions, such as phenols, cresols (*p*-cresol, *o*-cresol), and furans [35], as the concentration of those compounds in the distillation fractions increases with increasing boiling temperature as reported elsewhere [35,66,70,72].

The gasoline, light-kerosene, and kerosene-like fuel densities were 0.9146, 09191, and 0.9816 g/mL. The gasoline-like fuel density (fractions (40 °C < $T^{Boiling}$ < 175 °C), higher, but close to the density of distillation fraction of 0.8733 g/mL ($T^{Boiling}$ < 140 °C) for jatropha curcas cake pyrolysis bio-oil reported by Majhi et al. [46]. This is probably due to the high lipids content between 14–18 (wt.%) and 10–10.9 (wt.%) fiber, thus producing a bio-oil similar to lipid-based pyrolysis organic liquid products [77,78]. The gasoline, light-kerosene, and kerosene-like fuel kinematic viscosities were 1.457, 3.106, and 4.040 mm^2/s, lower than the distillation fraction kinematic viscosity of 2.350 mm^2/s ($T^{Boiling}$ < 140 °C) for jatropha curcas cake pyrolysis bio-oil reported by Majhi et al. [46].

The acid value of gasoline, light-kerosene, and kerosene-like fuel fractions were 14.94, 61.08, and 64.78 mg KOH/g, lower than the distillation fraction acid value of 0.05 mg KOH/g ($T^{Boiling}$ < 140 °C) for jatropha curcas cake pyrolysis bio-oil distillation reported by Majhi et al. [46], the organic phases (distillates) acid values of 4.1 (100 °C < $T^{Boiling}$), 15.1 (100 °C < $T^{Boiling}$ < 180 °C), and 7.41 (180 °C < $T^{Boiling}$ < 250 °C) mg KOH/g, for corn Stover bio-oil atmospheric distillation reported by Capunitan and Capareda [48], the organic phases (distillates) acid values of 3.0 (80 °C < $T^{Boiling}$), 13.9 (80 °C < $T^{Boiling}$ < 160 °C), and 5.0

(160 °C < $T^{Boiling}$ < 230 °C) mg KOH/g, for corn Stover bio-oil vacuum distillation reported by Capunitan and Capareda [48], the acid values of 13.5 mg KOH/g ($T^{Boiling}$ = 192 °C) and 5.3 mg KOH/g ($T^{Boiling}$ = 220 °C) of distillation fractions F_3 and F_4 of $TGRP_1$, and the acid value of 11.1 mg KOH/g ($T^{Boiling}$ = 235 °C) of distillation fraction F_5 of $TGRP_2$, for tail-gas reactive pyrolysis of horse manure ($TGRP_1$), switch grass ($TGRP_2$), and eucalyptus ($TGRP_3$), reported by Elkasabi et al. [53].

The results reported by Elkasabi et al. [53], show that fractional distillation was not effective to diminish the acid values of TGRP bio-oil with initial high acid values, what does not agree with the results reported by Capunitan and Capareda [48], as well as those presented in Table 6, showing that the acid values of distillation fractions are lower than that of raw bio-oil, proving that distillation was effective.

3.4. FT-IR and GC-MS Analyses of Bio-Oil and Distillation Fractions

3.4.1. FT-IR Spectroscopy of Bio-Oil and Distillation Fractions

The FT-IR analysis of bio-oils obtained by pyrolysis of Açaí seeds pyrolysis at 450 °C and 1.0 atmosphere, in laboratory, bench, and pilot scales, summarized in Figure 10, and the infrared bands representing the functional groups in Table 7. The identification of absorption bands/peaks was performed according to previous studies [28,48,59,73,77–80,82]. The FT-IR spectroscopy of bio-oils identify the presence of hydrocarbons (alkanes, alkenes, and aromatic hydrocarbons) and oxygenates (phenols, cresols, carboxylic acids, alcohols, ethers, ketones, and furans).

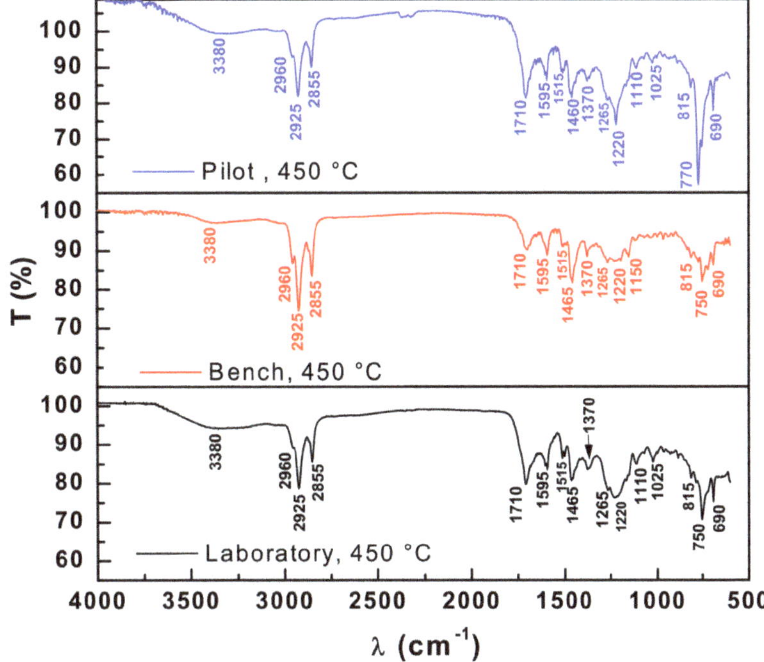

Figure 10. FT-IR of bio-oil obtained by pyrolysis of Açaí seeds pyrolysis at 450 °C and 1.0 atmosphere, in laboratory, bench, and pilot scales.

Table 7. Infrared bands identified in the bio-oils obtained by pyrolysis of Açaí seeds pyrolysis at 450 °C and 1.0 atmosphere, in laboratory, bench, and pilot scales.

Wave-Length (cm^{-1})	Functional Groups	Bio-Oil		
		450 °C Pilot	450 °C Bench	450 °C Laboratory
3380	O–H hydroxyl (polymeric association)	X	X	X
2960–2855	C-H aliphatics (alkanes)	X	X	X
1710	C=O carbonyl (carboxylic acids)	X	X	X
1595–1510	C=C (Aromatic)	X	X	X
1455–1465	-CH$_2$- angular deformation (methylene groups)	X	X	X
1370	CH$_3$ angular deformation (dimethyl groups)	X	X	X
1275–1020	C-O (esters, ethers, alcohols and phenols)	X	X	X
1225–1220	C-O (phenols)	X	X	X
1150	C-O (tertiary alcohol)	-	X	-
1110	C-O (secondary alcohol)	X	-	X
815–690	C=C (adjacent 3H aromatic rings)	X	X	X

Figure 11 illustrates the FT-IR analysis of bio-oil obtained by pyrolysis of Açaí seeds pyrolysis at 450 °C and 1.0 atmosphere, in pilot scale, and the distillation fractions (gasoline: 40–175 °C, light kerosene: 175–200 °C, and kerosene-like fraction: 200–215 °C). The absorption bands/peaks identified according to previous studies [28,48,59,73,77–80,82]. The FT-IR spectroscopy of bio-oil and distillation fraction identify the presence of hydrocarbons (alkanes, alkenes, and aromatic hydrocarbons) and oxygenates (phenols, cresols, carboxylic acids, alcohols, ethers, ketones, and furans).

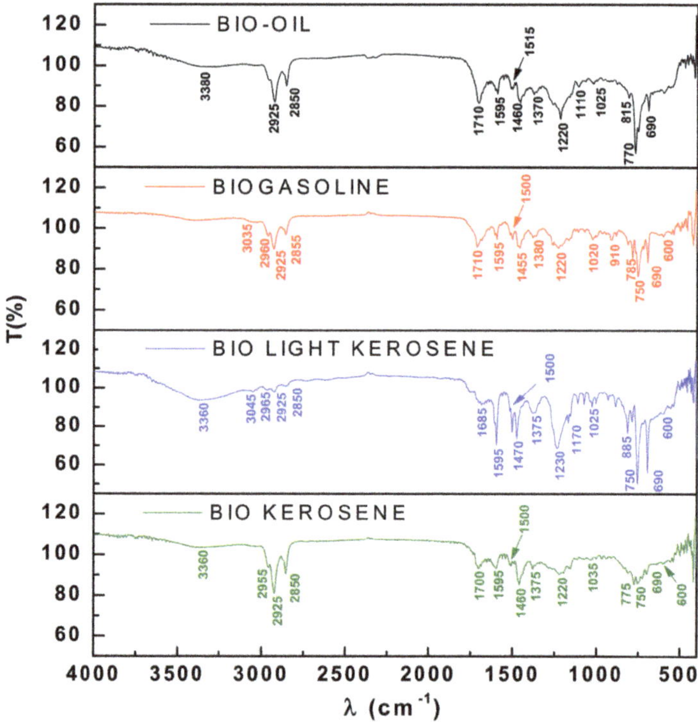

Figure 11. FT-IR of bio-oil obtained by pyrolysis of Açaí seeds pyrolysis at 450 °C and 1.0 atmosphere, in pilot scale, and the distillation fractions (gasoline: 40–175 °C, light kerosene: 175–200 °C, and kerosene-like fraction: 200–215 °C).

3.4.2. Chemical Compositional of Bio-Oil and Distillation Fractions by GC-MS

Chemical Compositional of Bio-Oils by GC-MS

The chromatograms of bio-oils obtained by pyrolysis of Açaí seeds pyrolysis at 450 °C and 1.0 atmosphere, in bench and pilot scales is shown in Figure 12. The peaks are concentrated between retention times of 8.0 and 22.0 min, with the highest one around 12.5 min for the bio-oil obtained pilot scale, while that obtained in bench scale, the peaks are concentrated between retention times of 7.0 and 18.0 min. The GC-MS identified hydrocarbons (alkanes, alkenes, aromatic hydrocarbons, and cycloalkenes) and oxygenates (esters, phenols, cresols, carboxylic acids, ketones, furans, and aldehydes) in bio-oils. The GC-MS identified 48.24 (area.%) hydrocarbons and 51.76 (area.%) oxygenates in the bio-oil in bench scale and 21.52 (area.%) hydrocarbons and 78.48 (area.%) oxygenates in the bio-oil in pilot scale, composed of 21.52 (area.%) hydrocarbons and 78.48 (area.%) oxygenates, as shown in Tables 8 and 9. The high acidity of both bio-oils, described in Table 4, is probably due to the presence of carboxylic acids, ketones, aldehydes, phenols, and cresols confer the high acidity of bio-oil.

The composition of bio-oil shows similarity to those reported in the literature [27,34,41,47,48,53,61], showing the presence of hydrocarbons, phenols, cresols, furans, carboxylic acids, and esters, among other classes of compounds [73]. The identification of hydrocarbons with carbon chain length between C_{11} and C_{15}, shows the presence of heavy gasoline compounds with C_{11} (C_5–C_{11}), light kerosene-like fractions (C_{11}–C_{12}), and light diesel-like fractions (C_{13}–C_{15}), according to Tables 8 and 9.

Chemical Compositional of Distillation Fractions by GC-MS

The chromatograms of bio-oil obtained in pilot scales and distillation fractions is shown in Figure 13. One observes that the spectrum of peaks is moving to the right, showing that distillation was effective to fractionate the bio-oil.

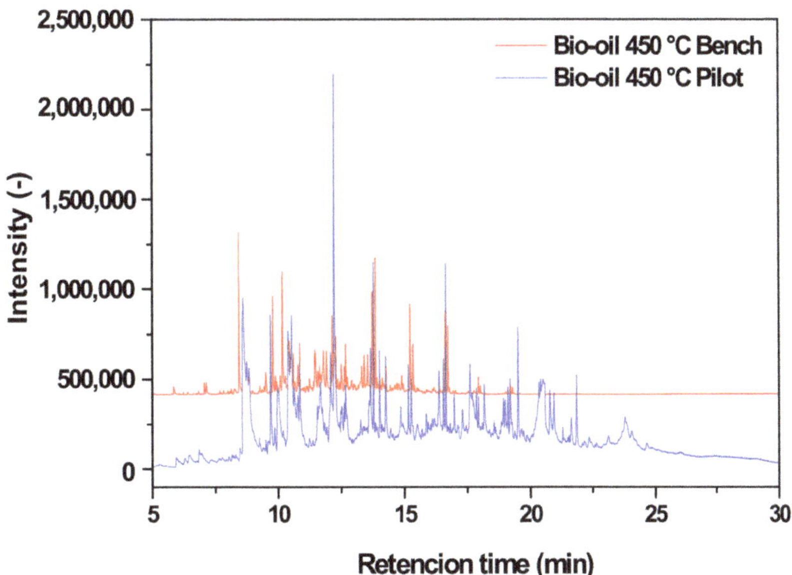

Figure 12. Chromatograms of bio-oils obtained in bench and pilot scales.

Table 8. Classes of compounds and summation of peak areas of hydrocarbons and oxygenates identified by CG-MS in bio-oil obtained in pilot scale.

Classes of Chemical Compounds	ωi (Area.%)
Alkanes	
Σ (Area.%)	7.521
Alkenes	
Σ (Area.%)	2.118
Cycloalkenes	
Σ (Area.%)	1.847
Aromatics	
Σ (Area.%)	10.038
Esters	
Σ (Area.%)	4.065
Carboxylic Acids	
Σ (Area.%)	8.523
Ketones	
Σ (Area.%)	3.533
Phenols	
Σ (Area.%)	35.167
Cresols	
Σ (Area.%)	20.526
Furans	
Σ (Area.%)	5.751
Aldehydes	
Σ (Area.%)	0.910

Table 9. Classes of compounds and summation of peak areas of hydrocarbons and oxygenates identified by CG-MS in bio-oil obtained in bench scale.

Classes of Chemical Compounds	ωi (Area.%)
Alkanes	
Σ (Area.%)	13.14
Alkenes	
Σ (Area.%)	25.50
Aromatics	
Σ (Area.%)	9.59
Alcohols	
Σ (Area.%)	0.43
Ketones	
Σ (Area.%)	4.43
Phenols	
Σ (Area.%)	30.87
Furans	
Σ (Area.%)	15.24
Aldehydes	
Σ (Area.%)	0.80

The GC-MS identified in gasoline-like fraction hydrocarbons and oxygenates. The gasoline-like fraction contains 64.0 (area.%) hydrocarbons [13.27 (area.%) alkenes, 9.41 (area.%) alkanes, and 41.32 (area.%) aromatic hydrocarbons] and 36.0 (area.%) oxygenates [5.50 (area.%) esters, 2.61 (area.%) ketones, 1.35 (area.%) phenols, 6.05 (area.%) alcohols, 13.24 (area.%) furans, and 7.25 (area.%) aldehydes]. The absence of carboxylic acids confers the low acidity of gasoline-like fraction, as summarized in Table 10.

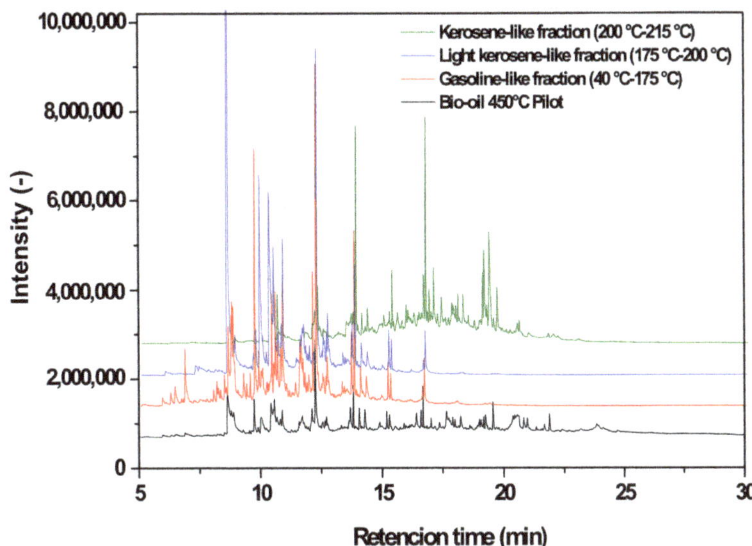

Figure 13. Chromatograms of bio-oil in pilot scale and distillation fractions.

Table 10. Classes of compounds and summation of peak areas of hydrocarbons and oxygenates identified by CG-MS in gasoline-like fraction (40–175 °C) after distillation of bio-oil obtained in pilot scale.

Classes of Chemical Compounds	ωi (Area.%)
Alkanes	
Σ (Area.%)	9.41
Alkenes	
Σ (Area.%)	13.27
Aromatics	
Σ (Area.%)	41.32
Alcohols	
Σ (Area.%)	6.05
Esters	
Σ (Area.%)	5.50
Ketones	
Σ (Area.%)	2.61
Phenols	
Σ (Area.%)	1.35
Furans	
Σ (Area.%)	13.24
Aldehydes	
Σ (Area.%)	7.25

The light kerosene-like fraction is composed of 66.67 (area.%) hydrocarbons [17.60 (area.%) alkenes, 32.65 (area.%) alkanes, and 16.42 (area.%) aromatic hydrocarbons] and 33.33 (area.%) oxygenates [6.16 (area.%) esters, 4.24 (area.%) ketones, 3.26 (area.%) carboxylic acids, 7.13 (area.%) phenols, 8.30 (area.%) alcohols, 2.39 (area.%) furans, and 1.86 (area.%) aldehydes]. The presence of carboxylic acids, ketones, furans, and phenols is associated to the high acidity of light kerosene-like fraction, as shown in Table 11.

Table 11. Classes of compounds and summation of peak areas of hydrocarbons and oxygenates identified by CG-MS in light-kerosene-like fraction (175–200 °C) after distillation of bio-oil obtained in pilot scale.

Classes of Chemical Compounds	ωi (Area.%)
Alkanes	
Σ (Area.%)	32.65
Alkenes	
Σ (Area.%)	17.60
Aromatics	
Σ (Area.%)	16.42
Esters	
Σ (Area.%)	6.16
Carboxylic Acids	
Σ (Area.%)	3.26
Ketones	
Σ (Area.%)	4.24
Phenols	
Σ (Area.%)	7.13
Alcohols	
Σ (Area.%)	8.30
Furans	
Σ (Area.%)	2.39
Aldehydes	
Σ (Area.%)	1.86

By the GC-MS analysis of kerosene-like fraction, hydrocarbons, and oxygenate were identified. The kerosene-like fraction contains 19.87 (area.%) hydrocarbons [2.79 (area.%) alkenes, 4.20 (area.%) alkanes, and 12.88 (area.%) aromatic hydrocarbons] and 81.13 (area.%) oxygenates [2.06 (area.%) esters, 0.80 (area.%) ethers, 3.50 (area.%) ketones, 60.79 (area.%) phenols, 0.96 (area.%) alcohols, 8.99 (area.%) furans, and 3.22 (area.%) aldehydes), as shown in Table 12. The presence of ketones, furans, ethers, esters, aldehydes, and phenols confer the high acidity of kerosene-like fraction, as summarized in Table 12. Finally, the content of hydrocarbons within gasoline (40 °C < $T^{Boiling}$ < 175 °C), light kerosene (175 °C < $T^{Boiling}$ < 200 °C), and kerosene-like fraction (200 °C < $T^{Boiling}$ < 215 °C) are higher than those reported in the literature [17–19,46–48,66,70,72], proving that distillation was not only effective to diminish the acidity, but also to concentrate hydrocarbons. The classes of compounds, summation of peak areas, CAS number, and retention times of chemical compounds identified by CG-MS in bio-oil obtained by pyrolysis of Açaí seeds at 450 °C and 1.0 atmosphere, in pilot scale, as well as in distillation fractions as summarized in details in Supplementary Tables S1–S4.

Table 12. Classes of compounds and summation of peak areas of hydrocarbons and oxygenates identified by CG-MS in kerosene-like fraction (200–215 °C) after distillation of bio-oil obtained in pilot scale.

Classes of Chemical Compounds	ωi (Area.%)
Alkanes	
Σ (Area.%)	4.20
Alkenes	
Σ (Area.%)	2.79
Aromatics	
Σ (Area.%)	12.88
Alcohols	
Σ (Area.%)	0.96

Table 12. *Cont.*

Classes of Chemical Compounds	ωi (Area.%)
Ethers	
Σ (Area.%)	0.80
Ketones	
Σ (Area.%)	3.50
Phenols	
Σ (Area.%)	60.79
Esters	
Σ (Area.%)	2.06
Furans	
Σ (Area.%)	8.99
Aldehydes	
Σ (Area.%)	3.22

4. Conclusions

The yields of bio-oil decrease with increasing reactor volume, while that of coke and gas increase. The yield of distillation fractions (gasoline, light kerosene, and kerosene-like like fractions), 77.61 (wt.%), is higher but according than those reported in the literature for both atmospheric and vacuum conditions [17–19,21,41,46–48,53,66,70,72]. The acid values of distillation fractions increase with increasing boiling temperature. However, the acidity of gasoline-like fraction is much lower than that of raw bio-oil (70.26 mg KOH/g). The same behavior was observed for the densities, kinematic viscosities, and refractive indexes of gasoline, light kerosene, and kerosene-like like fractions with increasing boiling temperature.

The FT-IR analysis of bio-oil and distillation fraction identify the presence of hydrocarbons (alkanes, alkenes, and aromatic hydrocarbons) and oxygenates (phenols, cresols, carboxylic acids, alcohols, ethers, ketones, and furans). The bio-oil is composed of 21.52 (area.%) hydrocarbons and 78.48 (area.%) oxygenates. The presence of carboxylic acids, as well as phenols and cresols, is associated to the high acidity of bio-oil.

The gasoline-like fraction is composed by 64.0 (area.%) hydrocarbons and 36.0 (area.%) oxygenates, while light kerosene-like fraction by 66.67 (area.%) hydrocarbons and 33.33 (area.%) oxygenates, and kerosene-like fraction by 19.87 (area.%) hydrocarbons and 81.13 (area.%) oxygenates. The content of hydrocarbons within the distillation fractions are higher than those reported in the literature [17–19,46–48,66,70,72], showing that distillation was effective not only to diminish the acidity, but also to concentrate hydrocarbons.

Supplementary Materials: The following are available online at https://www.mdpi.com/article/10.3390/en14133713/s1, Table S1: Classes of compounds, summation of peak areas, CAS number, and retention times of chemical compounds identified by CG-MS in bio-oil obtained by pyrolysis of Açaí seeds at 450 °C and 1.0 atmosphere, in pilot scale., Table S2: Classes of compounds, summation of peak areas, CAS number, and retention times of chemical compounds identified by CG-MS in gasoline-like fraction (40–175 °C) after distillation of bio-oil obtained by pyrolysis of Açaí seeds at 450 °C and 1.0 atmosphere, in pilot scale., Table S3: Classes of compounds, summation of peak areas, CAS number, and retention times of chemical compounds identified by CG-MS in light kerosene-like fraction (175–200 °C) after distillation of bio-oil obtained by pyrolysis of Açaí seeds at 450 °C and 1.0 atmosphere, in pilot scale., Table S4: Classes of compounds, summation of peak areas, CAS number, and retention times of chemical compounds identified by CG-MS in kerosene-like fraction (200–215 °C) after distillation of bio-oil obtained by pyrolysis of Açaí seeds at 450 °C and 1.0 atmosphere, in pilot scale.

Author Contributions: Investigation, Methodology, Formal analysis and Writing original draft preparation, D.A.R.d.C.; Formal analysis and Software, M.C.S.; Formal analysis, Software, and Visualization, H.d.S.A.; Investigation and Validation, H.J.d.S.R.; Investigation and Methodology, L.H.H.G.; Chemical analysis, S.J.B.; Chemical analysis, L.P.B.; Resources and Chemical analysis, S.D.J.;

Co-supervision and Resources, L.E.P.B.; Supervision, Conceptualization, and Data curation, N.T.M. All authors have read and agreed to the published version of the manuscript.

Funding: This research was partially funded by CNPq-Brazil, grant number: 207325/2014-6.

Informed Consent Statement: Not applicable.

Data Availability Statement: Data available in a publicly accessible repository that does not issue DOIs (https://drive.google.com/file/d/1mOxMKWsXkZDLVfnnsrWc6NsHLheW4u1p/view, accessed on 12 May 2021).

Acknowledgments: I would like to acknowledge and dedicate this research in memory to >Hélio da Silva Almeida, Professor at the Faculty of Sanitary and Environmental Engineering/UFPa, who passed away in 13 March 2021. His contagious joy, dedication, intelligence, honesty, seriousness, and kindness will always be remembered in our hearts.

Conflicts of Interest: The authors declare no conflict of interest.

References

1. Scherwinski-Pereira, J.E.; Guedes, R.D.S.; Da Silva, R.A.; Fermino, P.C.P.; Luis, Z.G.; Freitas, E.D.O. Somatic embryogenesis and plant regeneration in açaí palm (Euterpe oleracea). *Plant Cell Tissue Organ Cult. (PCTOC)* **2012**, *109*, 501–508. [CrossRef]
2. Schauss, A.G.; Wu, X.; Prior, R.L.; Ou, B.; Patel, D.; Huang, D.; Kababick, J.P. Phytochemical and Nutrient Composition of the Freeze-Dried Amazonian Palm Berry, Euterpe oleraceaeMart. (Acai). *J. Agric. Food Chem.* **2006**, *54*, 8598–8603. [CrossRef] [PubMed]
3. Sabbe, S.; Verbeke, W.; Deliza, R.; Matta, V.; Van Damme, P. Effect of a health claim and personal characteristics on consumer acceptance of fruit juices with different concentrations of açaí (Euterpe oleracea Mart.). *Appetite* **2009**, *53*, 84–92. [CrossRef] [PubMed]
4. Pacheco-Palencia, L.A.; Duncan, C.E.; Talcott, S.T. Phytochemical composition and thermal stability of two commercial açai species, Euterpe oleracea and Euterpe precatoria. *Food Chem.* **2009**, *115*, 1199–1205. [CrossRef]
5. Brondizio, E.S.; Safar, C.A.; Siqueira, A.D. The urban market of Açaí fruit (Euterpe oleracea Mart.) and rural land use change: Ethnographic insights into the role of price and land tenure constraining agricultural choices in the Amazon estuary. *Urban Ecosyst.* **2002**, *6*, 67–97. [CrossRef]
6. Dos Santos Bentes, E.; Oyama Homma, A.K.; Nunes dos Santos, C.A. Exportações de Polpa de Açaí do Estado do Pará: Situação Atual e Perspectivas. In: Anais Congresso da Sociedade Brasileira de Economia, Administração e Sociologia Rural, 55, Santa Maria, RS-Brazil. 2017. Available online: https://www.researchgate.net/publication/319465735_Exportacoes_de_Polpa_de_Acai_do_Estado_do_Para_Situacao_Atual_e_Perspectivas (accessed on 5 August 2020).
7. Almeida, A.V.D.C.; Melo, I.M.; Pinheiro, I.S.; Freitas, J.F. Appreciation of acai core of a pulp producer from Ananindeua/PA: Proposal of reverse channel structure oriented by NPSW and reverse logistics. *Rev. Gestão Produção Operações Sist.* **2017**, *12*, 59–83. [CrossRef]
8. Townsend, C.R.; de Lucena Costa, N.; de Araújo Pereira, R.G.; Clóvis, C. Diesel Senger. Características Químico-Bromatológica do Caroço de Açaí. COMUNICADO TÉCNICO N° 193 (CT/193), EMBRAPA-CPAF Rondônia, ago./01, 1–5. ISSN 0103-9458. Available online: https://ainfo.cnptia.embrapa.br/digital/bitstream/item/100242/1/Cot193-acai.pdf (accessed on 5 August 2020).
9. Fioravanti, C. Açaí: Do pé para o lanche. *Rev. Pesqui. Fapesp.* **2013**, *203*, 64–68. Available online: http://revistapesquisa.fapesp.br/2013/01/11/folheie-a-edicao-203/ (accessed on 5 August 2020).
10. De Santana, A.C.; De Santana, Á.L.; De Santana, Á.L.; Dos Santos, M.A.S.; De Oliveira, C.M. Análise discriminante múltipla do mercado varejista de açaí em Belém do Pará. *Rev. Bras. Frutic.* **2014**, *36*, 532–541. [CrossRef]
11. Pessoa, J.D.C.; Silva, P.V.D. Effect of temperature and storage on açaí (Euterpe oleracea) fruit water uptake: Simulation of fruit transportation and pre-processing. *Fruits* **2007**, *62*, 295–302. [CrossRef]
12. Cordeiro, M.A. Estudo da Hidrólise Enzimática do Caroço de açaí (Euterpe Oleracea, Mart) para a Produção de Etanol. Dissertação de Mestrado, Programa de Pós-Graduação em Engenharia Química, UFPA-Brazil. Marcio de Andrade Cordeiro. 2016. Available online: https://ppgeq.propesp.ufpa.br/ARQUIVOS/dissertacoes/2016/M%C3%A1rcio%20de%20Andrade%20Cordeiro/M%C3%A1rcio_Disserta%C3%A7%C3%A3o_Defesa.pdf (accessed on 5 August 2020).
13. Da Fonseca, T.R.; de Amorim Silva, T.; Alecrim, M.M.; da Cruz Filho, R.F.; Teixeira, M.F. Cultivation and nutritional studies of an edible mushroom from North Brazil. *Afr. J. Microbiol. Res.* **2015**, *9*, 1814–1822.
14. Kabacknik, A.; Roger, H. Determinação do Poder Calorífico do Caroço do açaí em três Distintas Umidades. In Proceedings of the 38th Congresso Brasileiro de Química, São Luiz, MA, Brazil, 21–25 Septembar 1998.
15. Altman, R.F.A. *O Caroço de açaí (Euterpe Oleracea, Mart)*; Boletim Técnico do Instituto Agronômico do Norte: Belém, Brasil, 1956; Volume 31, pp. 109–111. Available online: https://www.bdpa.cnptia.embrapa.br/consulta/busca?b=ad&biblioteca=vazio&busca=autoria:"ALTMAN,%20R.%20F.%20A." (accessed on 5 August 2020).
16. Özçimen, D.; Ersoy-Meriçboyu, A. Characterization of biochar and bio-oil samples obtained from carbonization of various biomass materials. *Renew. Energy* **2010**, *35*, 1319–1324. [CrossRef]

17. Adjaye, J.D.; Sharma, R.K.; Bakhshi, N.N. Characterization and stability analysis of wood-derived bio-oil. *Fuel Process. Technol.* **1992**, *31*, 241–256. [CrossRef]

18. Carazza, F.; Rezende, M.E.A.; Pasa, V.M.D.; Lessa, A. Fractionation of wood tar. *Proc. Adv. Thermochem. Biomass Convers.* **1994**, *2*, 465.

19. Adjaye, J.; Bakhshi, N. Production of hydrocarbons by catalytic upgrading of a fast pyrolysis bio-oil. Part I: Conversion over various catalysts. *Fuel Process. Technol.* **1995**, *45*, 161–183. [CrossRef]

20. Xu, B.J.; Lu, N. Experimental research on the bio-oil derived from biomass pyrolysis liquefaction. *Trans. Chin. Soc. Agric. Eng.* **1999**, *15*, 177–181.

21. Boucher, M.E.; Chaala, A.; Roy, C. Bio-oils obtained by vacuum pyrolysis of softwood bark as a liquid fuel for gas turbines. Part I: Properties of bio-oil and its blends with methanol and a pyrolytic aqueous phase. *Biomass Bioenergy* **2000**, *19*, 337–350. [CrossRef]

22. Oasmaa, A.; Kuoppala, E.; Gust, S.; Solantausta, Y. Fast Pyrolysis of Forestry Residue. 1. Effect of Extractives on Phase Separation of Pyrolysis Liquids. *Energy Fuels* **2003**, *17*, 1–12. [CrossRef]

23. Czernik, S.; Bridgwater, A.V. Overview of applications of biomass fast pyrolysis oil. *Energy Fuels* **2004**, *18*, 590–598. [CrossRef]

24. Mohan, D.; Pittman, C.U., Jr.; Steelee, P.H. Pyrolysis of wood/biomass for bio-oil: A critical review. *Energy Fuels* **2006**, *20*, 848–889. [CrossRef]

25. Yu, F.; Deng, S.; Chen, P.; Liu, Y.; Wan, Y.; Olson, A.; Kittelson, D.; Ruan, R. Physical and Chemical Properties of Bio-Oils from Microwave Pyrolysis of Corn Stover. *Appl. Biochem. Biotecnol.* **2007**, *136–140*, 957–970. [CrossRef]

26. Zhang, Q.; Chang, J.; Wang, T.; Xu, Y. Review of biomass pyrolysis oil properties and upgrading research. *Energy Convers. Manag.* **2007**, *48*, 87–92. [CrossRef]

27. Boateng, A.A.; Mullen, C.A.; Goldberg, N.; Hicks, K.B. Production of bio-oil from alfalfa stems by fluidized-bed fast pyrolysis. *Ind. Eng. Chem. Res.* **2008**, *47*, 4115–4122. [CrossRef]

28. Lu, Q.; Yang, X.L.; Zhu, X.F. Analysis on chemical and physical properties of bio-oil pyrolyzed from rice husk. *J. Anal. Appl. Pyrolysis* **2008**, *82*, 191–198. [CrossRef]

29. Junming, X.; Jianchun, J.; Yunjuan, S.; Yanju, L. Bio-oil upgrading by means of ethyl ester production in reactive distillation to remove water and to improve storage and fuel characteristics. *Biomass Bioenergy* **2008**, *32*, 1056–1061. [CrossRef]

30. Guo, Z.; Wang, S.; Zhu, Y.; Luo, Z.; Cen, K. Separation of acid compounds for refining biomass pyrolysis oil. *J. Fuel Chem. Technol.* **2009**, *7*, 49–52. [CrossRef]

31. Vispute, T.P.; Huber, G.W. Production of hydrogen, alkanes and polyols by aqueous phase processing of wood-derived pyrolysis oils. *Green Chem.* **2009**, *11*, 1433–1445. [CrossRef]

32. Song, Q.-H.; Nie, J.-Q.; Ren, M.-G.; Guo, Q.-X. Effective Phase Separation of Biomass Pyrolysis Oils by Adding Aqueous Salt Solutions. *Energy Fuels* **2009**, *23*, 3307–3312. [CrossRef]

33. Wang, S.; Gu, Y.; Liu, Q.; Yao, Y.; Guo, Z.; Luo, Z.; Cen, K. Separation of bio-oil by molecular distillation. *Fuel Process. Technol.* **2009**, *90*, 738–745. [CrossRef]

34. Oasmaa, A.; Elliott, D.C.; Korhonen, J. Acidity of Biomass Fast Pyrolysis Bio-oils. *Energy Fuels* **2010**, *24*, 6548–6554. [CrossRef]

35. De Castro, D.A.R. Processo de Produção de Bio-Óleo e Bio-Adsorventes via Pirólise das Sementes do Açaí (Euterpe oleraceae, Mart). Ph.D. Thesis, PRODERNA, UFPa, Belém, Brazil, 2019.

36. Guo, X.; Wang, S.; Guo, Z.; Liu, Q.; Luo, Z.; Cen, K. Pyrolysis characteristics of bio-oil fractions separated by molecular distillation. *Appl. Energy* **2010**, *87*, 2892–2898. [CrossRef]

37. Guo, Z.; Wang, S.; Gu, Y.; Xu, G.; Li, X.; Luo, Z. Separation characteristics of biomass pyrolysis oil in molecular distillation. *Sep. Purif.* **2010**, *76*, 52–57. [CrossRef]

38. Suota, M.J.; Simionatto, E.L.; Scharf, D.R.; Meier, H.F.; Wiggers, V.R. Esterification, Distillation, and Chemical Characterization of Bio-Oil and Its Fractions. *Energy Fuels* **2019**, *33*, 9886–9894. [CrossRef]

39. Nam, H.; Choi, J.; Capareda, S.C. Comparative study of vacuum and fractional distillation using pyrolytic microalgae (*Nannochloropsis oculata*) bio-oil. *Algal Res.* **2016**, *17*, 87–96. [CrossRef]

40. Christensen, E.D.; Chupka, G.M.; Smurthwaite, J.L.T.; Alleman, T.L.; Lisa, K.; Franz, J.A.; Elliott, D.C.; Mc Cormick, R.L. Analysis of oxygenated compounds in hydrotreated biomass fast pyrolysis oil distillate fractions. *Energy Fuels* **2011**, *25*, 5462–5471. [CrossRef]

41. Zheng, J.-L.; Wei, Q. Improving the quality of fast pyrolysis bio-oil by reduced pressure distillation. *Biomass Bioenergy* **2011**, *35*, 1804–1810. [CrossRef]

42. Pollard, A.; Rover, M.; Brown, R.C. Characterization of bio-oil recovered as stage fractions with unique chemical and physical properties. *J. Anal. Appl. Pyrolysis* **2012**, *93*, 129–138. [CrossRef]

43. Shah, A.; Darr, M.J.; Dalluge, D.; Medic, D.; Webster, K.; Brown, R.C. Physicochemical properties of bio-oil and biochar produced by fast pyrolysis of stored single-pass corn stover and cobs. *Bioresour. Technol.* **2012**, *125*, 348–352. [CrossRef] [PubMed]

44. Imam, T.; Capareda, S. Characterization of bio-oil, syngas and bio-char from switch grass pyrolysis at various temperatures. *J. Anal. Appl. Pyrolysis* **2012**, *93*, 170–177. [CrossRef]

45. Xiu, S.; Shahbazi, A. Bio-oil production and upgrading research: A review. *Renew. Sustain. Energy Rev.* **2012**, *16*, 4406–4414. [CrossRef]

46. Majhi, A.; Sharma, Y.K.; Naik, D.V. Blending optimization of Hempel distilled bio-oil with commercial diesel. *Fuel* **2012**, *96*, 264–269. [CrossRef]

47. Zhang, X.-S.; Yang, G.-X.; Jiang, H.; Liu, W.-J.; Ding, H.-S. Mass production of chemicals from biomass-derived oil by directly atmospheric distillation coupled with co-pyrolysis. *Sci. Rep.* **2013**, *3*, srep01120. [CrossRef] [PubMed]
48. Capunitan, J.A.; Capareda, S.C. Characterization and separation of corn stover bio-oil by fractional distillation. *Fuel* **2013**, *112*, 60–73. [CrossRef]
49. Sun, Y.; Gao, B.; Yao, Y.; Fang, J.; Zhang, M.; Zhou, Y.; Chen, H.; Yang, L. Effects of feedstock type, production method, and pyrolysis temperature on biochar and hydrochar properties. *Chem. Eng. J.* **2014**, *240*, 574–578. [CrossRef]
50. Yang, H.; Yao, J.; Chen, G.; Ma, W.; Yan, B.; Qi, Y. Overview of Upgrading of Pyrolysis Oil of Biomass. *Energy Procedia* **2014**, *61*, 1306–1309. [CrossRef]
51. Gooty, A.T.; Li, D.; Berruti, F.; Briens, C. Kraft-lignin pyrolysis and fractional condensation of its bio-oil vapors. *J. Anal. Appl. Pyrolysis* **2014**, *106*, 33–40. [CrossRef]
52. Biradar, C.H.; Subramanian, K.A.; Dastidar, M.G. Production and fuel upgrading of pyrolysis bio-oil Jatropha Curcas de-oiled seed cake. *Fuel* **2014**, *119*, 81–89. [CrossRef]
53. Elkasabi, Y.; Mullen, C.A.; Boateng, A.A. Distillation and isolation of commodity chemicals from bio-oil made by tail-gas reactive pyrolysis. *Sustain. Chem. Eng.* **2014**, *2*, 2042–2052. [CrossRef]
54. Gooty, A.T.; Li, D.; Briens, C.; Berruti, F. Fractional condensation of bio-oil vapors produced from birch bark pyrolysis. *Sep. Purif. Technol.* **2014**, *124*, 81–88. [CrossRef]
55. Wang, S.; Cai, Q.; Wang, X.; Zhang, L.; Wang, Y.; Luo, Z. Biogasoline Production from the Co-cracking of the Distilled Fraction of Bio-oil and Ethanol. *Energy Fuels* **2013**, *28*, 115–122. [CrossRef]
56. Papari, S.; Hawboldt, K. A review on the pyrolysis of woody biomass to bio-oil: Focus on kinetic models. *Renew. Sustain. Energy Rev.* **2015**, *52*, 1580–1595. [CrossRef]
57. Kambo, H.S.; Dutta, A. A comparative review of biochar and hydro-char in terms of production, physico-chemical properties and applications. *Renew. Sustain. Energy Rev.* **2015**, *45*, 359–378. [CrossRef]
58. Kumar, S.; Lange, J.-P.; Van Rossum, G.; Kersten, S.R. Bio-oil fractionation by temperature-swing extraction: Principle and application. *Biomass Bioenergy* **2015**, *83*, 96–104. [CrossRef]
59. Elkasabi, Y.; Boateng, A.A.; Jackson, M.A. Upgrading of bio-oil distillation bottoms into biorenewable calcined coke. *Biomass Bioenergy* **2015**, *81*, 415–423. [CrossRef]
60. Elkasabi, Y.; Mullen, C.A.; Jackson, M.A.; Boateng, A.A. Characterization of fast-pyrolysis bio-oil distillation residues and their potential applications. *J. Anal. Appl. Pyrolysis* **2015**, *114*, 179–186. [CrossRef]
61. Kanaujia, P.K.; Naik, D.V.; Tripathi, D.; Singh, R.; Poddar, M.K.; Siva Kumar Konathala, L.N.; Sharma, Y.K. Pyrolysis of Jatropha Curcas seed cake followed by optimization of liquid–liquid extraction procedure for the obtained bio-oil. *Anal. Appl. Pyrolysis* **2016**, *118*, 202–224. [CrossRef]
62. Kan, T.; Strezov, V.; Evans, T.J. Lignocellulosic biomass pyrolysis: A review of product properties and effects of pyrolysis parameters. *Renew. Sustain. Energy Rev.* **2016**, *57*, 1126–1140. [CrossRef]
63. Cai, J.; Banks, S.W.; Yang, Y.; Darbar, S.; Bridgwater, T. Viscosity of Aged Bio-oils from Fast Pyrolysis of Beech Wood and Miscanthus: Shear Rate and Temperature Dependence. *Energy Fuels* **2016**, *30*, 4999–5004. [CrossRef]
64. Zheng, Y.; Wang, F.; Yang, X.; Huang, Y.; Liu, C.; Zheng, Z.; Gu, J. Study on aromatics production via the catalytic pyrolysis vapor upgrading of biomass using metal-loaded modified H-ZSM-5. *J. Anal. Appl. Pyrolysis* **2017**, *126*, 169–179. [CrossRef]
65. Johansson, A.-C.; Iisa, K.; Sandström, L.; Ben, H.; Pilath, H.; Deutch, S.; Wiinikka, H.; Öhrman, O.G. Fractional condensation of pyrolysis vapors produced from Nordic feedstocks in cyclone pyrolysis. *J. Anal. Appl. Pyrolysis* **2017**, *123*, 244–254. [CrossRef]
66. Kuo, H.P.; Hou, B.R.; Huang, A.N. The influence of the gas fluidization velocity on the properties of bio-oils from fluidized bed pyrolizer with in-line distillation. *Appl. Energy* **2017**, *194*, 279–286. [CrossRef]
67. Guedes, R.E.; Luna, A.S.; Torres, A.R. Operating parameters for bio-oil production in biomass pyrolysis: A review. *J. Anal. Appl. Pyrolysis* **2018**, *129*, 134–149. [CrossRef]
68. Dhyani, V.; Bhaskar, T. A comprehensive review on the pyrolysis of lignocellulosic biomass. *Renew. Energy* **2018**, *129*, 695–716. [CrossRef]
69. Cai, W.; Liu, R.; He, Y.; Chai, M.; Cai, J. Bio-oil production from fast pyrolysis of rice husk in a commercial-scale plant with a downdraft circulating fluidized bed reactor. *Fuel Process. Technol.* **2018**, *171*, 308–317. [CrossRef]
70. Huang, A.-N.; Hsu, C.-P.; Hou, B.-R.; Kuo, H.-P. Production and separation of rice husk pyrolysis bio-oils from a fractional distillation column connected fluidized bed reactor. *Powder Technol.* **2018**, *323*, 588–593. [CrossRef]
71. Rahman, S.; Helleur, R.; MacQuarrie, S.; Papari, S.; Hawboldt, K. Upgrading and isolation of low molecular weight compounds from bark and softwood bio-oils through vacuum distillation. *Sep. Purif. Technol.* **2018**, *194*, 123–129. [CrossRef]
72. Yuan, X.; Sun, M.; Wang, C.; Zhu, X. Full temperature range study of rice husk bio-oil distillation: Distillation characteristics and product distribution. *Sep. Purif. Technol.* **2021**, *263*, 118382. [CrossRef]
73. De Castro, D.A.R.; Ribeiro, H.J.D.S.; Ferreira, C.C.; Cordeiro, M.D.A.; Guerreiro, L.H.H.; Pereira, A.M.; Dos Santos, W.G.; Santos, M.C.; De Carvalho, F.B.; Junior, J.O.C.S.; et al. Fractional Distillation of Bio-Oil Produced by Pyrolysis of Açaí (*Euterpe oleracea*) Seeds. In *Fractionation*; Ibrahim, H.A.-H., Ed.; IntechOpen: London, UK, 2019. [CrossRef]
74. De Andrade Cordeiro, M.; de Almeida, O.; de Castro, D.A.; da Silva Ribeiro, H.J.; Machado, N.T. Produção de Etanol através da Hidrólise Enzimática do Caroço de Açaí (Euterpe oleracea, Mart.). *Rev. Bras. Energ. Renov.* **2019**, *8*, 122–152.
75. *Acid-Insoluble Lignin in Wood and Pulp*; Tappi Method T 222 Om-06; Tappi Press: Atlanta, GA, USA, 2006.

76. Buffiere, P.; Loisel, D. *Dosage des fibres Van Soest*; Weened, Laboratoire de Biotechnologie de l'Environnement, INRA: Narbonne, France, 2007; pp. 1–14.
77. Almeida, H.D.S.; Corrêa, O.; Eid, J.; Ribeiro, H.; De Castro, D.; Pereira, M.; Pereira, L.; Aâncio, A.D.A.; Santos, M.; Da Mota, S.; et al. Performance of thermochemical conversion of fat, oils, and grease into kerosene-like hydrocarbons in different production scales. *J. Anal. Appl. Pyrolysis* **2016**, *120*, 126–143. [CrossRef]
78. Da Mota, S.A.; Mancio, A.A.; Lhamas, D.E.; De Abreu, D.H.; Da Silva, M.S.; Dos Santos, W.G.; De Castro, D.A.; De Oliveira, R.M.; Araújo, M.E.; Borges, L.E.; et al. Production of green diesel by thermal catalytic cracking of crude palm oil (Elaeis guineensis Jacq) in a pilot plant. *J. Anal. Appl. Pyrolysis* **2014**, *110*, 1–11. [CrossRef]
79. Ferreira, C.C.; Costa, E.C.; de Castro, D.A.; Pereira, M.S.; Mâncio, A.A.; Santos, M.C.; Lhamas, D.E.; Da Mota, S.A.; Leão, A.C.; Duvoisin, S., Jr.; et al. Deacidification of organic liquid products by fractional distillation in laboratory and pilot scales. *J. Anal. Appl. Pyrolysis* **2017**, *127*, 468–489. [CrossRef]
80. Seshadri, K.S.; Cronauer, D.C. Characterization of coal-derived liquids by 13C N.M.R. and FT-IR Spectroscopy. *Fuel* **1983**, *62*, 1436–1444. [CrossRef]
81. Abnisa, F.; Arami-Niya, A.; Daud, W.M.; Sahu, J.N. Characterization of Bio-oil and Bio-char from Pyrolysis of Palm Oil Wastes. *BioEnergy Res.* **2013**, *6*, 830–840. [CrossRef]
82. Tanneru, S.K.; Parapati, D.R.; Steele, P.H. Pretreatment of bio-oil followed by upgrading via esterification to boiler fuel. *Energy* **2014**, *73*, 214–220. [CrossRef]
83. Garcia-Perez, M.; Chaala, A.; Roy, C. Vacuum pyrolysis of sugarcane bagasse. *J. Anal. Appl. Pyrolysis* **2002**, *65*, 111–136. [CrossRef]
84. Nolte, M.W.; Liberatore, M.W. Viscosity of Biomass Pyrolysis Oils from Various Feedstocks. *Energy Fuels* **2010**, *24*, 6601–6608. [CrossRef]
85. Ba, T.; Chaala, A.; Garcia-Perez, M.; Rodrigue, D.; Roy, C. Colloidal properties of bio-oils obtained by vacuum pyrolysis of softwood bark. Characterization of water-soluble and water-insoluble fractions. *Energy Fuels* **2004**, *18*, 704–712. [CrossRef]

Article

Performance Optimisation of Fuel Pellets Comprising Pepper Stem and Coffee Grounds through Mixing Ratios and Torrefaction

Sunyong Park [1], Hui-Rim Jeong [2], Yun-A Shin [2], Seok-Jun Kim [1], Young-Min Ju [3], Kwang-Cheol Oh [4], La-Hoon Cho [1] and DaeHyun Kim [1,2,*]

[1] Department of Interdisciplinary Program in Smart Agriculture, Kangwon National University, Hyoja 2 Dong 192-1, Chuncheon-si 200-010, Korea; psy0712@kangwon.ac.kr (S.P.); ksj91@kangwon.ac.kr (S.-J.K.); jjola1991@kangwon.ac.kr (L.-H.C.)

[2] Department of Biosystems Engineering, Kangwon National University, Hyoja 2 Dong 192-1, Chuncheon-si 200-701, Korea; jhr524@naver.com (H.-R.J.); saltlyspy@naver.com (Y.-A.S.)

[3] Division of Wood Chemistry, Department of Forest Products, National Institute of Forest Science, Seoul 02455, Korea; jym889@korea.kr

[4] Green Materials & Processes R&D Group, Korea Institute of Industrial Technology, 55, Jongga-ro, Jung-gu, Ulsan 44413, Korea; okc@kitech.re.kr

* Correspondence: daekim@kangwon.ac.kr

Abstract: Agricultural by-products have several disadvantages as fuel, such as low calorific values and high ash contents. To address these disadvantages, this study examined the mixing of agricultural by-products and spent coffee grounds, for use as a solid fuel, and the improvement of fuel characteristics through torrefaction. Pepper stems and spent coffee grounds were first dried to moisture contents of <15% and then combined, with mixing ratios varying from 9:1 to 6:4. Fuel pellets were produced from these mixtures using a commercial pelletiser, evaluated against various standards, and classified as grade A, B, or Bio-SRF. The optimal ratio of pepper stems to spent coffee grounds was determined to be 8:2. The pellets were torrefied to improve their fuel characteristics. Different torrefaction temperatures improved the mass yields of the pellets to between 50.87% and 88.27%. The calorific value increased from 19.9% to 26.8% at 290 °C. The optimal torrefaction temperature for coffee ground pellets was 230 °C, while for other pellets, it was 250 °C. This study provides basic information on the potential enhancement of agricultural by-products for fuel applications.

Keywords: torrefaction; agricultural by-products; mixing ratios; solid fuel; pellet evaluation

Citation: Park, S.; Jeong, H.-R.; Shin, Y.-A.; Kim, S.-J.; Ju, Y.-M.; Oh, K.-C.; Cho, L.-H.; Kim, D. Performance Optimisation of Fuel Pellets Comprising Pepper Stem and Coffee Grounds through Mixing Ratios and Torrefaction. *Energies* **2021**, *14*, 4667. https://doi.org/10.3390/en14154667

Academic Editors: Andrea Di Carlo and Elisa Savuto

Received: 22 June 2021
Accepted: 23 July 2021
Published: 1 August 2021

Publisher's Note: MDPI stays neutral with regard to jurisdictional claims in published maps and institutional affiliations.

1. Introduction

In recent years, significant efforts have been made to reduce greenhouse gases in the atmosphere. Based on the Paris Agreement signed at the 2015 United Nations Conference on Climate Change (COP 21), which obligates all countries to reduce greenhouse gas emissions, the Korean government set a goal of achieving a 37% reduction by 2030 [1,2]. To meet this goal, the current government established the Renewable Energy 3020 Implementation Plan, which aimed to generate 20% of the total power supply from new and renewable energy sources by 2030, with 5% of this power generated from biomass, such as wood pellets. Biomass is a carbon-neutral fuel due to its carbon cycle, so it does not contribute to carbon emissions by itself [3]. It is advantageous because it can be utilised in existing coal-based power generation facilities, replacing solid fossil fuels, and has gained increasing attention in the field of energy utilisation [4]. Furthermore, unlike fossil fuels that cannot be recycled once imported and consumed, wood-based biomass is an existing domestic resource which has natural cyclical characteristics that allow its reproduction through the agricultural and forestry industries [5]. Various studies are being conducted to use the biomass. Menardo et al. (2012) conducted a study to produce methane through anaerobic digestion by mixing agricultural by-products and organic waste [6]. Ramírez-López et al. (2003) conducted a study on the treatment of polluted air by using five agricultural by-products, such as peanut shells, rice husks, and coconut shells, thrown away in South America as biofilters [7]. Fediuk (2016) reported that the Russian Far East improved the physical and mechanical properties by adding fibers to concrete as binders [8]. However, from the perspective of biosolid fuel, interest in wood pellets has increased because they facilitate the use of biomass. In 2019, the local production of wood pellets in the Republic of Korea (ROK) was approximately 243,000 tons, while the imported quantity was 2,566,000 tons [9]. This difference in quantities results from difficulties in procuring the raw materials for pellet production and the high cost of pelletising. To supply stable raw materials for producing wood pellets, various domestic agricultural by-products can be utilised [10].

Currently, the domestic biomass potential of the ROK is estimated to be 5141.39 PJ/year, while the supply potential of the current technology is 971.34 PJ/year [11]. Rice straw (approximately 93.97 PJ/year) and rice husk (approximately 16.97 PJ TOE/year) comprise the largest portion of agricultural biomass (Figure 1) and have their own applications including compost, animal feed, and livestock rugs [12]. After rice straw and chaff, pepper stem is the next most prominent biomass. Most other agricultural by-products are either burned or discarded. This is because the transportation and collection of agricultural by-products incur additional costs and storage is difficult, owing to their low calorific value and hygroscopicity.

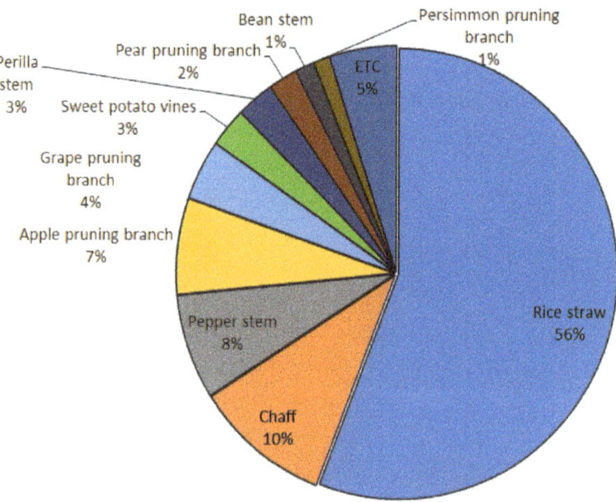

Figure 1. Annual agricultural by-product potential in Korea.

Coffee consumption is increasing worldwide. In 2019/20, it was estimated that approximately 8303 million kg was consumed, up 0.3% from the previous year [13]. ROK's consumption of coffee in 2016 averaged 377 cups per person; accordingly, large amounts of coffee waste (270,000 tons/year) are discharged as garbage every year. According to a previous study, the coffee grounds remaining after extraction contain 15 wt % of residual oil; therefore, the calorific value (24.5 MJ/kg) of waste coffee grounds exceeds that of wood pellets (18.36 MJ/kg) [14]. In addition, a previous study reported that waste coffee grounds have excellent adsorption abilities [15] and can be used as wood adhesives. However, this study found that if pelletising was performed using only waste coffee grounds, the lignin content was significantly lower than that of wood, and successful moulding required the use of an adhesive [16]. Park et al. (2020) showed the possibility of fuel applications by mixing spent coffee grounds and agricultural by-products in a 5:5 ratio, but they did not suggest an exact mixing ratio [17]. Ungureanu et al. (2018) reported that the optimal moisture content of pellets was 5–15%, and that pellets with >20% moisture may become unstable [18]. Samuelsson et al. (2012) reported that the moisture content of sawdust was the dominant factor determining the bulk density of the mixture and consequently the required pelletiser motor current; the optimal moisture content was reportedly 11–13% [19]. Liu et al. reported that a ratio of 3:2 is the most optimal when pellets are made by mixing bamboo and rice husks [20].

Against this backdrop, in this study, we were consider mixing and a torrefaction process for the effective utilisation of unused agricultural by-products and spent coffee ground as a possible fuel. In this study, unused agricultural by-product and spent coffee grounds, pepper stems, and waste coffee grounds were combined with different mixing ratios to produce six types of pellets. Pure pepper stem pellets and spent coffee ground pellets were also evaluated as controls. These pellets were evaluated based on the 'Non-woody biomass' standards of the Korea Energy Appliances Industry Association (KEAA) [21]. After evaluation, an optimal mixing ratio was determined. Then, the pellets were subjected to torrefaction to improve their fuel characteristics, and their mass reduction was measured. Subsequently, elemental analysis and calorific value measurements were performed, and the properties of the torrefied agro-pellets were evaluated.

2. Materials and Methods

In this study, red pepper stems and coffee waste were pelletised. The pepper stems were collected from Yangju-si in Gyeonggi Province, Yesan-gun in Chungcheongnam Province, and Gumi-si in Gyeongsangbuk Province. The spent coffee grounds were collected from a café in Kangwon National University, Korea. The pepper stem and spent coffee grounds were naturally dried to a moisture content of <15%. The moisture content of pepper stem and spent coffee waste was approximately 8% and 14%, respectively. Then, the elemental compositions of the pepper stems from each source were analysed (Table 1).

Table 1. Elemental analysis of pepper stems from each of the three sources.

	C	H	O	N	S	Cl
Yangju-si	42.70	5.69	49.89	1.72	0.14	0.25
Yesan-gun	45.24	5.60	47.24	1.92	0.15	0.44
Gumi	43.57	5.98	47.84	2.61	0.18	0.86

2.1. Production of Pellets

The pepper stems were pulverised to less than 2.36 mm in size using a shredder (SH01-2540, Woojin Techtool, Seoul, Korea), while the spent coffee grounds were already less than 0.4 mm. A pelletiser (SP-75, Geumgang ENG, Gwangju, Korea) was used to pelletise the collected by-products (Figure 2). The specifications of the pelletiser are listed in Table 2. The mixing ratios of pepper stems and coffee waste were varied from 9:1 to 6:4 in mass fraction. For evaluation, approximately 7–10 kg of each variety of pellet were produced. The single-material pepper stem and coffee waste pellets were designated as PP and CP, respectively. The mixed pellets were named PACB, where A and B represent the relative proportions of pepper stems and coffee waste, respectively. For example, the pellets with a pepper stem and coffee waste mix ratio of 9:1 were named P9C1.

Figure 2. Pelletiser.

Table 2. Specifications of pelletiser.

Description (Unit)	Specification
Inner matrix-diameter (mm)	200
Dices (n)	108
Dice diameter (mm)	6
Length of dices (mm)	34
Press rollers (n)	2
Nominal power of press motor (kW)	5.5

The pepper stems were pulverised to less than 2.36 mm in size, while the spent coffee grounds were already less than 0.6 mm (Table 3). A pelletiser (SP-75, Geumgang ENG, Korea) was used to pelletise the collected by-products (Figure 2). The specifications of the pelletiser are listed as Table 2. The mixing ratios of pepper stems and coffee waste were varied from 9:1 to 6:4 in mass fraction. For evaluation, approximately 7–10 kg of each variety of pellet were produced. The single-material pepper stem and coffee waste pellets were designated as PP and CP, respectively. The mixed pellets were named PACB, where A and B represent the relative proportions of pepper stems and coffee waste, respectively. For example, the pellets with a pepper stem and coffee waste mix ratio of 9:1 were named P9C1.

Table 3. Particle size analysis of the samples.

Sieve Size (μm)	Particle Distribution (%)		Particle Size Range (μm)
	Pepper Stem	Spent Coffee Grounds	
2360	6.36	-	2360–600
600	60.82	51.14	425–600
425	17.40	39.77	75–425
75	10.24	9.09	<75

2.2. Evaluation of Pellets

The pellets were evaluated according to the KEAA standards. The evaluation indicators were size; bulk density; moisture content; ash content; fine particles; durability; gross calorific value; and the contents of chlorine, sulphur, nitrogen, and eight kinds of heavy metals (As, Cd, Cr, Cu, Pb, Hg, Ni, and Zn).

2.3. Size

The length and diameter of the pellets were measured using Vernier callipers.

2.4. Bulk Density

A 1 L container was filled with pellets to approximately 20 cm above the top edge. Then, the container was dropped thrice onto a rigid flat wooden floor, and any pellets remaining above the top of the container were removed with a flat object and weighed. This process was repeated twice. The bulk density, expressed in kg/m^3, was calculated using Equation (1), and the resulting value was rounded off to the first digit.

$$BD_{ar} = \frac{m_{pc} - m_c}{V} \tag{1}$$

where BD_{ar}: Pellet bulk density on wet basis (kg/m^3). m_{pc}: Weight of container with pellet (kg). m_c: Weight of empty container (kg). V: Volume of empty container (m^3).

2.5. Moisture Content

In accordance with ISO 18134-2:2017 [22], the pellets were dried at 105 °C for 3 days and weighed before and after drying. Then, the moisture content was calculated from the difference in weight using Equation (2):

$$MC = \frac{\left(m_o - m_f\right)}{m_o} \tag{2}$$

where MC: moisture content of pellet (%), m_o: weight of pellet before drying (kg), m_f: weight of pellet after drying (kg).

2.6. Ash Content

Ash content measurements were conducted in accordance with the ISO 18122:2015 [23] using a proximate analyser (PrepASH229, Precisa, Switzerland).

2.7. Fine Particles

A fine particle is defined as a particle with a diameter of <3.15 mm. In accordance with ISO 18846:2016 [24], pellets weighing 50 g or more were selected and weighed to two decimal digits. The pellets were placed in a metal sieve with a mesh size of 3.15 mm, which was mounted on a shaker and filtered. The mass of the pellets remaining in the sieve was measured twice. The mean of the values, calculated using Equation (3), was rounded up to 0.1%.

$$F = \frac{m_a - m_o}{m_e} \times 100 \tag{3}$$

where F: Fine particle content (%), m_a: pellet weight before sieving (g), m_o: pellet weight after sieving (g).

2.8. Durability

In accordance with ISO 17831-1 [25], 500 ± 50 g of pellets were poured into a durability tester, and 500 rotation tests were conducted at the rate of 50 ± 2 rev/min. After the test, the pellets were filtered using a sieve with a mesh size of 3.15 mm, and the weight of the pellets remaining in the sieve was measured twice. The mean of the values, calculated using Equation (4), was rounded up to 0.1%:

$$DU = \frac{m_a}{m_b} \times 100 \tag{4}$$

where DU: durability (%), m_a: pellet weight post-sieving after durability test (g), m_b: pellet weight prior to sieving before durability test (g).

2.9. Calorific Value

The calorific value was measured through combustion using an automated calorimeter (6400, Parr, Moline, IL, USA). The calorimeter was calibrated using benzoic acid as a standard substance. The pellets were dried in a dryer, and their calorific value was measured thrice. The average calorific value of the samples was rounded off to the first digit.

2.10. Elemental Analysis

In accordance with the quality test method for solid fuel products stipulated by the Ministry of Environment [26], elemental analysis was conducted to determine the contents of the elements C, H, N, and O by an elemental analyser (FlashEA 1112, Thermo Fisher Scientific, Waltham, MA, USA).

2.11. Sulphur and Chlorine

The sulphur and chlorine contents were measured in accordance with ASTM D7359 through combustion ion chromatography (AQF-2100H, Mitsubishi Chemical Analytech Co., LTD, Tokyo, Japan).

2.12. Heavy Metals

The Hg content was measured using an Hg analyser (Hydra II C, Teledyne Leeman Labs, Hudson, NH, USA). The contents of the other metals were measured using inductively coupled plasma optical emission spectroscopy (ICP-OES; OPTIMA 7300 DV, PerkinElmer, Waltham, MA, USA) after pre-treatment by microwave heating.

2.13. Torrefaction Experiments

Torrefaction experiments were conducted by placing 12 ± 2 g of the pellets into a prototype capsule with a height of 100 mm and a diameter of 24 mm. The capsule was sealed with heat-resistant tape to reduce external effects during the experiments. The experiments were performed using an electric furnace (N7/H/B410, Nabertherm GmbH, Lower Saxony, Germany). To isolate the effect of temperature, the duration of each experiment was set to 30 min and the process temperature was varied from 210 to 290 °C in increments of 20 °C. After the experiments, all samples were cooled for 30 min to prevent rapid reactions with oxygen. After cooling, the mass reduction was measured. Based on the mass reduction, the mass yield was calculated using Equation (5):

$$MY = m_{torrefied}/m_{raw} \times 100 \tag{5}$$

where MY: mass yield (%), $m_{torrefied}$: mass after torrefaction (g), m_{raw}: mass before torrefaction (g).

To determine the microstructures of the samples, scanning electron microscopy (SEM) was conducted using a field-emission scanning electron microscope (S-4800, Hitachi, Tokyo, Japan).

2.14. Fuel Characteristic Analysis

2.14.1. Elemental Analysis and Van Krevelen Diagram

The elemental composition of a pellet can vary based on the torrefaction conditions, such as time, temperature, and atmospheric conditions. The H, C, and O contents of a pellet significantly affect its combustion characteristics. Van Krevelen diagrams show the ratios of these three elements. To investigate the changes in element composition, elemental analysis (EA3000, Eurovector, Pavia, Italy) was performed, and Van Krevelen diagrams were plotted from the results of the analysis.

2.14.2. Calorific Value and Energy Yield

To account for changes in both calorific value and mass loss, the energy yield was calculated using Equation (6):

$$EY = HHV_{torrefied} / HHV_{raw} \times MY \tag{6}$$

where EY: energy yield (%), $HHV_{torrefied}$: higher heating value after torrefaction (MJ/kg), HHV_{raw}: higher heating value before torrefaction (MJ/kg).

3. Results

3.1. Mixed Pellets

Agricultural by-product pellets were produced. With increases in the quantity of coffee waste, the observed colours of the mixed pellets became darker. In length, the CP pellets were shorter than the other pellets.

3.2. Evaluation of Pellets

The pellets were classified as A-grade or B-grade 'Non-wood pellets', based on the standards stipulated by KEAA, and as Bio-SRF grade based on the 'Act on the Promotion of Saving and Recycling of Resources' of the Ministry of Environment (Table 4) [27]. The criteria for A-grade and B-grade non-wood pellets are more extensive than those for the Bio-SRF standards. Therefore, in this study, the grades of the pellets were categorised according to their quality in the order of A, B, and Bio-SRF.

Table 4. Standards for A-grade and B-grade pellets and Bio-SRF.

Properties	Unit	A-Grade Pellet *	B-Grade Pellet *	Bio-SRF **
Length	mm	$\leq$40	$\leq$50	$\leq$100
Diameter	mm	6–10	$\leq$25	$\leq$50
Bulk density	kg/m^3	$\geq$600	$\geq$600	-
Moisture content	%	$\leq$12	$\leq$15	$\leq$10
Ash content	%	$\leq$15.0	$\leq$15.0	$\leq$15.0
Fine particle	%	$\leq$2.0	$\leq$3.0	-
Durability	%	$\geq$97.5	$\geq$96.0	-
Gross Calorific value	MJ/kg	$\geq$16.5	$\geq$16.5	$\geq$12.56
Chlorine	%	$\leq$0.1	$\leq$0.3	$\leq$0.5
Sulphur	%	$\leq$0.2	$\leq$0.3	$\leq$0.6
Biomass	%	-	-	$\geq$95
Nitrogen	%	$\leq$1.5	$\leq$2.0	-
Arsenic	mg/kg	$\leq$1.0	$\leq$1.0	$\leq$5.0
Cadmium	mg/kg	$\leq$0.5	$\leq$0.5	$\leq$5.0
Chromium	mg/kg	$\leq$10	$\leq$10	$\leq$70.0
Copper	mg/kg	$\leq$20	$\leq$20	-
Lead	mg/kg	$\leq$10	$\leq$10	$\leq$100
Mercury	mg/kg	$\leq$0.1	$\leq$0.1	$\leq$0.6
Nickel	mg/kg	$\leq$10	$\leq$10	-
Zinc	mg/kg	$\leq$100	$\leq$100	-

* A- and B-grade: Criteria according to quality standards for non-wood pellets. ** Bio-SRF: Based on the Bio-SRF standard for moulding according to Article 10 of Annex 1 of the Enforcement Regulation of the Act on the Promotion of Saving and Recycling of Resources.

3.3. Size

All pellets had a 6 mm diameter, as required by the A-grade standard in diameter. Expect for the CP pellets, the length of pellets was 21.3 $\pm$ 0.9 mm, and there were no significant variations with the different mixing ratios. However, the average lengths of the CP pellets were approximately 10 mm, while in some cases, the length was <10 mm.

3.4. Bulk Density

The bulk density varied with the mixing ratio. The bulk densities increased from PP to P8C2 but decreased from P7C3 to CP. All pellets had bulk densities above 600 kg/m^3 except CP, which had a bulk density of 563 kg/m^3 and therefore did not satisfy the A- or B-grade criterion.

3.5. Moisture Content

Excluding P6C4 and CP, the moisture content in the pellets was observed to be within 10–12%, satisfying the A-grade criterion. The moisture contents of P6C4 and CP were approximately 13.3% and 20.4%, respectively. Therefore, we conclude that an increased proportion of coffee grounds resulted in a higher moisture content, owing to the hygroscopic properties of coffee waste.

3.6. Ash Content

All pellets met the A-grade criterion for ash content. This analysis showed that a higher proportion of pepper stem corresponded to increased observed ash contents.

3.7. Fine Particles

The proportion of fine particles was determined to be less than 0.2% in all pellets, excluding CP (2%). Therefore, all pellets were determined to be of A-grade, except for CP, which was categorised as B-grade.

3.8. Durability

The durability varied only slightly with changes in the coffee waste mixing ratio and was observed to be over 98% for all pellets except CP. In particular, P9C1 showed an increase in durability compared with PP. A previous study (Park et al., 2020b), reported a durability of 98.6% for PP, similar to the results obtained in this study, although it was slightly lower. The same study reported a durability of 85.9% for CP, which was approximately 3.3 percentage points lower than the value obtained in this study. Neither of the pellets could be graded as A or B.

3.9. Gross Calorific Value

All pellets had higher gross calorific values than those specified by the grading criteria. In particular, the observed value for CP was 23.14 MJ/kg, which is similar to that of sub-bituminous coal. The gross calorific value increased as the proportion of coffee waste increased.

3.10. Nitrogen, Sulphur, and Chlorine Content

The N and S contents of pellets have important implications for the generation of oxides of these elements (NO_x and SO_x). P6C4 and CP were identified as Bio-SRF in grade, while the others met the B-grade criterion. All pellets except PP had less than 0.1% S and 0.3% Cl. PP showed the highest S content, but no significant differences were observed between the pellets. The Cl content in PP was observed to be higher than that suggested by the B-grade criterion. Thus, PP was determined to be of Bio-SRF. Mixed pellets (P9C1, P8C2, P7C3, and P6C4) had 0.1–0.3% Cl and were therefore identified as B-grade. Only CP could be classified as A-grade.

3.11. Heavy Metals

All pellets met the criterion for heavy metal content. CP exhibited the highest Cu content; therefore, the mixed pellets also showed higher Cu than PP. The Zn content of the pellets ranged from 20% to 30% but was less than 10% for CP.

3.12. Evaluation Summary

All evaluation results presented above are summarised in Table 5. PP conformed to neither grade A nor grade B because of its high Cl content. P6C4 did not meet the requirements for grades A or B because of its N content, while CP had excessively high moisture and N contents and therefore could not be graded as A or B. Furthermore, because of its lower durability, CP was graded as Bio-SRF, which lacks a specific durability criterion. With increasing coffee waste content, the Cl content of the pellets decreased while the calorific value increased. P6C4 was found to be easily contaminated with fungi and was therefore unsuitable for storage. The durability of PP was similar to that of pellets evaluated in a previous study [13]. CP exhibited a durability of 89.2%, which is higher than the durability of 85.9% obtained in a previous study. The Cu content of the pellets varied with the different sources of pepper stems, which was likely due to the particular soil conditions and rainfall patterns of the different areas. Based on these evaluations, both PP and P6C4 were categorised as Bio-SRF grade, while P9C1, P8C2, and P7C3 were categorised as B-grade. Only CP could not meet any of these standards. The pellets with the optimum mixing ratios of pepper stems to spent coffee grounds were P8C2 and P7C3. Both pellet types were categorised as B-grade. P8C2 showed a higher bulk density and durability and a lower moisture content than P7C3, while P7C3 showed a higher gross calorific value and lower Cl and N contents. Based on all evaluations, P9C1, P8C2, and P7C3 showed similar characteristics. Among them, P7C3 had the highest calorific value. However, considering transportation and storage in warehouses, P8C2 was considered to have the optimal characteristics owing to its maximal durability.

Table 5. Evaluation of agricultural by-product pellets.

Properties	Unit	PP	P9C1	P8C2	P7C3	P6C4	CP
Length	mm	21.9 (A)	20.0 (A)	22.6 (A)	20.8 (A)	21.1 (A)	10.0 (A)
Diameter	mm	6.0 (A)	6.0 (A)	6.0 (A)	6.0 (A)	6.0 (A)	6.0 (A)
Bulk density	kg m^{-3}	640 (A)	651 (A)	659 (A)	628 (A)	623 (A)	563 (Bio-SRF) *
Moisture content	%	10.9 (A)	10.3 (A)	10.4 (A)	11.7 (A)	13.3 (B)	20.4 (OUT)
Ash content	%	6.2 (A)	5.9 (A)	5.7 (A)	5.2 (A)	4.4 (A)	1.5 (A)
Fine particle	%	0.1 (A)	0.2 (A)	0.2 (A)	0.2 (A)	0.1 (A)	2.4 (B)
Durability	%	98.9 (A)	99.4 (A)	98.4 (A)	98.2 (A)	98.6 (A)	89.2 (Bio-SRF) *
Gross calorific value	MJ kg^{-1}	16.67 (A)	19.21 (A)	19.43 (A)	19.85 (A)	20.35 (A)	23.15 (A)
Chlorine	%	0.36 (Bio-SRF) *	0.29 (B)	0.27 (B)	0.24 (B)	0.17 (B)	0.02 (A)
Sulphur	%	0.1 (A)	0.09 (A)	0.09 (A)	0.07 (A)	0.06 (A)	0.08 (A)
Nitrogen	%	1.9 (B)-	1.7 (B)-	1.8 (B)	1.7 (B)	2.8 (Bio-SRF) *	3.0 (Bio-SRF) *
As	mg kg^{-1}	0.4 (A)	<0.05 (A)	<0.05 (A)	<0.05 (A)	<0.05 (A)	<0.05 (A)
Cd	mg kg^{-1}	0.1 (A)	0.2 (A)	0.2 (A)	<0.1 (A)	0.1 (A)	<0.1 (A)
Cr	mg kg^{-1}	2 (A)	5 (A)	5 (A)	3 (A)	7 (A)	1 (A)

Table 5. *Cont.*

Properties	Unit	PP	P9C1	P8C2	P7C3	P6C4	CP
Cu	mg kg^{-1}	<0.05 (A)	8 (A)	9 (A)	6 (A)	9 (A)	17 (A)
Pb	mg kg^{-1}	<0.05 (A)	<0.05 (A)	<0.05 (A)	<0.05 (A)	<0.05 (A)	<0.05 (A)
Hg	mg kg^{-1}	0.05 (A)	0.03 (A)	0.02 (A)	<0.01 (A)	<0.01 (A)	<0.01 (A)
Ni	mg kg^{-1}	<0.05 (A)	1 (A)	1 (A)	1 (A)	1 (A)	1 (A)
Zn	mg kg^{-1}	24 (A)	28 (A)	29 (A)	20 (A)	23 (A)	9 (A)
Grade		Bio-SRF	B	B	B	Bio-SRF	OUT

OUT: Out of grade. * Grade according to 'Solid Fuel Quality Standards' specified by Ministry of Environment.

3.13. Torrefaction

Torrefaction was conducted to improve the fuel properties, such as the calorific value. Higher processing temperatures corresponded to darker pellet colours and higher mass losses. At low (210 °C and 230 °C) and intermediate temperatures (250 °C), mixed pellets showed lower mass losses than the single-material pellets (PP and CP), regardless of the mixing ratio. The difference in mass losses between the different pellets was less than 3%. However, in the high-temperature range (270 °C and 290 °C), it was observed that a higher proportion of spent coffee grounds corresponded to a lower mass loss. The mass yields of PP were found to be 69.16% and 50.87% at 270 °C and 290 °C, but those of CP were 75.34% and 60.81%, respectively (Table 6).

Table 6. Mass yield after torrefaction (%).

	210 °C	230 °C	250 °C	270 °C	290 °C
PP	88.27 (0.13)	84.87 (0.50)	81.85 (1.18)	69.16 (2.73)	50.87 (0.85)
P9C1	87.63 (0.12)	87.05 (0.28)	82.58 (2.02)	70.51 (7.44)	51.63 (1.10)
P8C2	88.33 (0.45)	86.98 (0.75)	84.13 (0.84)	71.66 (3.28)	52.37 (0.93)
P7C3	88.05 (0.33)	85.21 (0.31)	83.05 (0.62)	71.54 (3.36)	57.21 (0.44)
P6C4	86.90 (0.34)	84.28 (0.26)	83.60 (1.14)	72.35 (1.68)	59.67 (1.64)
CP	87.72 (0.50)	84.67 (0.82)	81.07 (0.62)	75.34 (3.70)	60.81 (1.11)

Standard deviation in parentheses.

SEM imaging was conducted on P8C2 (identified as the optimal composition), on the two single-material pellets (PP and CP), and on torrefied pellets of these three compositions. However, the CP pellet was easily broken up, so SEM could not be performed. The SEM images are shown in Figure 3. The images are magnified by factors of 300, 1.5 K, and 3 K to demonstrate the microstructures of the pellet cross-sections. Many linear fibre structures were observed in PP. The PP treated at 290 °C exhibited a more broken down linear fibre structure compared to the non-torrefied PP. In P8C2, unlike PP, it seems that the spent coffee grounds were attached to linear fibre structures. In P8C2 treated at 290 °C, no linear fibres were visible, and some parts had collapsed. It was determined that these changes were due to the thermal decomposition of the attached spent coffee grounds during torrefaction.

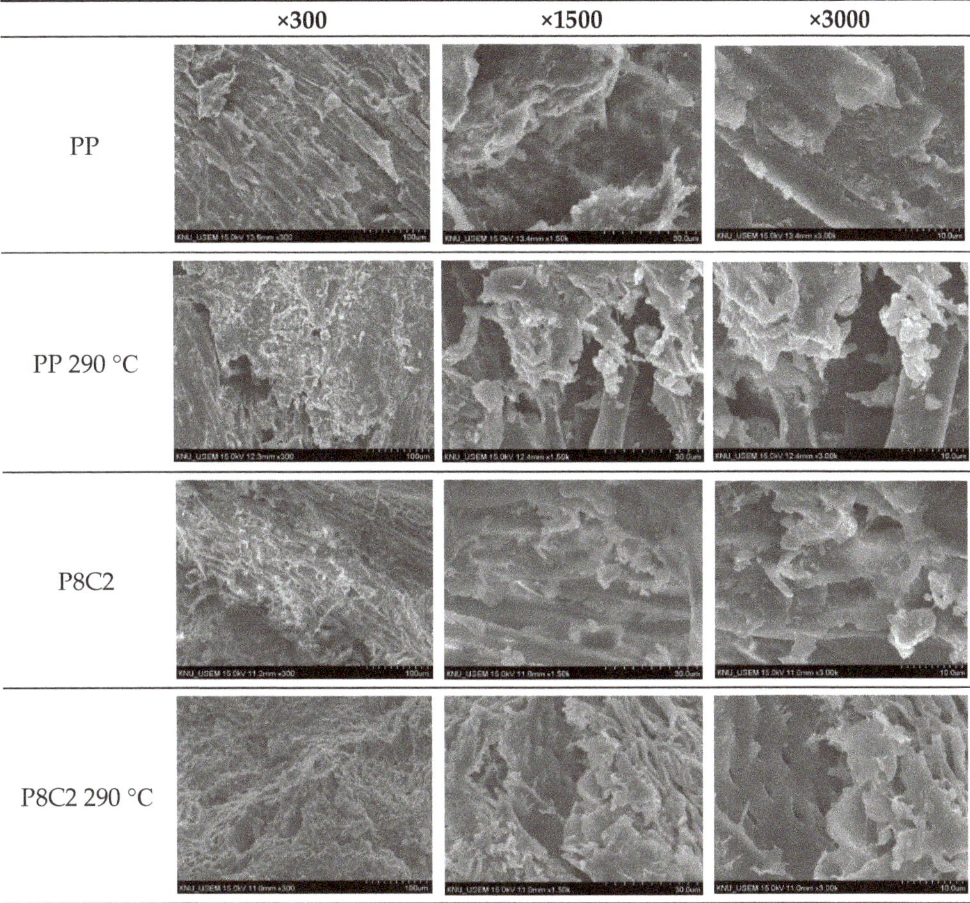

Figure 3. SEM images of untreated and torrefied pellets.

3.14. Fuel Characteristics and Van Krevelen Diagram

The torrefaction-induced changes in the elemental composition of each of the pellets were significant, as shown in Figure 4. C, O, and H exhibited the largest variations, as these elements are commonly associated with combustion characteristics. With higher torrefaction temperatures, the proportion of C in the biomass was increased, but the proportions of O and H were decreased. The N concentration did not exhibit significant differences with changes in the processing temperature. As a result of the changes in the elemental composition, all pellets showed decreases in the O/C and H/C ratios, as shown in Figure 5. The H/C and O/C ratios of all pellets were improved by torrefaction. Higher C contents and lower O contents were observed for pellets with a higher proportion of waste coffee grounds.

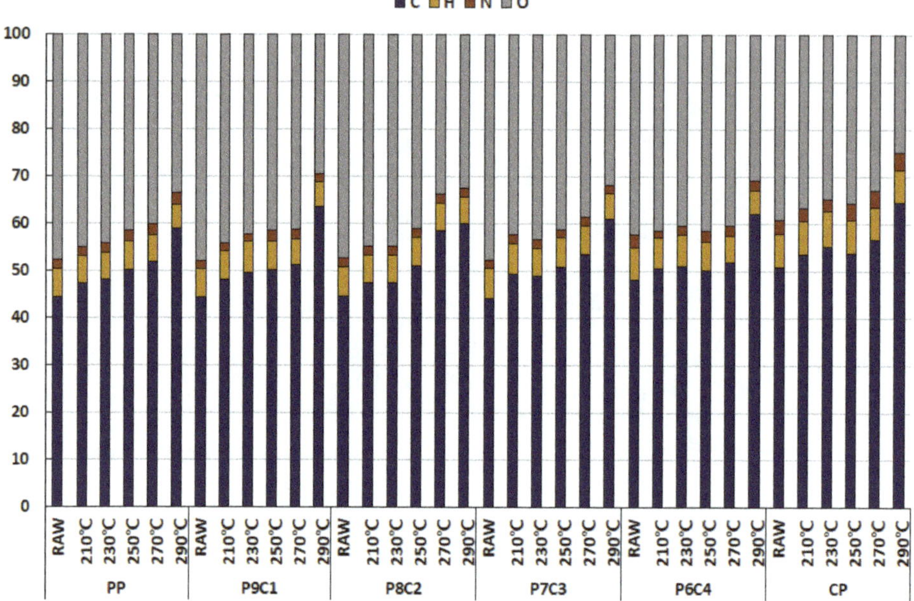

Figure 4. Element component ratio of each torrefied pellet.

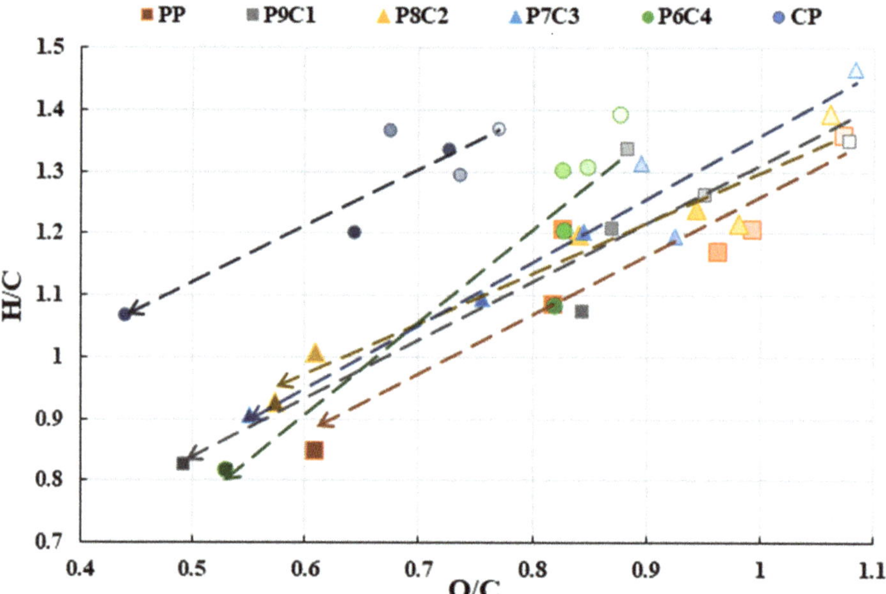

Figure 5. Torrefaction-induced changes in H/C and O/C for each pellet.

3.15. Calorific Value and Energy Yield

The calorific values increased with increasing torrefaction temperature, as shown in Figure 6. Relative to the raw (non-torrefied) pellets, pellets torrefied at a relatively low temperature (210–230 °C) showed a slight increase in calorific value, while CP showed a slight decrease at 230 °C. According to a previous study, this is possibly because the removal of the oil component of CP during torrefaction reduces its calorific value [28]. P9C1 showed the highest increase in calorific value (26.8%), while CP showed the lowest increase (19.9%) compared to that of the raw pellet. At temperatures below 250 °C, pellets with a higher proportion of waste coffee grounds showed higher calorific values, similar to the trend observed for the raw pellets. However, in the high-temperature range (270–290 °C), a different tendency was observed. For the mixed pellets, the variation in calorific value exhibited a U-shaped trend. The calorific value increased as the proportion of spent coffee grounds increased. Furthermore, P7C3 showed the highest calorific value of the pellets torrefied at 270 °C, but P6C4 showed a lower calorific value than P7C3. Moreover, P8C2 showed the highest calorific value after torrefaction at 290 °C. The calorific value of pellets torrefied at this temperature decreased with increases in the proportion of spent coffee grounds.

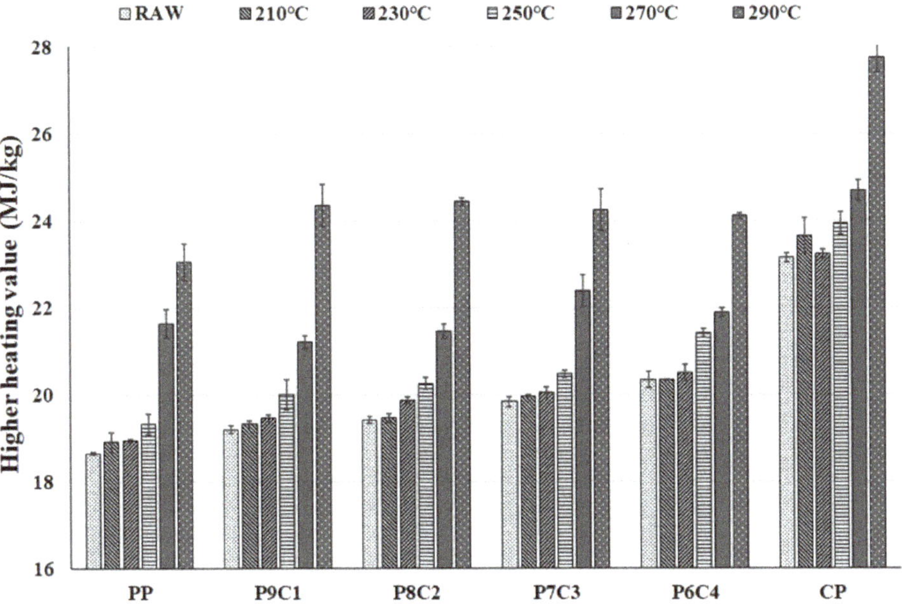

Figure 6. Variation in higher heating value with different torrefaction temperatures.

In terms of energy yield (Figure 7), all pellets showed an approximate 10% decrease in yield when torrefied at 210 °C, relative to the raw (non-torrefied) pellets. This caused a significant decrease in mass compared to the increase in the calorific value due to moisture evaporation. All pellets torrefied at 250 °C showed energy yields of more than 80%. In particular, pellets torrefied at 290 °C exhibited a significant decrease in mass compared to those torrefied at other temperatures. However, except for temperatures of 290 °C, no clear tendency was observed between the mass loss and the proportion of coffee grounds. As the energy yield is the product of the mass yield and the calorific value, considerably high energy yields were observed in some cases, even though the mass yield was low (e.g., P6C4 torrefied at 230 °C and 250 °C).

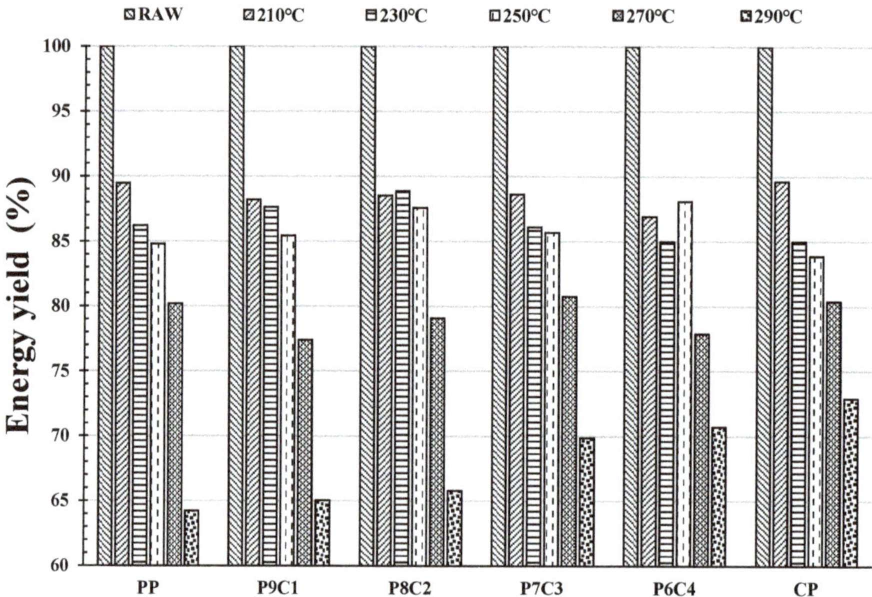

Figure 7. Changes in the energy yield of different pellets, with different torrefaction temperatures.

To determine the optimal conditions, the energy yield was used to reflect the increases in both mass yield and calorific value. An energy yield of approximately 85% was selected as the optimal condition because it indicated a mass yield of over 80%, as well as a significant increase in the calorific value.

4. Discussion and Conclusions

To determine the optimal mixture ratio of fuel pellets comprising agricultural by-product pepper stems and spent coffee grounds, and to improve the fuel quality through torrefaction, this study evaluated mixed pellets using established fuel quality standards and improved the fuel quality of the pellets through torrefaction. Regardless of the mixing ratio, the Cl and N contents of the pellets were determined to be the main factors hindering fuel performance. Through various evaluation criteria, P8C2 was determined to be the optimal mixing ratio. In torrefaction tests, higher mass yields were correlated with higher concentrations of pepper stems at relatively low torrefaction temperatures of 210–230 °C. However, this trend was reversed at temperatures above 250 °C. For the evaluation of fuel characteristics, the calorific value and element component ratio of each mixing pellet was determined. Based on energy yield, 230 °C was selected as the optimal torrefaction temperature for CP, while 250 °C was optimal for the others. It was determined that the calorific value of CP decreased at higher torrefaction temperatures as a result of the decomposition of volatile matter.

This study showed that the properties of single-material pellets (PP and CP) can be improved upon by mixing the two materials at a ratio of 8:2. The high calorific value of coffee grounds can be complemented by the relatively low hygroscopicity of agricultural by-products. Compared with previous study [17], the moisture content of CP can be improved. In addition, it was confirmed that the mixing of agricultural by-products and spent coffee grounds can be the way of improving the disadvantages of existing agricultural by-products in various standards. With another previous study [29], mixing with spent coffee ground showed more longer length. In durability, a higher mixing ratio of spent coffee ground showed less durability in previous study, but, in this study, there were significant difference in mixing. It was observed that their fuel characteristics could be optimised through torrefaction at 250 °C. This study provides baseline data for research on using agricultural by-products as solid fuels. Based on this study, further studies on the mixing of various agricultural by-products and spent coffee grounds, pilot-scale torrefaction experiments, and methods to decrease Cl and N contents should be conducted.

Author Contributions: Conception and design of study: S.P., S.-J.K., D.K.; acquisition of data: S.P., S.-J.K., H.-R.J., Y.-A.S.; analysis and/or interpretation of data: S.P., S.-J.K., L.-H.C., Y.-M.J., K.-C.O.; drafting the manuscript: S.P., S.-J.K., H.-R.J., Y.-A.S.; revising the manuscript critically for important intellectual content: S.P., S.-J.K., H.-R.J., Y.-A.S., D.K.; approval of the version of the manuscript to be published: S.P., H.-R.J., Y.-A.S., S.-J.K., Y.-M.J., K.-C.O., L.-H.C. and D.K. All authors have read and agreed to the published version of the manuscript.

Funding: This study was carried out with the support of R&D Program for Forest Science Technology (Project No. "FTIS 2021352B10-2123-AC03") provided by Korea Forest Service (Korea Forestry Promotion Institute).

Institutional Review Board Statement: Not applicable.

Informed Consent Statement: Not applicable.

Conflicts of Interest: The authors declare no conflict of interest.

References

1. Lee, Y.; Jeon, C. The Biomass Pre-treatment Effect on the Combustion Characteristics of Coal and Biomass Blends. *Trans. Korean Hydrog. New Energy Soc.* **2018**, *29*, 81–89.
2. Yoon, I.J. Issues and Prospects of the Paris Agreement. *Han Yang Law Rev.* **2017**, *28*, 113–144.
3. Jeong, J.S.; Kim, G.M.; Jeong, H.J.; Kim, G.B.; Jeon, C.H. A Study on the Improved the Hydrophobicity of Torrefied Biomass. *Trans. Korean Hydrog. New Energy Soc.* **2019**, *30*, 49–57.
4. Park, H.M.; Mock, C.S.; Ryu, C.K.; Choi, S.C. Combustion characteristics of single biomass pellets in a hot gas flow. *J. Korean Soc. Power Syst. Eng.* **2019**, *23*, 91–104. [CrossRef]
5. Yoo, K.H. Technology Development Trend using Woody biomass. *Konetic Rep.* **2016**, *72*, 1–13.
6. Menardo, S.; Balsari, P. An Analysis of the Energy Potential of Anaerobic Digestion of Agricultural By-Products and Organic Waste. *Bioenergy Res.* **2012**, *5*, 759–767. [CrossRef]
7. Ramírez-López, E.; Corona-Hernández, J.; Dendooven, L.; Rangel, P.; Thalasso, F. Characterization of five agricultural by-products as potential biofilter carriers. *Bioresour. Technol.* **2003**, *88*, 259–263. [CrossRef]
8. Fediuk, R. High-strength fibrous concrete of Russian Far East natural materials. *IOP Conf. Ser. Mater. Sci. Eng.* **2016**, *116*, 1–5. [CrossRef]
9. Forest Biomass Energy Association Production and Import of Wood pellet. Available online: http://www.biomassenergy.kr/ (accessed on 23 February 2020).
10. Oh, D.G.; Kim, Y.H.; Son, H.S. Fuel Properities of Spent Coffee Bean by Torrefaction. *New Renew. Energy* **2013**, *9*, 29–35. [CrossRef]
11. Ministry of Food, Agriculture, F. and F. 377 Cups of Coffee per Year per Adult in Korea. Available online: https://www.mafra.go.kr/mafra/293/subview.do%3Bjsessionid=ptxeoPi-fKyzHO9WlW3TrDv6.inst11?enc=Zm5jdDF8QEB8JTJGYmJzJTJGbWFmcmElMkY2OCUyRjMxNDQ2MCUyRmFydGNsVmlldy5kbyUzRg%3D%3D (accessed on 23 February 2020).
12. Lee, C.G.; Lee, S.Y.; Joo, S.Y.; Cho, L.H.; Park, S.Y.; Lee, S.H.; Oh, K.C.; Kim, D.H. A Study on Agricultural by-products for Biomass-to-energy Conversion and Korean Collecting Model. *New Renew. Energy* **2017**, *13*, 27–35. [CrossRef]
13. ICO. International Coffee Organization Coffee Development Report (2020). In *The Value of Coffee–Sustainability, Inclusiveness and Resilience of the Coffee Global Value Chain*; ICO: London, UK, 2020; p. 108.
14. Lee, S.G.; Lee, K.H. A Method for Making Coffee-Meal Pellets Fuel and Coffee-Meal Pellets Thereby. Korea Patent 1010332120000, 2011.
15. Lee, H.-S.; Kang, J.-W.; Yang, W.-H.; Zong, M.-S. A study on Preparation of Adsorbent from Coffee Grounds and Removal of Trichloroethylene in Water Treadtment. *Korean J. Environ. Health Soc.* **1998**, *24*, 20–31.
16. Kim, S.; Ha, J. A Study on the Characteristics of Coffee Ground (CG)-RDF by Using Different Drying Method. *Korea Acad. Ind. Coop. Soc.* **2019**, *20*, 451–457.
17. Park, S.Y.; Kim, S.J.; Oh, K.C.; Cho, L.; Kim, M.J.; Jeong, I.S.; Lee, C.G.; Kim, D.H. Investigation of agro-byproduct pellet properties and improvement in pellet quality through mixing. *Energy* **2020**, *190*, 116380. [CrossRef]
18. Ungureanu, N.; Vladut, V.; Voicu, G.; Dinca, M.; Zabava, B. Influence of biomass moisture content on pellet properties—Review. *Eng. Rural Dev.* **2018**, *17*, 1876–1883. [CrossRef]
19. Samuelsson, R.; Larsson, S.H.; Thyrel, M.; Lestander, T.A. Moisture content and storage time influence the binding mechanisms in biofuel wood pellets. *Appl. Energy* **2012**, *99*, 109–115. [CrossRef]

20. Liu, Z.; Liu, X.; Fei, B.; Jiang, Z.; Cai, Z.; Yu, Y. The properties of pellets from mixing bamboo and rice straw. *Renew. Energy* **2013**, *55*, 1–5. [CrossRef]

21. Association Korea Energy Appliances Industry. *Non-Wood Biomass*; Association Korea Energy Appliances Industry: Ansan, Koera, 2018.

22. International Organization for Standardization. *ISO 18134-2:2017 Solid Biofuels-Determination of Moisture Content—Oven Dry Method—Part 2: Total Moisture—Simplified method No Title*; ISO: Geneva, Switzerland, 2017.

23. International Organization for Standardization. *ISO 18122:2015 Solid Biofuels—Determination of Ash Content*; ISO: Geneva, Switzerland, 2015.

24. International Organization for Standardization. *ISO 18846:2016 Solid Biofuels—Determination of Fines Content in Quantities of Pellets*; ISO: Geneva, Switzerland, 2016.

25. International Organization for Standardization. *ISO 17831-1:2015 Solid Biofuels—Determination of Mechanical Durability of Pellets and Briquettes—Part 1: Pellets*; ISO: Geneva, Switzerland, 2015.

26. Ministry of Environment. *Quality Test and Analysis Method of Solid Fuel Product*; Ministry of Environment: Seoul, Koera, 2014.

27. Ministry of Environment. *Act on the Promotion of Saving and Recycling of Resources*; Ministry of Environment: Seoul, Koera, 2019.

28. Park, S.; Kim, S.J.; Oh, K.C.; Cho, L.H.; Kim, M.J.; Jeong, I.S.; Lee, C.G.; Kim, D.H. Characteristic analysis of torrefied pellets: Determining optimal torrefaction conditions for agri-byproduct. *Energies* **2020**, *13*, 423. [CrossRef]

29. Woo, D.G.; Kim, S.H.; Kim, T.H. Solid fuel characteristics of pellets comprising spent coffee grounds and wood powder. *Energies* **2021**, *14*, 371. [CrossRef]

Process Analysis of Main Organic Compounds Dissolved in Aqueous Phase by Hydrothermal Processing of Açaí (*Euterpe oleraceae*, Mart.) Seeds: Influence of Process Temperature, Biomass-to-Water Ratio, and Production Scales

Conceição de Maria Sales da Silva [1], Douglas Alberto Rocha de Castro [1], Marcelo Costa Santos [1], Hélio da Silva Almeida [1,2], Maja Schultze [3], Ulf Lüder [4], Thomas Hoffmann [3] and Nélio Teixeira Machado [1,2,3,*]

[1] Graduate Program of Natural Resources Engineering of Amazon, Rua Corrêa N° 1, Campus Profissional-UFPA, Belém 66075-110, Pará, Brazil; concisales@ufpa.br (C.d.M.S.d.S.); douglascastro87@hotmail.com (D.A.R.d.C.); marceloenqui@yahoo.com (M.C.S.); helioalmeida@ufpa.br (H.d.S.A.)

[2] Faculty of Sanitary and Environmental Engineering, Rua Corrêa N° 1, Campus Profissional-UFPA, Belém 66075-900, Pará, Brazil

[3] Leibnitz-Institüt für Agrartechnik Potsdam-Bornin e.V, Department of Postharvest Technology, Max-Eyth-Alee 100, 14469 Potsdam, Germany; mschultze@atb-potsdam.de (M.S.); THoffmann@atb-potsdam.de (T.H.)

[4] SunCoal Industries GmbH, Rudolf-Diesel-Straße 15, 14974 Ludwigsfelde, Germany; info@suncoal.com

* Correspondence: machado@ufpa.br; Tel.: +55-91-984620325

Citation: da Silva, C.d.M.S.; de Castro, D.A.R.; Santos, M.C.; Almeida, H.d.S.; Schultze, M.; Lüder, U.; Hoffmann, T.; Machado, N.T. Process Analysis of Main Organic Compounds Dissolved in Aqueous Phase by Hydrothermal Processing of Açaí (*Euterpe oleraceae*, Mart.) Seeds: Influence of Process Temperature, Biomass-to-Water Ratio, and Production Scales. *Energies* **2021**, *14*, 5608. https://doi.org/10.3390/en14185608

Academic Editors: Andrea Di Carlo and Elisa Savuto

Received: 10 June 2021
Accepted: 25 July 2021
Published: 7 September 2021

Publisher's Note: MDPI stays neutral with regard to jurisdictional claims in published maps and institutional affiliations.

Abstract: This work aims to systematically investigate the influence of process temperature, biomass-to-water ratio, and production scales (laboratory and pilot) on the chemical composition of aqueous and gaseous phases and mass production of chemicals by hydrothermal processing of Açaí (*Euterpe oleraceae*, Mart.) seeds. The hydrothermal carbonization was carried out at 175, 200, 225, and 250 °C at 2 °C/min and a biomass-to-water ratio of 1:10; at 250 °C at 2 °C/min and biomass-to-water ratios of 1:10, 1:15, and 1:20 in technical scale; and at 200, 225, and 250 °C at 2 °C/min and a biomass-to-water ratio of 1:10 in laboratory scale. The elemental composition (C, H, N, S) in the solid phase was determined to compute the HHV. The chemical composition of the aqueous phase was determined by GC and HPLC and the volumetric composition of the gaseous phase using an infrared gas analyzer. For the experiments in the pilot test scale with a constant biomass-to-water ratio of 1:10, the yields of solid, liquid, and gaseous phases varied between 53.39 and 37.01% (wt.), 46.61 and 59.19% (wt.), and 0.00 and 3.80% (wt.), respectively. The yield of solids shows a smooth exponential decay with temperature, while that of liquid and gaseous phases showed a smooth growth. By varying the biomass-to-water ratios, the yields of solid, liquid, and gaseous reaction products varied between 53.39 and 32.09% (wt.), 46.61 and 67.28% (wt.), and 0.00 and 0.634% (wt.), respectively. The yield of solids decreased exponentially with increasing water-to-biomass ratio, and that of the liquid phase increased in a sigmoid fashion. For a constant biomass-to-water ratio, the concentrations of furfural and HMF decreased drastically with increasing temperature, reaching a minimum at 250 °C, while that of phenols increased. In addition, the concentrations of CH_3COOH and total carboxylic acids increased, reaching a maximum concentration at 250 °C. For constant process temperature, the concentrations of aromatics varied smoothly with temperature. The concentrations of furfural, HMF, and catechol decreased with temperature, while that of phenols increased. The concentrations of CH_3COOH and total carboxylic acids decreased exponentially with temperature. Finally, for the experiments with varying water-to-biomass ratios, the productions of chemicals (furfural, HMF, phenols, cathecol, and acetic acid) in the aqueous phase is highly dependent on the biomass-to-water ratio. For the experiments at the laboratory scale with a constant biomass-to-water ratio of 1:10, the yields of solids ranged between 55.9 and 51.1% (wt.), showing not only a linear decay with temperature but also a lower degradation grade. The chemical composition of main organic compounds (furfural, HMF, phenols, catechol, and acetic acid) dissolved in the aqueous phase in laboratory-scale study showed the same behavior as those obtained in the pilot-scale study.

Keywords: Açaí seeds; hydrothermal carbonization; hot compressed water; process analysis; HMF; furfural; acetic acid; mass production

1. Introduction

Açaí (*Euterpe oleracea* Mart.) is a palm native to the Brazilian Amazon [1]. It has an abundant occurrence in the Amazon estuary floodplains [2,3]. The Açaí fruits in nature have great economic importance for the agroindustry, as well as extractive activities of rural communities of the Brazilian Amazonian state of Pará [4]. The fruit is a small, dark-purple, berry-like fruit, almost spherical, weighing between 2.6 to 3.0 g [5]. The fruit has a diameter around 10.0 and 20.0 mm [5], containing a large core seed/kernel that occupies between 85 and 95% (vol./vol.) of its volume [3,6].

When the pulp and skin are processed/extracted with warm water to produce a thick, purple-colored juice [3,6], a waste is generated [7–11], namely the Açaí seeds, a rich lignin-cellulosic residue with great potential for energetic use [8,9,12]. In the crop years 2016–2017, around 1200–1274 million tons of Açaí fruits were produced in Brazil;–the state of Pará is the main producer (94%), generating large amounts of solid residues consisting of seeds and fibers [4–9].

The residue of Açaí fruit (seeds + fibers) processing has a fibrous outer layer, containing 46.51% (wt.) cellulose and 30.31% (wt.) lignin [11]; this residue (seeds + fibers) is composed of 36.13% (wt.) cellulose, 47.92% (wt.) lignin, 1.57% (wt.) ash, and 16.64% (wt.) extractives [11], representing an important renewable biomass source for energetic applications [8,9,12].

The thermo-chemical transformation of lignin–cellulose-rich biomass with H_2O in the sub-critical or supercritical state is a promising technique, and the literature reports several studies on the subject [13–49]. Li et al. [13] applied statistical methods to investigate the role of process conditions (temperature, reaction time, biomass-to-water ratio) and chemical raw material characteristics on the physical–chemical properties of hydrothermal carbonization products (solid, liquid, and gas). In addition, Li et al. [13] reported that the most commonly cited hydrothermal carbonization product parameter was solid (hydro-char) yield (71%), while little attention has been paid to carbon-related information on the liquid and gaseous phases (<18%), which includes the analysis of chemical composition.

In fact, only a few studies systematically investigated the composition of the main chemical compounds, such as aromatic-ring compounds (furfural, HMF, phenols, cresols, catechol, guaiacol, etc.) [14,18,39–41], carboxylic acids ($HCOOH$, CH_3COOH, CH_3CH_2COOH, $CH_3CH(OH)COOH$, $CH_3C(O)CH_2CH_2COOH$, etc.) [14,18,39,40,43], alcohols (CH_3OH, CH_3CH_2OH) [14], sugars (glucose, xylose, galactose, fructose, sucrose, mannosan, levoglucosan, etc.) [18,40,43], and BTEX (benzene, toluene, xylenes, and ethylbenzene) [41]. In these studies, the above compounds were dissolved in the process water by hydrothermal carbonization of biomass, including the hydrothermal processing of corn stover [14,18]; Tahoe mix, pinyon/juniper, loblolly pine wood, sugar bagasse, and rice hulls [18]; wheat straw, wheat straw digestate, poplar, garapa, massaranduba, pine wood, α-Cellulose, and D-(+)-xylose [39]; wheat straw, poplar, and α-Cellulose [40]; wheat straw, wheat straw digestate, poplar, garapa, massaranduba, and pine wood [41]; Tahoe mix [43]; and biodegassed pellets [49]. Recently, Poerschmann et al. [44] investigated the distribution of medium molar mass compounds dissolved in process water by hydrothermal carbonization of glucose, fructose, and xylose by GC-MS and IC, identifying more than 50 compounds, the most abundant being carboxylic acids (formic, acetic, glycolic, lactic, and levulinic acids) and aromatic-ring compounds (furfural, 5-(Hydroxymethyl)-2-furfural).

Another process parameter affecting the physical–chemical properties of hydrothermal carbonization products (solid, liquid, and gas) is the water-to-biomass ratio [19,25–30]. In recent years only a few studies have investigated the influence of the water-to-biomass ratio from the hydrothermal processing of biomass, including tomato-peel-waste [26], olive stone [27], microalgae [28], sawdust [29], banana peels [30], wood chips [25], and corn stalks [19], but no study has examined their influence consistently [13], particularly the composition of the main chemical compounds dissolved in the process water.

Açaí (*Euterpe oleracea*, Mart.) is the only fruit species whose seed's centesimal and elemental composition is completely different from wood biomass (e.g., poplar, garapa, massaranduba, Tahoe mix, pinyon/juniper, loblolly pine, and pine wood) [18,39–41,43], agriculture residues of cereal grains (corn stover, corn stalk, rice hulls, and wheat straw), or agriculture residues of sugar cane (sugar bagasse) [18]. Although hydrothermal treatment has been applied to enhance enzymatic hydrolysis of Açaí seeds in aqueous H_2SO_4 at 121 °C [47], no systematic study has investigated the influence of temperature, biomass-to-water ratio, and production scales (laboratory and pilot) on the chemical composition of aqueous and gaseous phases, as well as on the mass production of chemicals by hydrothermal carbonization of Açaí seeds on a technical scale.

This work aims to investigate the influence of process temperature, biomass-to-water ratio, and production scales (laboratory and pilot) on the chemical composition of the main chemical compounds, such as aromatic-ring compounds, carboxylic acids, and alcohols, dissolved in process water, the gaseous phase composition, and mass production of chemicals by hydrothermal processing of Açaí (*Euterpe oleracea*, Mart.) seeds on a technical scale.

2. Materials and Methods

2.1. Methodology

The process flow sheet shown in Figure 1 summarizes the applied methodology, described in a logical sequence of ideas, chemical methods, and procedures, to analyze the chemical composition of main compounds in the liquid phase by HTC of Açaí seeds.

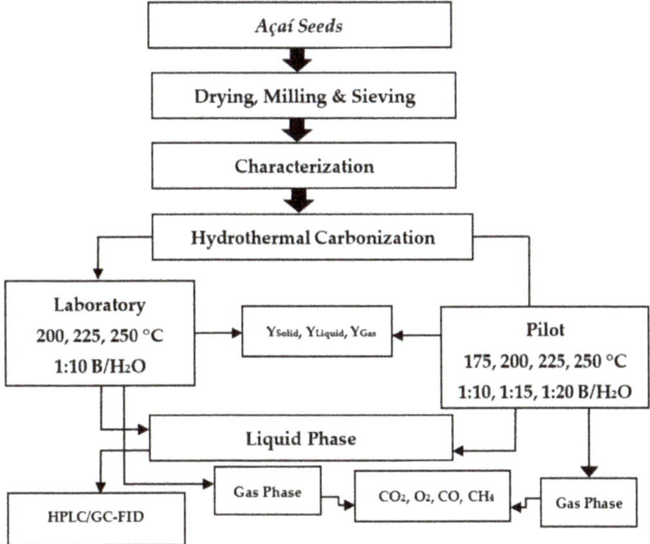

Figure 1. Process flow sheet for hydrothermal processing of Açaí (*Euterpe oleracea*, Mart.) seeds with hot compressed H_2O in laboratory and pilot scales.

Initially, the Açaí seeds were collected. Afterward, they were subjected to pre-treatments of drying, followed by milling and sieving. The HTC was carried out at laboratory and pilot scales to investigate the influence of temperature (175, 200, 225, 250 °C) and biomass-to-water ratio (1:10, 1:15, 1:20) on the yields of hydro-char, H_2O, and gas, as well as on the chemical composition in the liquid phase. Finally, the influence of production scales on the composition of the main compounds (furfural, HMF, phenols, catechol, guaiacol, HCOOH, etc.) in the liquid phase was investigated. The composition in the gas phase (CH_4, CO_2, O_2, CO) was determined by infrared spectroscopy, and that of the liquid phase was determined by HPLC and GC-FID.

2.2. Materials, Pre-Treatment, and Characterization of Açaí Seeds in Nature

The charges of Açaí seeds in nature obtained in a small Açaí store of in the city of Belém-Pará, Brazil [12]. The seeds were dried at 105 °C, ground using a knife cutting mill, and sieved using an 18 mesh sieve [12]. Afterward, the seeds underwent physical–chemistry characterization for moisture (AOAC 935.29), volatile matter (ASTM D 3175-07), ash (ASTM D 3174-04), fixed carbon (ASTM D6316-09), lipids (AOAC 963.15), proteins (AOAC 991.20) [12], and fibers according to the literature [50] and for insoluble lignin by the modified method of Klason [51]. The elemental composition (C, H, N, S) of Açaí seeds in the solid phase was also determined. Figure 2 shows the Açaí seeds after drying, milling, and sieving (seeds + fibers (a); comminuted seeds + fibers (b); and sieved seeds + fibers (c)). The HTC experiments were carried out with sieved seeds + fibers, as shown in Figure 2c.

Figure 2. Material after drying, milling, and sieving process of Açaí seeds (seeds + fibers (**a**); comminuted seeds + fibers (**b**); and sieved seeds (**c**)).

2.3. Experimental Apparatus and Procedures

2.3.1. Experimental Apparatus and Procedures in Pilot Scale

Pilot-scale apparatus is illustrated in Figure 3, described in details elsewhere [14].

The hydrothermal processing of dried Açaí seeds was carried out with hot compressed water at 175, 200, 225, and 250 °C for 240 min with a biomass-to-water ratio of 1:10 and at 250 °C for 240 min with biomass-to-water ratios of 1:10, 1:15, and 1:20, as described in detail elsewhere [14].

2.3.2. Experimental Apparatus and Procedures in Laboratory Scale

The laboratory-scale apparatus is illustrated in Figure 4. The laboratory-scale cylindrical stirred tank stainless steel reactor is 1.0 L (Parr, USA, Model: 4577), with internal diameters of 9.525 cm and 15.748 cm and a weight of 7.257 kg. The reactor contains a mechanical agitation system with a stirrer motor of $\frac{1}{4}$ hp, 1.81 Nm torque, and 02 impellers (ID = 5.08 cm) with 6 blades, a ceramic movable heater of 2800 W, a modular controller (Parr, Moline, IL, USA, Model: 4848), and 02 type J thermocouples inside a thermos well and operates at maximum conditions of 345 bar and 500 °C. The hydrothermal processing of dried Açaí seeds was carried out with hot compressed water at 200, 225, and 250 °C for 240 min, with a biomass-to-water ratio of 1:10. Initially, the solid moisture was determined, and the mass of water was computed based on the dried solid. A mass of 73.50 g of solid with 25% (wt.) moisture was added to 532.82 g of distilled and deionized water and was introduced in the reactor. The operating temperatures (200, 225, and 250 °C) set for a heating rate of 2.0 °C/min. The reaction time was computed from the time the reactor reached the setpoint temperature (τ_0). Afterwards, the reactor was cooled until ambient temperature. The reaction products, including a moist dewatered solid phase and a liquid phase, were determined gravimetrically. Then, the moist solid phase was dried at 105 °C for 24 h. The volume of the gas and its composition were determined, as described elsewhere [14]. The samples of moist dewatered solid, liquid, and dried solid phases were stored for physicochemical analysis.

Figure 3. Pilot-scale stirred tank stainless steel reactor of 18.92 P L (Parr, USA, Model: 4555).

Figure 4. Pilot-scale stirred tank stainless steel reactor of 1.0 L (Parr, Moline, IL, USA, Model: 4577).

2.4. Compositional Analysis of Reaction Products

2.4.1. Aqueous Phase

The chemical analysis of volatile low-chain length carboxylic acids (R-COOH, with $R_1=CH_3$, $R_2=C_2H_5$, $R_3=C_3H_7$, $R_4=C_4H_9$) and alcohols (R-OH, with $R_1=CH_3$, $R_2=C_2H_5$, $R_3=C_3H_8$), as well as selective cellulose/hemicellulose-derived compounds, was done by hydrothermal processing of biomass and then identified by GC while the selective cellulose/hemicellulose-derived phenolic (phenol, cresol, catechol, and guaiacol) and aldehyde (furfural, HMF) compounds were identified by HPLC. The equipment specifications (GC-FID, HPLC), and operating conditions, including the GC-FID temperature program, as well as HPLC procedures, are described in detail elsewhere [14].

2.4.2. Gaseous Phase

The volume of gas, degassed at 25 °C and 1.0 atmosphere, was measured with a gas flow meter, while an infrared gas analyzer was used to determine the volumetric composition of gaseous products [14]. The equipment's specifications and procedures are described in detail elsewhere [14].

2.5. Steady-State Material Balance by Hydrothermal Carbonization

The yields of reaction products (solid, liquid, and gaseous phases) were determined by applying the mass conservation principle within the stirred tank reactor, operating in batch mode in a closed thermodynamic system, and the equations are described in detail elsewhere [14].

3. Results

3.1. Hydrothermal Processing of Açaí Seeds

3.1.1. Material Balances, Operating Conditions, and Yields of Reaction Products
Influence of Temperature on the Yields of Reaction Products

The material balances, operating conditions, and yields of reaction products by hydrothermal processing of Açaí seeds *in nature* at 175, 200, 225, and 250 °C, with an increase of 2 °C/min for 240 min, and a biomass-to-water ratio of 1:10, at a pilot scale, are summarized in Table 1.

Figure 5 shows the effect of temperature on the yields of reaction products by hydrothermal processing of Açaí seeds in nature with hot, compressed H_2O heated by 2 °C/min to 175, 200, 225, 250 °C for 240 min with a biomass-to-water ratio of 1:10. The exponential function applied to regress the yields of reaction products correlated well with the experimental data for both the solid and liquid phases, with r^2 (R-Squared) between 0.996 and 0.996. The yield of dry solids shows a smooth first-order exponential decay behavior, while that of the liquid and gaseous phases shows a smooth first-order exponential growth. At 175 °C, hydrothermal carbonization takes place, as the main reaction product is a solid [15]. At 200 °C hydrothermal liquefaction occurs, and the main reaction products are liquids [15]. The hydro-char yields were similar to those reported in the literature for rye straw [16], eucalyptus leaves [17], corn stover [14], sugarcane bagasse [18], and corn stalk [19].

Table 1. Mass balances, processes, operating conditions, and yields of dry solid, liquid, and gaseous products by hydrothermal processing of Açaí seeds with hot compressed H_2O at 175, 200, 225, 250 °C, 2 °C/min, 240 min, and biomass-to-water ratio of 1:10, at a pilot scale.

	Temperature (°C)			
Process Parameters	**175**	**200**	**225**	**250**
Mass of Açaí Seeds [g]	300.00	299.82	299.98	300.16
Mass of H_2O [g]	2997.60	3000.20	3001.30	2999.90
Mechanical Stirrer Speed [rpm]	90	90	90	90
Initial Temperature [°C]	30	30	30	30
Heating Rate [°C/min]	2	2	2	2
Process Time [min]	240	240	240	240
Mass of Slurry [g]	3252.20	3240.20	3216.50	3167.40
Volume of Gas [mL], T = 25 °C, P = 1 atm	0	5290	5590	7470
Mass of Gas [g]	0	7.564	8.231	11.408
Process Loss (I) [g]	45.40	59.82	84.78	132.66
Input Mass of Slurry (Pressing) [g]	3252.20	3240.20	3216.50	3161.70
Process Loss (II) [g]	0.00	0.00	0.00	5.70
Mass of Liquid Phase [g]	2638.53	2615.56	2637.97	2556.96
Mass of Moist Hydro-char [g]	588.10	587.37	557.61	591.29
Process Loss (III) [g]	25.57	37.27	20.92	13.41
Mass of Dry Hydro-char [g]	160.16	118.53	113.052	111.092
(Mass of Liquid Phase + Σ Process Loss + Mass of Moist Hydro-char − Mass of Dry Hydro-char − Mass of Gas) [g]	3137.44	3173.926	3179.997	3177.52
Process Loss (I + II + III) [g]	70.97	97.09	105.70	151.77
Mass of Liquid$_{Reaction}$ [g]	139.84	173.726	178.697	177.62
Yield of Hydro-char [wt.%]	53.39	39.534	37.686	37.011
Yield of Liquid Phase [wt.%]	46.61	57.943	59.570	59.188
Yield of Gas [wt.%]	0.000	2.523	2.744	3.801

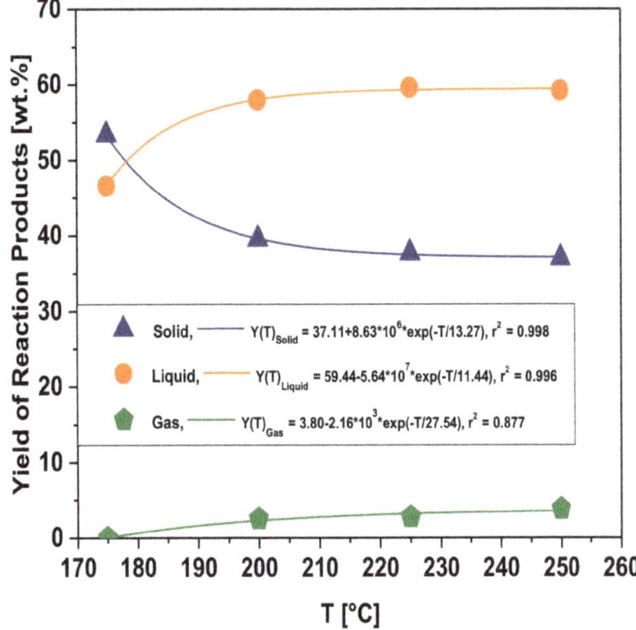

Figure 5. Effect of temperature on the yields of reaction products by hydrothermal processing of Açaí seeds *in nature*.

The centesimal composition of Açaí (*Euterpe oleracea* Mart.) seeds [12] was found to be 40.29% (wt.) cellulose, 5.5% (wt.) hemi-cellulose, 4.0% (wt.) lignin, 29.79% fibers, as well as 6.25% (wt.) protein, 0.61% (wt.) lipid, 10.15% (wt.) moisture, 0.5% (wt.) volatile matter,

0.83% (wt.) fixed carbon, and 0.15% (wt.) ash. The centesimal composition of Açaí fibers *in nature* is 41.37% (wt.) cellulose, 11.54% (wt.) hemi-cellulose, 40.25% (wt.) lignin, 1.96% ash, and 8.88% moisture, as reported by Tavares et al. [7].

Based on the facts that 61.25% (wt.) cellulose and 15% (wt.) lignin decompose at 200 °C over 24 h with a 1:10 biomass-to-water ratio (according to Falco et al. [16]), that 5.0% (wt.) lipids and 72% (wt.) proteins decompose at 350 °C over 90 min with a 1:5.7 biomass-to-water ratio (as reported by Teri et al. [20]), that hemicellulose degrades 65% (wt.) by a severity factor between 4.5 and 5.0 (as reported by Borrero-López et al. [21]), and that approximately 50% (wt.) cotton fibers decompose at 210 °C over 8.0 h with a 1:41 biomass-to-water ratio (as reported by Zhang et al. [22]), one may perform a centesimal mass balance to compute the approximate theoretical mass degradation of Açaí seeds at 200 °C, 2 °C/min, 240 min, and a biomass-to-water ratio of 1:10, using the following equation, computed in Table 2. The $Y_{\text{Hydro-char/Cellulose}}$ is computed by multiplying the percentage of cellulose in the centesimal composition of Açaí seeds by the fraction of non-degraded cellulose ($Y_{\text{Hydro-char/Cellulose}} = 40.29\% \times (100 - 61.25)/100 = 15.6124\%$), as summarized in Table 2.

$$Y_{\text{Hydro-char}} = \sum Y_{\text{Hydro-char/Cellulose}} + Y_{\text{Hydro-char/Lignin}} + Y_{\text{Hydro-char/Hemi-Cellulose}} + Y_{\text{Hydro-char/Proteins}} + Y_{\text{Hydro-char/Lipids}} + Y_{\text{Hydro-char/Fibers}} \quad (1)$$

The yield of hydro-char computed using Equation (1) is very close to the experimental value of 39.534% (wt.), showing a deviation of 2.51%, if the yield of $Y_{\text{Hydro-char/Fibers}}$ is computed based on the decomposition of cellulose, lignin, and hemi-cellulose ($Y_{\text{Hydro-char/Fibers}} = 16.17\%$), and 5.68%, if the yield of $Y_{\text{Hydro-char/Fibers}}$ is computed based on the decomposition of fibers ($Y_{\text{Hydro-char/Fibers}} = 14.895\%$). In addition, the computed amount of Açaí seeds fibers (29.79% (wt.)) degraded at 200 °C, 240 min, and the biomass-to-water ratio of 1:10 was 16.17% (wt.), giving a degradation percentage of 54.28%, very close to the results reported by Zhang et al. [22] who stated that approximately 50% (wt.) cotton fibers decompose at 210 °C over 8.0 h with a 1:41 biomass-to-water ratio.

Table 2. Computation of hydro-char yield. Centesimal composition of Açaí seeds and Açaí fibers from nature [7,12], decomposition cellulose and lignin [16], decomposition of lipids and proteins [20], decomposition of hemicellulose [21], decomposition of cotton fibers [22].

		Decomposition					
Centesimal Composition	**[wt.%]**	**Cellulose [wt.%]**	**Lignin [wt.%]**	**Hemi-Cellulose [wt.%]**	**Proteins [wt.%]**	**Lipids [wt.%]**	**Fibers [wt.%]**
		61.25	15.00	65.00	72.00	5.00	50.00
Açaí Seeds [12]		-	-	-	-	-	-
Cellulose	40.29	15.6124	-	-	-	-	-
Lignin	4.00	-	3.40	-	-	-	-
Hemi-cellulose	5.50	-	-	1.925	-	-	-
Proteins	6.25	-	-	-	1.82	-	-
Lipids	0.61	-	-	-	-	0.5795	-
Fibers	29.79	-	-	-	-	-	14.895
Moisture	10.15	-	-	-	-	-	-
Volatile matter	0.50	-	-	-	-	-	-
Fixed carbon	0.83	-	-	-	-	-	-
Ash	0.15	-	-	-	-	-	-
$Y_{\text{Hydro-char/Seeds}}$ **[wt.%]**	37.287	15.6124	3.400	1.925	1.820	0.5795	14.895
Açaí Fibers [7]		-	-	-	-	-	-
Cellulose	41.37	16.080	-	-	-	-	-
Lignin	40.25	-	34.213	-	-	-	-
Hemi-cellulose	11.54	-	-	4.039	-	-	-
Moisture	8.88	-	-	-	-	-	-
Ash	1.96	-	-	-	-	-	-
$Y_{\text{Hydro-char/Fibers}}$ **[wt.%]**	16.170	4.775	10.192	1.203	-	-	-

By analyzing the thermal decomposition behavior of cellulose and lignin reported by Falco et al. [16], one observes that the decomposition of cellulose is almost constant between 200 °C and 240 °C (38.75→37.00), showing that degradation/de-polymerization of cellulose is less intense [19], while lignin loses only 7.5% of its initial mass (85%→77.5%), thus making it possible to explain the small differences for the solid phase yields between 200 °C and 250 °C.

The material balances, operating conditions, and yields of reaction products by hydrothermal processing of Açaí seeds from nature at 200, 225, and 250 °C heated at 2 °C/min over 240 min with a biomass-to-water ratio of 1:10, at laboratory scale, are summarized in Table 3.

Table 3. Mass balances, processes, operating conditions, and yields of solid, liquid, and gaseous products by hydrothermal processing of Açaí seeds with hot compressed H_2O at 200, 225, 250 °C, 2 °C/min, 240 min, and biomass-to-water ratio of 1:10, in laboratory scale.

	Temperature [°C]		
Process Parameters	**200**	**225**	**250**
Mass of Açaí Seeds [g]	73.50	73.50	73.50
Moisture Content of Açaí Seeds [wt.%]	25.23	25.23	25.23
Mass of H_2O in Açaí Seeds [g]	18.54	18.54	18.54
Mass of H_2O [g]	531.01	531.01	531.01
Mechanical Stirrer Speed [rpm]	90	90	90
Initial Temperature [°C]	30	30	30
Heating Rate [°C/min]	2	2	2
Process Time [min]	240	240	240
Volume of Gas [mL], T = 25 °C, P = 1 atm	1395	2475	3215
CO_2 [vol.%]	48.60	44.80	35.80
CH_4 [vol.%]	1.90	1.20	0.60
O_2 [vol.%]	9.20	9.90	12.10
$100 - \Sigma\,(CO_2 + CH_4 + O_2)$ [vol.%]	40.30	44.10	51.50
Mass of Gas [g]	1.94	3.57	4.70
Mass of Dry Hydro-char [g]	30.77	29.43	28.16
Mass of Liquid$_{Reaction}$ [g]	34.72	36.01	36.59
Yield of Hydro-char [wt.%]	41.86	40.04	38.31
Yield of Liquid Phase [wt.%]	47.24	48.99	49.78
Yield of Gas [wt.%]	2.64	4.86	6.39

Figure 6 compares the yields of solid-phase products as a function of temperature by hydrothermal processing of Açaí seeds from nature with hot compressed H_2O at 175, 200, 225, and 250 °C, heated 2 °C/minovery 240 min, and biomass-to-water ratio of 1:10, in laboratory and pilot scales. One can observe, for the temperature interval 200–250 °C, that the yield of solid-phase products varied between 39.53 and 37.01% (wt.) in pilot scale, while that in the laboratory varied between 41.86 and 38.31% (wt.), showing deviations between 5.56 and 3.39%. The yield of solid-phase products at laboratory and pilot scales are very close, showing that production scales had little effect on hydro-char yield. A linear function was applied to regress the yield of hydro-char at a laboratory scale, correlating very well to the experimental data, with an r^2 (R-Squared) of 0.999.

Influence of Biomass-to-Water Ratio on the Yields of Reaction Products

Table 4 describes in detail the material balances, operating conditions, and yields of reaction products by hydrothermal processing of Açaí seeds *in nature* at 250 °C, 2 °C/min, 240 min, and biomass-to-water ratios of 1:10, 1:15, and 1:20 pilot scale.

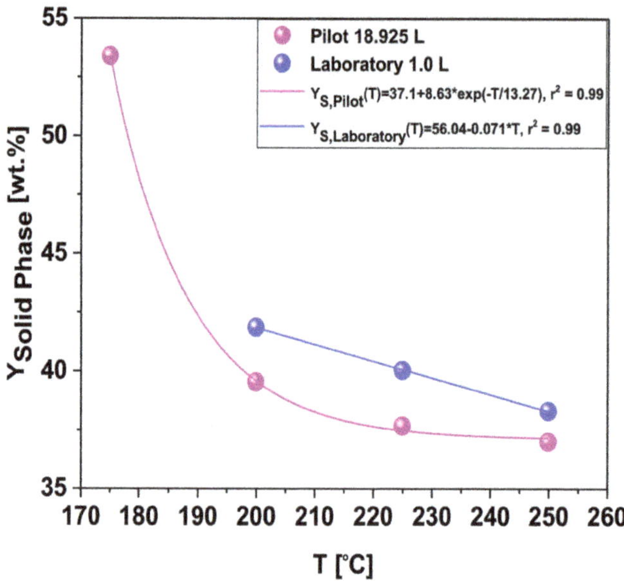

Figure 6. Influence of temperature on the solid phase yield by hydrothermal processing of Açaí seeds *in nature* at laboratory and pilot scales.

Table 4. Mass balances, processes, operating conditions, and yields of solid, liquid, and gaseous phases by hydrothermal processing of Açaí seeds with hot compressed H_2O at 250 °C, 2 °C/min, 240 min, and biomass-to-water ratios of 1:10, 1:15, and 1:20, in pilot scale.

	250 °C		
	Biomass/H_2O [-]		
Process Parameters	**1:10**	**1:15**	**1:20**
Mass of Açaí Seeds [g]	300.16	300.28	300.07
Mass of H_2O [g]	2999.90	4502.90	6000.80
Mechanical Stirrer Speed [rpm]	90	90	90
Initial Temperature [°C]	30	30	30
Heating Rate [°C/min]	2	2	2
Process Time [min]	240	240	240
Mass of Slurry [g]	3167.40	4696.50	6217.40
Volume of Gas [mL], T = 25 °C, P = 1 atm	7470	1240	1225
Mass of Gas [g]	11.408	1.905	1.863
Process Loss (I) [g]	132.66	106.68	83.47
Input Mass of Slurry (Pressing) [g]	3161.70	4696.50	6209.30
Process Loss (II) [g]	5.70	0.00	8.10
Mass of Liquid Phase [g]	2556.96	4077.05	5663.60
Mass of Moist Hydro-char [g]	591.29	585.83	518.45
Process Loss (III) [g]	13.41	33.62	35.35
Mass of Dry Hydro-char [g]	111.092	102.25	96.302
(Mass of Liquid Phase + Σ Process Loss + Mass of Moist Hydro-char − Mass of Dry Hydro-char − Mass of Gas) [g]	3177.52	5000.795	6210.805
Process Loss (I + II + III) [g]	151.77	140.30	126.92
Mass of Liquid$_{Reaction}$ [g]	177.62	196.125	202.535
Yield of Hydro-char [wt.%]	37.011	34.051	32.093
Yield of Liquid Phase [wt.%]	59.188	65.315	67.286
Yield of Gas [wt.%]	3.801	0.634	0.621

The effect of the H_2O-to-biomass ratio on the yields of reaction products (solid, liquid, and gas) by hydrothermal processing of Açaí seeds *in nature* is illustrated in Figure 7, and comparison of hydro-char yields with similar data reported in the literature is shown in Figure 8. At 250 °C, hydrothermal liquefaction is dominant, as the main reaction products formed are liquids [15]. The yields of reaction products, illustrated in Figure 7, were regressed using a dose–response function, showing an r^2 (R-Squared) between 0.97 and 0.99. The yields of hydro-char and gas decrease with H_2O-to-water ratios, while that of the liquid phase increases. With an increased H_2O-to-biomass ratio, the amount of reaction media (hot compressed H_2O) increases, increasing the number of hydroxonium ions (H_3O^+) and hydroxide ions (OH^-) dissociated within the reaction system, thus improving the catalyses of chemical reactions such as hydrolysis and organic compound degradation (e.g., depolymerization, fragmentation) without the aid of a catalyst [23]. In fact, according to the literature [24,25], increasing the H_2O-to-biomass ratio causes a great impact on hydrolysis reactions by hydrothermal processing of biomass.

A compilation of similar data on the effect of the H_2O-to-biomass ratio over hydro-char yields is illustrated in Figure 8. The behavior of the hydro-char yield is similar, showing a decrease in the hydro-char yield as the H_2O-to-biomass ratio increases. The data for Açaí seeds, tomato-peel-waste [26], olive stone [27], and corn stalk [19] were regressed using a dose–response function, showing an r^2 between 0.941 and 0.969. The experimental data not only resemble similar data reported in the literature for tomato-peel-waste [26], olive stone [27], microalgae [28], sawdust [29], banana peels [30], and wood chips [25] but are close to those of corn stalk [19] carried out at 250 °C and 4.0 h.

By analyzing Figure 8, one can observe that temperature has a combined effect on the hydro-char yield with varying H_2O-to-biomass ratios. At higher temperatures (250 °C), the effect of H_2O-to-biomass is more intense, playing an important role in hydro-char yield. For low–medium hydrothermal processing temperatures, the effect of H_2O-to-biomass on hydro-char yield is secondary, as reported by [26].

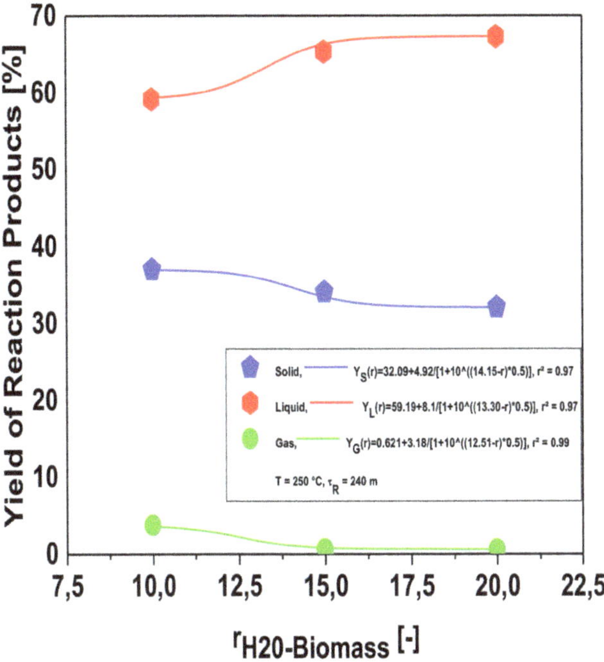

Figure 7. Effect of H_2O-to-biomass ratio on the yields of reaction products.

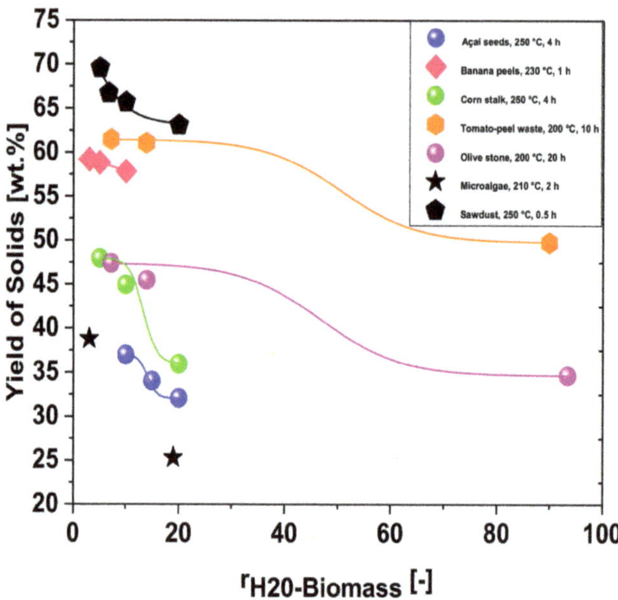

Figure 8. Comparison of hydro-char yield with similar data in the literature [19,26–29].

3.1.2. Chemical Composition of Gas Reaction Products
Influence of Temperature on the Chemical Composition of Gas Reaction Products

The volume of gas degassed at 25 °C and 1.0 atmosphere by hydrothermal processing of Açaí seeds with hot compressed H_2O at 175, 200, 225, and 250 °C, heated by 2 °C/min over 240 min, and biomass-to-water ratio of 1:10, at pilot scale, is shown in Table 5.

Table 5. Volume of gas and composition of gas products at 25 °C and 1.0 atmosphere by hydrothermal processing of Açaí seeds with hot compressed H_2O at 175, 200, 225, 250 °C, 2 °C/min, 240 min, and biomass-to-water ratio of 1:10, in pilot scale.

	Temperature [°C]			
Composition [vol.%]	**175**	**200**	**200**	**200**
CO_2 [vol.%]	0.00	44.40	51.30	60.60
CH_4 [vol.%]	0.00	0.40	0.50	1.30
O_2 [vol.%]	0.00	1.0	0.30	0.00
$100 - \Sigma\,(CO_2 + CH_4 + O_2)$ [vol.%]	0.00	54.20	47.90	38.10
Volume of Gas [mL]	0.00	5290	5590	7470
Volume of CO_2 [mL]	0.00	2348.76	2867.67	4526.82
Volume of CH_4 [mL]	0.00	21.16	27.95	97.11
Volume of O_2 [mL]	0.00	52.90	16.77	0.00
$V_{100-\Sigma\,(CO2+CH4+O2)} \approx V_{CO}$ [mL]	0.00	2867.18	2677.61	2846.07
Mass of Gas[g]	0.00	7.564	8.231	11.408
Mass of CO_2 [g]	0.00	4.191	5.117	8.078
Mass of CH_4 [g]	0.00	0.014	0.018	0.063
Mass of O_2 [g]	0.00	0.068	0.022	0.000
Mass of CO [g]	0.00	3.291	3.074	3.267
Composition of Gas [mol.%]				
Y^{CO2}	0.00	0.441410	0.510306	0.603507
Y^{CH4}	0.00	0.004055	0.004936	0.012943
Y^{O2}	0.00	0.009848	0.003017	0.000000
$Y^{100-\Sigma\,(CO2+CH4+O2)} \approx Y^{CO}$	0.00	0.544687	0.481741	0.383550

Figure 9 illustrates the effect of process temperature on the volume of gas degassed at 25 °C and 1.0 atmosphere and the volumetric composition of gaseous products shown in Figure 10. The volume of gas increases exponentially as the process temperature increases, and the same behavior was reported for the hydrothermal carbonization of corn stover by Machado et al. [14]. Similar studies reported that the volume of gaseous products increases with temperature [18,31,32].

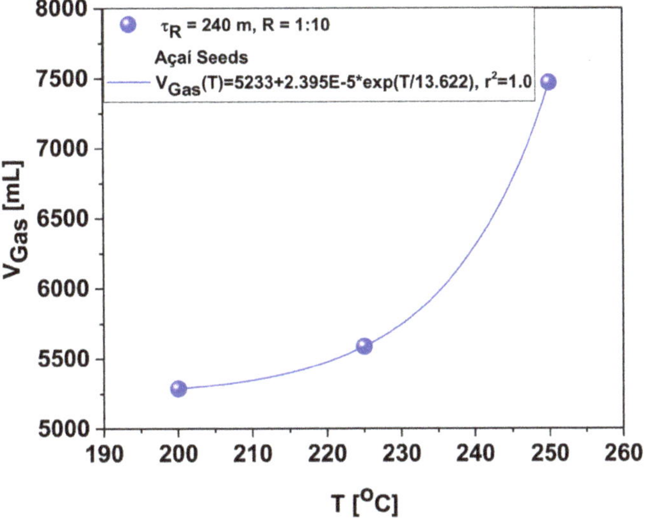

Figure 9. Effect of process temperature on the volume of gas degassed at 25 °C and 1.0 atmosphere.

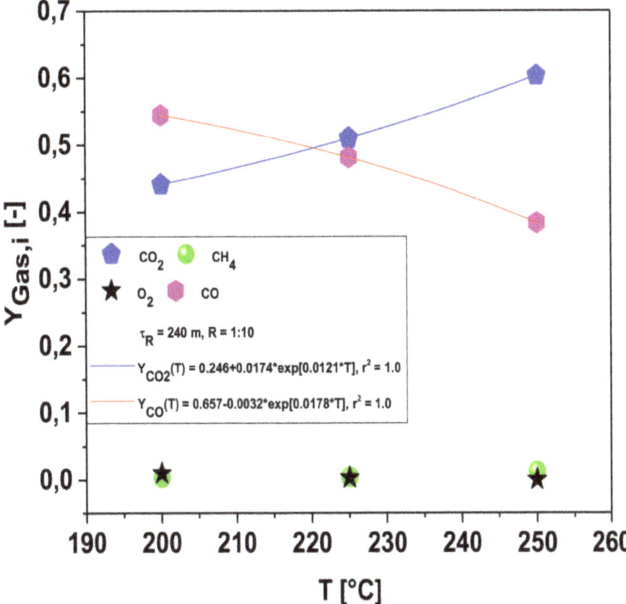

Figure 10. Effect of process temperature on the chemical composition of gas reaction products, expressed in mole fractions.

The infrared gas analyzer identified the presence of CO_2, O_2, CH_4, and CO, which was computed by difference [14], as summarized in Table 5, with CO_2 being the most abundant gaseous species produced. This is in accordance with similar studies on the evaluation of gaseous products and compositions by hydrothermal processing of biomass [14,18,31,33,34]. The presence of high volumetric concentrations of CO_2 in the gaseous phase indicates that decarboxylation is probably one of the dominant reaction mechanisms/pathways by hydrothermal processing of Açaí seeds *in nature*, according to Li et al. [35]. In fact, according to the literature [36], by hydrothermal processing of biomass, decarboxylation takes place, yielding CO_2, but other sources can also produce CO_2, including the decomposition of HCOOH, produced during the hydrothermal degradation of cellulose, and until condensation reactions occur.

The effect of temperature on the chemical composition of gas reaction products is shown in Figure 10. The mole fraction of CO shows a smooth exponential decay behavior, and the mole fraction of CO_2 shows a smooth exponential growth. An increase in CO_2 concentration in the gaseous phase by hydrothermal processing of biomass may be explained by analogy to the mild torrefaction process of biomass, as reported by Wannapeera et al. [37]. By increasing the processing temperature, the oxygen functional groups in the Açaí seeds are decomposed, resulting not only in higher amounts of gas formed but also in higher yields of CO_2.

Influence of H_2O-to-Biomass Ratio on the Volume of Gas Reaction Products

The effect of H_2O-to-biomass ratio on the volume of gas degassed at 25 °C and 1.0 atmosphere by hydrothermal of Açaí seeds from nature with hot compressed H_2O at 250 °C, heated by 2 °C/min over 240 min, and biomass-to-water ratios of 1:10, 1:15, and 1:20, at a pilot scale, illustrated in Table 6 and Figure 11. By increasing the H_2O-to-Biomass ratio, the volume of gas depletes, indicating that hydrolysis may be the dominant reaction mechanism [24,38].

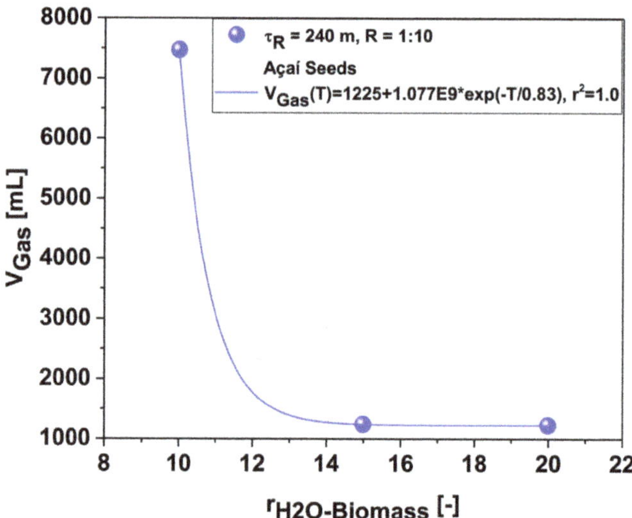

Figure 11. Effect of H_2O-to-biomass on the volume of gas degassed at 25 °C and 1.0 atmosphere by hydrothermal processing of Açaí seeds with hot compressed H_2O.

Table 6. Volume of gas and composition of gas products at 25 °C and 1.0 atmosphere as a function of temperature by hydrothermal processing of Açaí seeds with hot compressed H_2O at 250 °C, 2 °C/min, 240 min, and biomass-to-water ratios of 1:10, 1:15, and 1:20, in pilot scale.

	250 °C		
	Biomass/H_2O [-]		
Composition [vol.%]	1:10	1:15	1:20
CO_2 [vol.%]	60.60	63.40	59.90
CH_4 [vol.%]	1.30	3.10	1.60
O_2 [vol.%]	0.00	0.00	0.00
$\Sigma\,(CO_2 + CH_4 + O_2)$	61.90	65.50	61.50
$100 - \Sigma\,(CO_2 + CH_4 + O_2)$ [vol.%]	38.10	33.50	38.50
Volume of Gas [mL]	-	-	-
Volume of CO_2 [mL]	4526.82	786.16	733.775
Volume of CH_4 [mL]	97.11	38.44	19.60
Volume of O_2 [mL]	0.00	00.00	00.00
$V_{100\,-\,\Sigma\,(CO2\,+\,CH4\,+\,O2)} \approx V_{CO}$ [mL]	2846.07	415.40	471.625

3.1.3. Chemical Composition of Organic Compounds in the Aqueous Phase

Effect of Temperature on the Chemical Composition of Organic Compounds in the Aqueous Phase

The effect of temperature on the concentration profile of aromatic-ring compounds (furfural, HMF, phenols, and catechol) and carboxylic acids (CH_3COOH, CH_3CH_2COOH) by hydrothermal processing of Açaí seeds is illustrated in Figures 12–14, and the data are summarized in Table 7.

Figure 12 compares the concentration of HMF in the aqueous phase as a function of temperature by hydrothermal processing of Açaí seeds *in nature* with hot compressed H_2O at 175, 200, 225, 250 °C, heated by 2 °C/min over 240 min, and biomass-to-water ratio of 1:10, at laboratory and pilot scales. We observed, for the temperature interval 175–250 °C, at pilot scale, that the concentration of HMF showed a Gaussian distribution. In the temperature interval 200–250 °C, the concentration of HMF decreased drastically, showing an exponential decay behavior at both pilot and laboratory scales, and was not detectable at 250 °C.

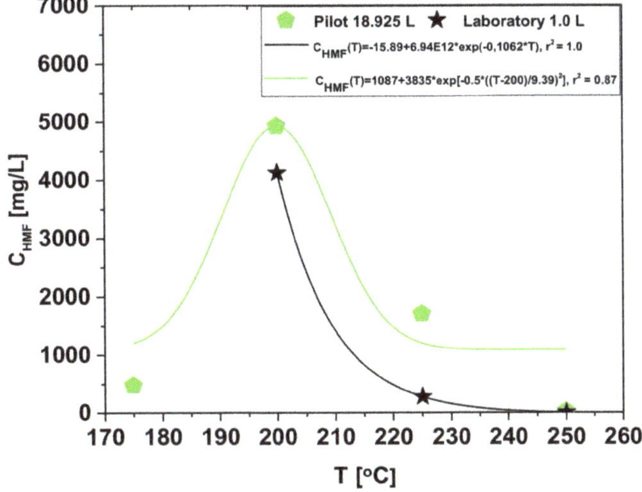

Figure 12. Effect of temperature on the concentration of HMF in the aqueous phase by hydrothermal processing of Açaí seeds from nature at laboratory and pilot scales.

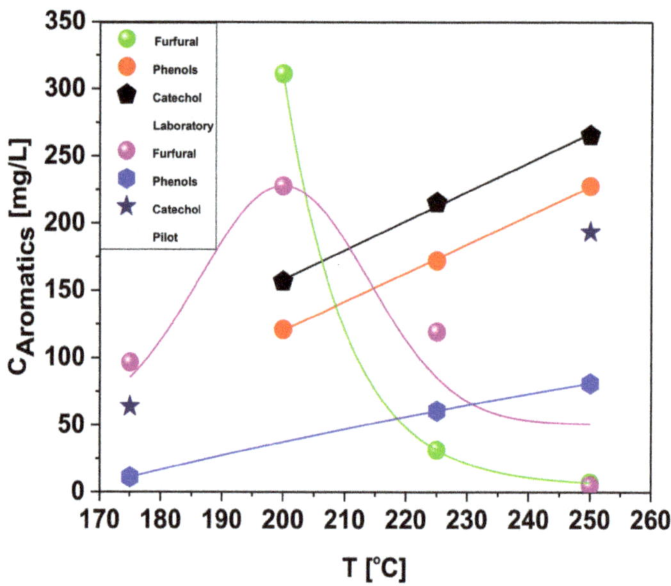

Figure 13. Effect of temperature on the concentration of aromatic compounds (phenols, furfural, and cathecol).

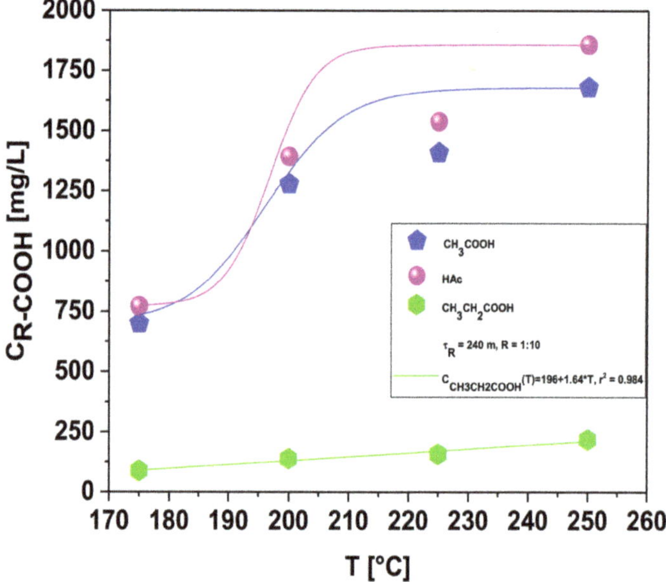

Figure 14. Effect of temperature on the concentration of carboxylic acids (CH$_3$COOH, CH$_3$CH$_2$COOH) and total carboxylic acids (HAc).

Table 7. Concentration of aromatics compounds (HMF, furfural, phenol, catechol), carboxylic acids (CH$_3$COOH, CH$_3$CH$_2$COOH), and total carboxylic acids (HAc) in the aqueous phase at 25 °C and 1.0 atmosphere by hydrothermal processing of Açaí seeds with hot compressed H$_2$O at 175, 200, 225, 250 °C, 2 °C/min, 240 min, and biomass-to-water ratio of 1:10, at pilot and laboratory scales.

	Temperature [°C]			
	Pilot			
Concentration of aromatics [mg/L]	**175**	**200**	**225**	**250**
HMF: CAS: 67-47-0	474.8	4923	1701	24.45
Furfural: CAS:98-01-1	96.83	227.8	119.5	5.05
Phenol: CAS: 108-95-2	11.34	-	60.81	81.24
Catechol: CAS: 120-80-9	63.86	-	-	193.8
Concentration of carboxylic acids [mg/L]	-	-	-	-
Acetic acid: CAS: 64-19-7	700	1280	1410	1680
Total acetic acid (HAc)	773.2	1394	1540	1859
Propionic Acid: CAS: 79-09-4	90	140	160	220
	Laboratory			
Concentration of aromatics [mg/L]	-	**200**	**225**	**250**
HMF: CAS: 67-47-0	-	4123	275	4.56
Furfural: CAS:98-01-1	-	311.5	31.82	7.28
Phenol: CAS: 108-95-2	-	121.2	172.5	227.99
Catechol: CAS: 120-80-9	-	156.84	215.4	365.61
Concentration of carboxylic acids [mg/L]	-	-	-	-
Acetic acid: CAS: 64-19-7	-	3420	3690	3810
Total acetic acid (HAc)	-	3797	3996	4286
Propionic Acid: CAS: 79-09-4	-	240	310	410

The concentration of HMF showed the same behavior in the temperature interval 200–250 °C at laboratory and pilot scales. However, there is a significant difference between concentration values, showing that production scales had a great effect on HMF concentration.

When the process temperature was increased, the concentrations of furfural, a byproduct of cellulose degradation, decreased exponentially in the temperature interval 200–250 °C and were present at very low concentrations at 250 °C, while the concentrations of phenols and catechol, products of furfural and HMF degradation, increased, as shown in Figure 13.

During hydrothermal processing of biomass, as cellulose hydrolyzes, it forms glucose, being transformed by isomerization reactions into fructose [38]. The decomposition of monosaccharides (glucose, fructose) produces volatile carboxylic acids, dissociating within the reaction media, thus producing hydroxonium ions (H$_3$O$^+$) and increasing the ionic product of reacting media, improving the degradation of biomass [38]. The monosaccharides (glucose, fructose) also undergo dehydration and fragmentation reactions, producing furfural-derived compounds (furfural, HMF), as well as acids and aldehydes [38]. Finally, as temperature increases, furfural-derived compounds (furfural, HMF) suffer degradation, producing acids, aldehydes, and phenols [38]. In this context, based on the reaction mechanism described by Sevilla and Fuertes [38], it is expected that, due to increasing process temperature, the concentrations of furfural and HMF will decrease, while those of catechol and phenols will increase. The results align with similar studies reported in the literature [14,18,39–41]. Jung et al. [42] studied the growth mechanism of hydro-char and the kinetic model of fructose degradation by hydrothermal carbonization, concluding that HMF degrades, forming hydro-char and H$_2$O (HMF $\rightarrow$ Hydro-char + H$_2$O), following first-order kinetics $\frac{d[\text{Hydro-char}]}{dt} = K \times [\text{HMF}]$. This is according to the results for hydro-char yields in Table 1; that is, the higher the concentration of HMF, the higher the yield of hydro-char.

Figure 14 shows that temperature has a great effect on concentrations of carboxylic acids (CH$_3$COOH, CH$_3$CH$_2$COOH) and total carboxylic acids (HAc) by hydrothermal processing of Açaí seeds with hot compressed H$_2$O at 175, 200, 225, 250 °C, heated by

2 °C/min over 240 min, and biomass-to-water ratio of 1:10, at pilot scale. The concentrations of carboxylic acids, particularly CH_3COOH, the most predominant one, and total carboxylic acids (HAc) increase strongly with temperature. With the hydrothermal processing of biomass, the monosaccharides (glucose, fructose) produced by hydrolysis are decomposed, forming volatile carboxylic acids, including acetic and propionic acids [38]. As reported by Hoekman et al. [18,43] and Machado et al. [14], the concentrations of acetic acid and total organic acids produced by hydrothermal processing of different biomass feedstocks increase with temperature. Poerschmann et al. [44] investigated the distribution of main medium molar mass compounds dissolved in process water by hydrothermal carbonization of glucose, fructose, and xylose at 180, 220, and 250 °C by GC-MS and IC, reporting acetic acid concentrations of 4560 and 3920 for the degradation of glucose and fructose, respectively, at 220 °C and 2.0 h.

It is known that monosaccharides (glucose, fructose) not only decompose, producing volatile carboxylic acids, but also undergo dehydration and fragmentation reactions, producing furfural-derived compounds (furfural, HMF). According to Kabyemela et al. [45], the reaction mechanism/pathway of cellobiose decomposition in sub- and supercritical H_2O (300 °C/25 MPa, 350 °C/25 MPa, 350 °C/40 MPa, and 400 °C/40 MPa), follows the sequence: hydrolysis of cellobiose to form glucose, followed by pyrolysis to form glycosyl-erythrose and glycosyl-glycol-aldehyde, which undergo hydrolysis to produce erythrose + glucose/fructose and glycol-aldehyde + glucose/fructose. tTat is, glucose- and fructose are intermediate-reaction products, being produced continuously along the hydrothermal process. However, Hoekman et al. [43] reported that concentrations of glucose/xylose and total sugars decrease with increasing process temperature (215, 235, 255, 275, 295 °C) from 1.02% (wt.) to 0.08% (wt.) and 1.41% (wt.) to 0.22% (wt.), respectively, and are not detected at 275 and 295 °C. Therefore, one may suppose that the degradation of monosaccharides (glucose, fructose) is not the only reaction mechanism to produce volatile carboxylic acids by hydrothermal processing of biomass as glucose, which according to Falco et al. [16], starts to be produced at 140 °C, reaches a maximum at 200 °C, and then begins to decompose.

Effect of Biomass-to-Water Ratio on the Chemical Composition of Organic Compounds in the Aqueous Phase

The effect of biomass-to-water ratio on the concentration profile of aromatic-ring compounds (furfural, HMF, phenols, and catechol) and carboxylic acids (CH_3COOH, CH_3CH_2COOH) by hydrothermal processing of Açaí seeds, illustrated in Figures 15 and 16, and the data are summarized in Table 8.

Table 8. Concentration of aromatics compounds (HMF, furfural, phenol, catechol), carboxylic acids (CH_3COOH, CH_3CH_2COOH), and total carboxylic acids (HAc) in the aqueous phase at 25 °C and 1.0 atmosphere by hydrothermal processing of Açaí seeds with hot compressed H_2O at 250 °C, 2 °C/min, 240 min, and biomass-to-water ratio of 1:10, in pilot scale.

	250 °C		
	Biomass/H_2O [-]		
Concentration of aromatics [mg/L]	**1:10**	**1:15**	**1:20**
HMF: CAS: 67-47-0	24.450	5.188	3.002
Furfural: CAS:98-01-1	5.054	2.972	2.194
Phenol: CAS: 108-95-2	81.24	81.78	89.33
Catechol: CAS: 120-80-9	195.6	193.8	185.9
Concentration of carboxylic acids [mg/L]	-	-	-
Acetic acid: CAS: 64-19-7	1680	1270	1070
Total acetic acid (HAc)	1859	1424	1070
Propionic Acid: CAS: 79-09-4	220	190	20

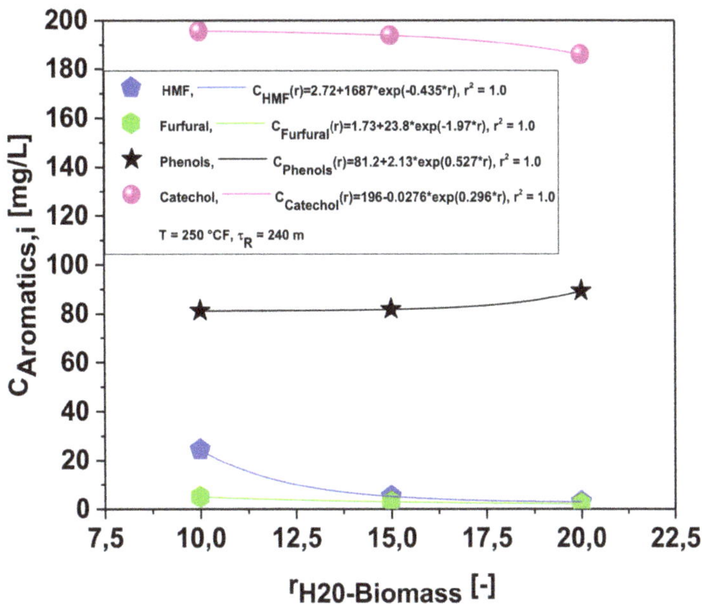

Figure 15. Effect of H_2O-to-biomass ratio on the concentration of aromatic (phenols, furfural, HMF, and catechol).

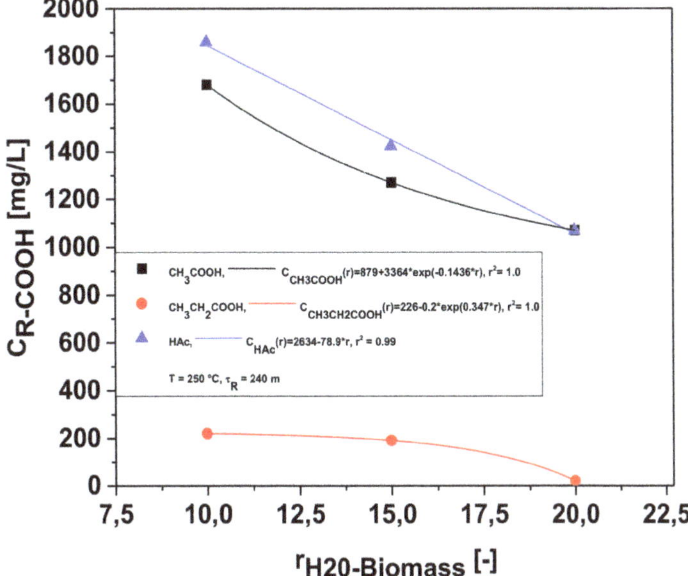

Figure 16. Effect of H_2O-to-biomass ratio on the concentration of carboxylic acids (CH_3COOH, CH_3CH_2COOH) and total carboxylic acids (HAc).

When the H_2O-to-biomass ratio was increased, the concentrations of furfural and HMF become very low and decrease smoothly, while that of phenols shows a smooth first-order exponential growth behavior, as shown in Figure 15. In addition, the carboxylic acids (CH_3COOH, CH_3CH_2COOH) and total carboxylic acids (HAc) also decrease as the H_2O-to-biomass ratio increases, as illustrated in Figure 16. The concentration is measured in mg/L, so increasing the H_2O-to-biomass ratio increases the volume of reaction, and it is hence expected to decrease the concentration of organic compounds dissolved in processed H_2O.

Effect of Biomass-to-Water Ratio on the Mass Production of Chemicals in the Aqueous Phase

By performing a mass balance by multiplying the concentration of main organic compounds dissolved in process water, described in Table 8, and the volume of process water (Mass of Liquid Phase + Σ Process Loss + Mass of Moist Hydro-char − Mass of Dry Hydro-char − Mass of Gas), described in Table 4, it can be shown that increasing the H_2O-to-biomass ratio has caused an increase in the mass production of chemicals, including phenols and catechol, products of furfural and HMF degradation, as well as acetic acid as shown in Figures 17 and 18.

According to the literature [24,25], increasing the H_2O-to-biomass ratio causes a great impact on the hydrolysis reactions by hydrothermal processing of the biomass so that the remaining cellulose in biomass is hydrolyzed, producing monosaccharides (glucose, fructose), and the decomposition of monosaccharides (glucose, fructose) produces volatile carboxylic acids, particularly acetic acid (Figure 17). As observed in Figure 17, the mass of furfural and HMF decreases while those of phenols and catechol increase, confirming that furfural and HMF were transformed into phenols and catechol.

It may be concluded that hydrolysis is probably the dominant reaction mechanism, but not the only one, by hydrothermal processing of Açaí seeds with hot compressed H_2O at 250 °C, heated by 2 °C/min over 240 min, as the biomass-to-water ratio increases from 1:10 to 1:20.

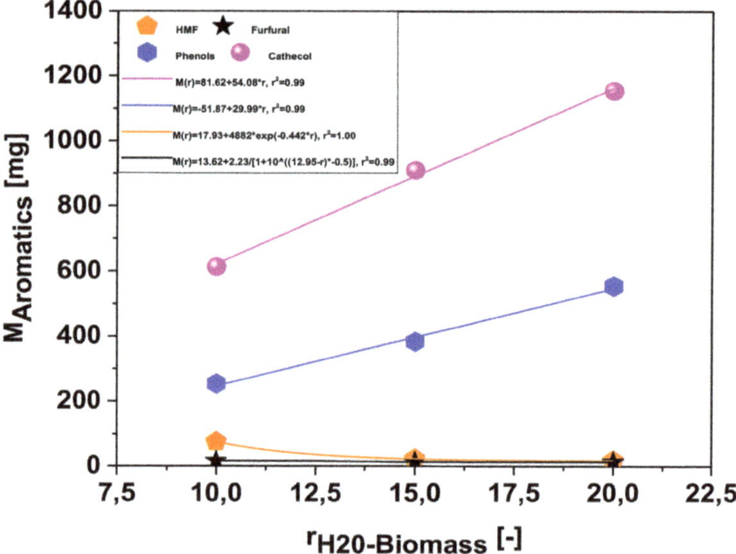

Figure 17. Effect of H_2O-to-biomass ratio on the mass production of acetic acid (CH_3COOH).

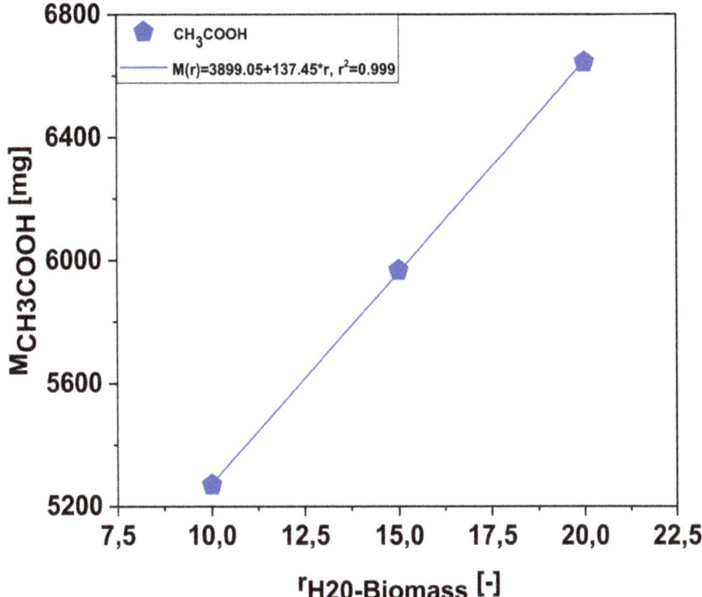

Figure 18. Effect of H_2O-to-biomass ratio on the mass production of aromatic compounds (phenols, catechol).

4. Conclusions

Based on the centesimal composition of Açaí (*Euterpe oleracea* Mart.) seeds, one may perform a centesimal mass balance to compute the approximate theoretical mass degradation of Açaí seeds at 200 °C, 2 °C/min, 240 min, and the biomass-to-water ratio of 1:10. The computed solid-phase yield was 38.562% (wt.), very close to the experimental value of 39.534% (wt.), showing a deviation of 2.51%.

The yields of hydro-char and gas decrease with H_2O-to-biomass ratios, while that of the liquid phase increases, showing that the H_2O-to-biomass ratio causes a great impact on hydrolysis reactions by hydrothermal processing of the biomass.

The presence of high volumetric concentrations of CO_2 in the gaseous phase indicates that decarboxylation is probably one of the dominant reaction mechanisms/pathways by hydrothermal processing of Açaí seeds from nature.

When the H_2O-to-biomass ratio is increased, the concentrations of furfural and HMF are very low and decrease smoothly, while that of phenols shows a smooth first-order exponential growth behavior. In addition, the carboxylic acids (CH_3COOH, CH_3CH_2COOH) and total carboxylic acids (HAc) also decrease as the H_2O-to-biomass ratio increases. By performing a mass balance, it can be shown that increasing the H_2O-to-biomass ratio caused an increase in the mass production of chemicals, including acetic acid, phenols, and catechol.

It may be concluded that hydrolysis is probably the dominant reaction mechanism, but not the only one, by hydrothermal processing of Açaí seeds with hot compressed H_2O at 250 °C, 2 °C/min, 240 min, as the biomass-to-water ratio increases from 1:10 to 1:20.

Author Contributions: The individual contributions of all the co-authors are provided as follows: C.d.M.S.d.S. contributed to formal analysis and writing—original draft preparation; D.A.R.d.C. contributed to formal analysis and writing—original draft preparation; M.C.S. contributed to formal analysis and software; H.d.S.A. contributed to formal analysis, software, and visualization; U.L. contributed to the investigation and validation; M.S. contributed to the investigation and methodology; T.H. contributed to resources and project administration; and N.T.M. contributed with supervision, conceptualization, and data curation. All authors have read and agreed to the published version of the manuscript.

Funding: This research was partially funded by CNPq-Brazil, grant number: 207325/2014-6.

Acknowledgments: We would like to acknowledge the Department of Postharvest Technology at Leibnitz-Institüt für Agrartechnik Potsdam-Bornin e.V for the opportunity to research at the Laboratory of Biochar, as well as for providing all the technical support (administrative, infrastructure, analytics) and materials to develop this research.

Conflicts of Interest: The authors declare no conflict of interest.

References

1. Lindolfo, M.M.; De Matos, G.S.B.; Pereira, W.V.D.S.; Fernandes, A.R. Productivity and nutrition of fertigated açaí palms according to boron fertilization. *Rev. Bras. Frutic.* **2020**, *42*, e601. [CrossRef]
2. Heinrich, M.; Dhanji, T.; Casselman, I. Açai (Euterpe oleracea. Mart.)—A phytochemical and pharmacologicalassessment of the species' health claims. *Phytochem. Lett.* **2011**, *4*, 10–21. [CrossRef]
3. Sabbe, S.; Verbeke, W.; Deliza, R.; Matta, V.; van Damme, P. Effect of a health claim and personal characteristics on consumer acceptance offruit juices with different concentrations of açaí (Euterpe oleracea Mart.). *Appetite* **2009**, *53*, 84–92. [CrossRef]
4. Del Pozo-Insfran, D.; Del Pozo-Insfran, D.; Brenes, C.H.; Brenes, C.H.; Talcott, S.T.; Talcott, S.T. Phytochemical Composition and Pigment Stability of Açai (Euterpe oleracea Mart.). *J. Agric. Food Chem.* **2004**, *52*, 1539–1545. [CrossRef]
5. Pessoa, J.D.C.; Silva, P.V.D.S.E. Effect of temperature and storage on açaí (Euterpe oleracea) fruit water uptake: Simulation of fruit transportation and pre-processing. *Fruits* **2007**, *62*, 295–302. [CrossRef]
6. Pompeu, D.; da Silva, E.M.; Rogez, H. Optimisation of the solvent extraction of phenolic antioxidants from fruits of Euterpe oleracea using Response Surface Methodology. *Bioresour. Technol.* **2009**, *100*, 6076–6082. [CrossRef] [PubMed]
7. Tavares, F.F.D.; de Almeida, M.D.C.; da Silva, J.A.P.; Araújo, L.L.; Cardozo, N.S.M.; Santana, R.M.C. Thermal treatment of açaí (Euterpe oleracea) fiber for composite reinforcement. *Polímeros* **2020**, *30*, e2020003. [CrossRef]
8. Bufalino, L.; Guimaraes, A.A.; de Silva, B.M.; de Souza, R.L.F.; de Melo, I.C.N.A.; de Oliveira, D.N.P.S.; Trugilho, P.F. Local variability of yield and physical properties of açaí waste and improvement of its energetic attributes by separation of lignocellulosic fibers and seeds. *J. Renew. Sustain. Energy* **2018**, *10*, 053102. [CrossRef]
9. De Lima, A.C.P.; Bastos, D.L.R.; Camarena, M.A.; Bon, E.P.S.; Cammarota, M.C.; Teixeira, R.S.S.; Gutarra, M.L.E. Physicochemical characterization of residual biomass (seed and fiber) from açaí (Euterpe oleracea) processing and assessment of the potential for energy production and bioproducts. *Biomass Convers. Biorefin.* **2019**, *11*, 925–935. [CrossRef]
10. Pessoa, J.D.C.; Arduin, M.; Martins, M.A.; De Carvalho, J.E.U. Characterization of açaí (E. oleracea) fruits and its processing residues. *Braz. Arch. Biol. Technol.* **2010**, *53*, 1451–1460. [CrossRef]
11. Barbosa, A.D.M.; Rebelo, V.S.M.; Martorano, L.G.; Giacon, V.M. Caracterização de partículas de açaí visando seu potencial uso na construção civil. *Matéria* **2019**, *24*. [CrossRef]
12. De Castro, D.A.R.; Ribeiro, H.J.D.S.; Ferreira, C.C.; Cordeiro, M.D.A.; Guerreiro, L.H.H.; Pereira, A.M.; dos Santos, W.; Santos, M.C.; De Carvalho, F.B.; Silva, J.O.C., Jr.; et al. Fractional Distillation of Bio-Oil Produced by Pyrolysis of Açaí (Euterpe oleracea) Seeds. *Fractionation* **2019**. [CrossRef]
13. Li, L.; Flora, J.R.; Caicedo, J.M.; Berge, N.D. Investigating the role of feedstock properties and process con-ditions on products formed during the hydrothermal carbonization of organics using regression techniques. *Bioresour. Technol.* **2015**, *187*, 263–274. [CrossRef]
14. Machado, N.; de Castro, D.; Santos, M.; Araújo, M.; Lüder, U.; Herklotz, L.; Werner, M.; Mumme, J.; Hoffmann, T. Process analysis of hydrothermal carbonization of corn Stover with subcritical H2O. *J. Supercrit. Fluids* **2018**, *136*, 110–122. [CrossRef]
15. Möller, M.; Nilges, P.; Harnisch, F.; Schröder, U. Subcritical Water as Reaction Environment: Fundamentals of Hydrothermal Biomass Transformation. *ChemSusChem* **2011**, *4*, 566–579. [CrossRef]
16. Falco, C.; Baccile, N.; Titirici, M. Morphological and structural differences between glucose, cellulose and lignocellulosic biomass derived hydrothermal carbons. *Green Chem.* **2011**, *13*, 3273–3281. [CrossRef]
17. Liu, Z.; Balasubramanian, R. Hydrothermal carbonization of waste biomass for energy generation. *Procedia Environ. Sci.* **2012**, *16*, 159–166. [CrossRef]
18. Hoekman, S.K.; Broch, A.; Robbins, C.; Zielinska, B.; Felix, L. Hydrothermal carbonization (HTC) of selected woody and herbaceous biomass feedstocks. *Biomass Convers. Biorefin.* **2012**, *3*, 113–126. [CrossRef]
19. Guo, S.; Dong, X.; Wu, T.; Zhu, C. Influence of reaction conditions and feedstock on hydrochar properties. *Energy Convers. Manag.* **2016**, *123*, 95–103. [CrossRef]
20. Teri, G.; Luo, L.; Savage, P. Hydrothermal Treatment of Protein, Polysaccharide, and Lipids Alone and in Mixtures. *Energy Fuels* **2014**, *28*, 7501–7509. [CrossRef]
21. Borrero-López, A.M.; Masson, E.; Celzard, A.; Fierro, V. Modelling the reactions of cellulose, hemicellulose and lignin submitted to hydrothermal treatment. *Ind. Crop. Prod.* **2018**, *124*, 919–930. [CrossRef]

22. Zhang, Y.; Hou, W.; Guo, H.; Shi, S.; Dai, J. Preparation and Characterization of Carbon Microspheres From Waste Cotton Textiles By Hydrothermal Carbonization. *J. Renew. Mater.* **2019**, *7*, 1309–1319. [CrossRef]
23. Öztürk, I.; Irmak, S.; Hesenov, A.; Erbatur, O. Hydrolysis of kenaf (Hibiscus cannabinus L.) stems by catalytical thermal treatment in subcritical water. *Biomass Bioenergy* **2010**, *34*, 1578–1585. [CrossRef]
24. Román, S.; Nabais, J.M.V.; Laginhas, C.; Ledesma, B.; González, J.F. Hydrothermal carbonization as an effective way of densi-fying the energy content of biomass. *Fuel Process. Technol.* **2012**, *103*, 78–83. [CrossRef]
25. Sermyagina, E.; Saari, J.; Kaikko, J.; Vakkilainen, E. Hydrothermal carbonization of coniferous biomass: Effect of process parameters on mass and energy yields. *J. Anal. Appl. Pyrolysis* **2015**, *113*, 551–556. [CrossRef]
26. Sabio, E.; Álvarez-Murillo, A.; Román, S.; Ledesma, B. Conversion of tomato-peel waste into solid fuel by hydrothermal car-bonization: Influence of the processing variables. *Waste Manag.* **2016**, *47*, 122–132. [CrossRef] [PubMed]
27. Álvarez-Murillo, A.; Roman, S.; Ledesma, B.; Sabio, E. Study of variables in energy densification of olive stone by hydrothermal carbonization. *J. Anal. Appl. Pyrolysis* **2015**, *113*, 307–314. [CrossRef]
28. Heilmann, S.M.; Davis, H.T.; Jader, L.R.; Lefebvre, P.A.; Sadowsky, M.; Schendel, F.J.; Von Keitz, M.G.; Valentas, K.J. Hydrothermal carbonization of microalgae. *Biomass Bioenergy* **2010**, *34*, 875–882. [CrossRef]
29. Oktaviananda, C.; Rahmawati, R.F.; Prasetya, A.; Purnomo, C.W.; Yuliansyah, A.T.; Cahyono, R.B. Effect of temperature and biomass-water ratio to yield and product characteristics of hydrothermal treatment of biomass. In Proceedings of the AIP Conference Proceedings, Bydgoszcz, Poland, 9–11 May 2018; AIP Publishing Center: Yogyakarta, Indonesia, 2017; p. 020029. [CrossRef]
30. Putra, H.; Damanhuri, E.; Dewi, K.; Pasek, A.D. Hydrothermal carbonization of biomass waste under low temperature condition. In Proceedings of the 2nd International Conference on Engineering and Technology for Sustainable Development (ICET4SD 2017), Yogyakarta, Indonesia, 13–14 September 2017; Volume 154, p. 01025. [CrossRef]
31. Castello, D.; Kruse, A.; Fiori, L. Biomass gasification in supercritical and subcritical water: The effect of the reactor material. *Chem. Eng. J.* **2013**, *228*, 535–544. [CrossRef]
32. Lucian, M.; Fiori, L. Hydrothermal Carbonization of Waste Biomass: Process Design, Modeling, Energy Efficiency and Cost Analysis. *Energies* **2017**, *10*, 211. [CrossRef]
33. Axel, F.; Felix, R.; Andrea, K. Experimental comparison of hydrothermal and vapothermal carbonization. *Fuel Process. Technol.* **2013**, *115*, 261–269.
34. Zhang, B.; Huang, H.J.; Ramaswamy, S. Reaction kinetics of the hydrothermal treatment of lignin. *Appl. Biochem. Biotechnol.* **2008**, *147*, 119–131. [CrossRef]
35. Li, L.; Diederick, R.; Flora, J.R.; Berge, N.D. Hydrothermal carbonization of food waste and associated packaging materials for energy source generation. *Waste Manag.* **2013**, *33*, 2478–2492. [CrossRef]
36. Funke, A.; Ziegler, F. Hydrothermal carbonization of biomass: A summary and discussion of chemical mechanisms for process engineering. *Biofuels Bioprod. Biorefin.* **2010**, *4*, 160–177. [CrossRef]
37. Wannapeera, J.; Fungtammsan, B.; Worasuwannarak, N. Effects of temperature and holding time during tor-refaction on the pyrolysis behaviors of woody biomass. *J. Anal. Appl. Pyrolysis* **2011**, *92*, 99–105. [CrossRef]
38. Sevilla, M.; Fuertes, A.B. The production of carbon materials by hydrothermal carbonization of cellulose. *Carbon* **2009**, *47*, 2281–2289. [CrossRef]
39. Becker, R.; Dorgerloh, U.; Paulke, E.; Mumme, J.; Nehls, I. Hydrothermal Carbonization of Biomass: Major Organic Components of the Aqueous Phase. *Chem. Eng. Technol.* **2014**, *37*, 511–518. [CrossRef]
40. Reza, M.T.; Wirth, B.; Lueder, U.; Werner, M. Behavior of selected hydrolyzed and dehydrated products during hydrothermal carbonization of biomass. *Bioresour. Technol.* **2014**, *169*, 352–361. [CrossRef] [PubMed]
41. Becker, R.; Dorgerloh, U.; Helmis, M.; Mumme, J.; Diakité, M.; Nehls, I. Hydrothermally carbonized plant materials: Patterns of volatile organic compounds detected by gas chromatography. *Bioresour. Technol.* **2013**, *130*, 621–628. [CrossRef]
42. Jung, D.; Zimmermann, M.; Kruse, A. Hydrothermal Carbonization of Fructose: Growth Mechanism and Kinetic Model. *ACS Sustain. Chem. Eng.* **2018**, *6*, 13877–13887. [CrossRef]
43. Hoekman, S.K.; Broch, A.; Robbins, C. Hydrothermal Carbonization (HTC) of Lignocellulosic Biomass. *Energy Fuels* **2011**, *25*, 1802–1810. [CrossRef]
44. Poerschmann, J.; Weiner, B.; Koehler, R.; Kopinke, F.D. Hydrothermal Carbonization of Glucose, Fructose, and Xylose-Identification of Organic Products with Medium Molecular Masses. *ACS Sustainable Chem. Eng.* **2017**, *5*, 6420–6428. [CrossRef]
45. Kabyemela, B.M.; Takigawa, M.; Adschiri, T.; Malaluan, R.M.; Arai, K. Mechanism and Kinetics of Cellobiose Decomposition in Sub- and Supercritical Water. *Ind. Eng. Chem. Res.* **1998**, *37*, 357–361. [CrossRef]
46. Román, S.; Libra, J.; Berge, N.; Sabio, E.; Ro, K.; Li, L.; Ledesma, B.; Álvarez, A.; Bae, S. Hydrothermal Carbonization: Modeling, Final Properties Design and Applications: A Review. *Energies* **2018**, *11*, 216. [CrossRef]
47. Oliveira, J.; Komesu, A.; Maciel Filho, R. Hydrothermal Pretreatment for Enhancing Enzymatic Hydrolysis of Seeds of Açaí (Euterpe oleracea) and Sugar Recovery. *Chem. Eng. Trans.* **2014**, *37*, 785–792.
48. Bauer, S.K.; Cheng, F.; Colosi, L.M. Evaluating the Impacts of ACP Management on the Energy Performance of Hydrothermal Liquefaction via Nutrient Recovery. *Energies* **2019**, *12*, 729. [CrossRef]

49. Shah, A.; Toor, S.; Nielsen, A.; Pedersen, T.; Rosendahl, L. Bio-Crude Production through Recycling of Pretreated Aqueous Phase via Activated Carbon. *Energies* **2021**, *14*, 3488. [CrossRef]
50. Standards, T. *Acid-Insoluble Lignin in Wood and Pulp Tappi Method T 222 Om-06*; Tappi Press: Atlanta, GA, USA, 2006.
51. Buffiere, P.; Loisel, D.; Van Soest, P.J. *Dosage des Fibres*; Inra-Lbe Laboratoire De Biotechnologie De L'environnement: Narbonne, France, 2007; pp. 1–14.

Article

Characterization of Bio-Adsorbents Produced by Hydrothermal Carbonization of Corn Stover: Application on the Adsorption of Acetic Acid from Aqueous Solutions

Maria Elizabeth Gemaque Costa [1], Fernanda Paula da Costa Assunção [2], Tiago Teribele [1], Lia Martins Pereira [1], Douglas Alberto Rocha de Castro [1], Marcelo Costa Santo [1], Carlos Emerson Ferreira da Costa [3], Maja Shultze [4], Thomas Hofmann [4] and Nélio Teixeira Machado [1,4,5,*]

[1] Graduate Program of Natural Resources Engineering of Amazon, Rua Augusto Corrêa N° 1, Campus Profissional-UFPA, Belém 66075-110, Brazil; gemaquebeth@yahoo.br (M.E.G.C.); teribele@globo.com (T.T.); liapereira@ufpa.br (L.M.P.); douglascastro87@hotmail.com (D.A.R.d.C.); marcelo.santos@ufra.edu.br (M.C.S.)
[2] Graduate Program of Civil Engineering, Rua Augusto Corrêa N° 1, Campus Profissional-UFPA, Belém 66075-110, Brazil; fernanda.assuncao.itec@gmail.com
[3] Graduate Program of Chemistry, Rua Augusto Corrêa N° 1, Campus Profissional-UFPA, Belém 66075-110, Brazil; emmerson@ufpa.br
[4] Department of Postharvest Technology, Leibnitz-Institüt für Agrartechnik Potsdam-Bornin e.V, Max-Eyth-Alee 100, 14469 Potsdam, Germany; mschultze@atb-potsdam.de (M.S.); THoffmann@atb-potsdam.de (T.H.)
[5] Faculty of Sanitary and Environmental Engineering, Rua Corrêa N° 1, Campus Profissional-UFPA, Belém 66075-900, Brazil
*   Correspondence: machado@ufpa.br; Tel.: +55-91-984-620-325

Citation: Costa, M.E.G.; da Costa Assunção, F.P.; Teribele, T.; Pereira, L.M.; de Castro, D.A.R.; Santo, M.C.; da Costa, C.E.F.; Shultze, M.; Hofmann, T.; Machado, N.T. Characterization of Bio-Adsorbents Produced by Hydrothermal Carbonization of Corn Stover: Application on the Adsorption of Acetic Acid from Aqueous Solutions. *Energies* **2021**, *14*, 8154. https://doi.org/10.3390/en14238154

Academic Editor: Andrea Di Carlo

Received: 15 October 2021
Accepted: 3 November 2021
Published: 5 December 2021

Publisher's Note: MDPI stays neutral with regard to jurisdictional claims in published maps and institutional affiliations.

Abstract: In this work, the influence of temperature on textural, morphological, and crystalline characterization of bio-adsorbents produced by hydrothermal carbonization (HTC) of corn stover was systematically investigated. HTC was conducted at 175, 200, 225, and 250 °C, 240 min, heating rate of 2.0 °C/min, and biomass-to-H_2O proportion of 1:10, using a reactor of 18.927 L. The textural, morphological, crystalline, and elemental characterization of hydro-chars was analyzed by TG/DTG/DTA, SEM, EDX, XRD, BET, and elemental analysis. With increasing process temperature, the carbon content increased and that of oxygen and hydrogen diminished, as indicated by elemental analysis (C, N, H, and S). TG/DTG analysis showed that higher temperatures favor the thermal stability of hydro-chars. The hydro-char obtained at 250 °C presented the highest thermal stability. SEM images of hydro-chars obtained at 175 and 200 °C indicated a rigid and well-organized fiber structure, demonstrating that temperature had almost no effect on the biomass structure. On the other hand, SEM images of hydro-chars obtained at 225 and 250 °C indicated that hydro-char structure consists of agglomerated micro-spheres and heterogeneous structures with nonuniform geometry (fragmentation), indicating that cellulose and hemi-cellulose were decomposed. EDX analysis showed that carbon content of hydro-chars increases and that of oxygen diminish, as process temperature increases. The diffractograms (XRD) identified the occurrence of peaks of higher intensity of graphite (C) as the temperature increased, as well as a decrease of peaks intensity for crystalline cellulose, demonstrating that higher temperatures favor the formation of crystalline-phase graphite (C). The BET analysis showed 4.35 m^2/g surface area, pore volume of 0.0186 cm^3/g, and average pore width of 17.08 µm. The solid phase product (bio-adsorbent) obtained by hydrothermal processing of corn stover at 250 °C, 240 min, and biomass/H_2O proportion of 1:10, was activated chemically with 2.0 M NaOH and 2.0 M HCl solutions to investigate the adsorption of CH_3COOH. The influence of initial acetic acid concentrations (1.0, 2.0, 3.0, and 4.0 mg/mL) was investigated. The kinetics of adsorption were investigated at different times (30, 60, 120, 240, 480, and 960 s). The adsorption isotherms showed that chemically activated hydro-chars were able to recover acetic acid from aqueous solutions. In addition, activation of hydro-char with NaOH was more effective than that with HCl.

Keywords: corn stover; hydrothermal process; hydrochar; adsorption; acetic acid; thermo-gravimetric analysis; scanning electron microscopy; X-ray diffraction; BET analysis

1. Introduction

Hydro-char is porous carbonaceous material with reactive, functionalized/aromatic surfaces [1]. These morphological and textural properties make hydrochar a potential adsorbent to remove/recover chemical contaminants from process water [1–5], until the process aqueous phase produced by hydrothermal carbonization of biomass [6]. Hydrochars differ from biochars due to its lower aromaticity, consisting of mostly alkyl moieties [7]. In recent years, the literature reports the application of hydro-chars as bio-adsorbents to selectively remove organic and inorganic compounds, as well as heavy metal, that is, the adsorption/sorption of aqueous contaminants onto hydro-chars [1–6,8–51].

Hydro-chars have been applied as adsorbents due to their capacity to remove and or selectively absorb polar and non-polar organic compounds, including acetic acid [6], bisphenol A, 17α-ethinyl estradiol and phenanthrene [22], 2-naphthol [11], pyrene [22]; phenolic compounds (phenol, guaiacol, vanillyl alcohol, and resorcinol) [31]; dyes, such as methylene blue [8,9,23,24,26,27,29,36,38–47], methyl orange [36,49], methylene green [48], and Congo red [11]; herbicides, such as fluridone, norflurazon [18], and isoproturon [19]; pharmaceuticals, such as triclosan, estrone, carbamazepine, acetaminophen [12], sulfamethoxazole, diclofenac, bezafibrate, carbamazepine, atrazine [16,51], tetracycline [20], diclofenac sodium, salicylic acid, and flurbiprofen [21]; alkaline (K^+, Na^+) and alkaline earth metals (Mg^{2+}, Ca^{2+}) [31]; heavy metals, such as lead [2–4,10,15,23–25,28,30,32,42], cadmium (II) [2,4,5,10,13,24,32,38], uranium(VI) [1,35], antimony [13], copper(II) [4,5,10,14,17,24,39,50], nickel [4], zinc [10], and chromium(II) [33]; and fertilizers such as phosphate [23,34], orthophosphate [50], and ammonium [34].

The chemical activation of hydro-chars was carried out by either modifying the reaction media composition (H_2O + *modifier*) or the solid phase reaction products (hydro-char + modifier) and has been intensively investigated in recent years [4,8–10,21,26–30,32,33,35,36,38–43,45–51]. The physical activation of hydrochars has also been investigated [24]. Changes in the mesoporous and surface properties of hydro-chars by modifying the reaction medium composition include the addition of H_3PO_4 (phosphoric acid) [30], $CH_2 = CH$-COOH (acrylic acid) [29,32], terminal amino hyper-branched polymer solutions [33], $C_2H_2(CO)_2O$ (maleic anhydride) followed by deprotonation of carboxyl groups with $NaHCO_3$ solution [35], $ZnCl_2$ (zinc chloride) [41], NaOH [48], HCl, NaOH, and NaCl [51]. In addition, hydro-char mesoporous and surface properties have been chemically activated with H_2O_2 [4,25], NaOH [6,8,9,26,27,29], KOH [10,28,47,50], H_3PO_4 [21], polyaminocarboxylated modified hydro-char [39], O_3/NaCl [42], HNO_3 [43], and until etherification, amination and protonation reaction [49].

The water process streams by hydrothermal carbonization of lignocellulosic materials is a complex mixture containing aromatic-ring compounds (furfural, HMF, phenols, cresols, catechol, and guaiacol) [52–54], carboxylic acids (formic acid, acetic acid, propionic acid, and lactic acid) [52–54], alcohols (methanol and ethanol) [52,53], and sugars [54], with high concentrations of volatiles carboxylic acids [52–54], as well as hazardous substances such as HMF (hydromethylfurfural), thus making its reuse, even as washing water not possible, so that application of separation processes is necessary to recover organic and inorganic compounds. In addition, Machado et al. [52] stated that complex chemical composition of water process streams by HTC poses a hard separation task, because of the huge differences in chemical structure, as well as thermal and physical properties such as boiling point. In fact, many of those compounds present boiling points higher than H_2O, so that application of separation processes (except adsorption [6,31]), such as distillation and evaporation are not effective [52].

Despite some studies on the adsorption/sorption of organic compounds within hydrochar produced by hydrothermal carbonization of biomass, activated chemically with NaOH [6,8,9,26,27,29], only a few have investigated the uptake of organic compounds from aqueous process streams from hydrothermal carbonization/liquefaction [6,31], as summarized synthetically below. In addition, until now, no study has investigated the adsorption of acetic acid in hydro-char activated chemically with HCl.

Machado et al. [6], studied the adsorption of acetic acid form H_2O solutions on hydro-chars produced by hydrothermal carbonization of corn stover at 200, 225, and 250 °C, 240 min, heating rate of 2 °C/min, biomass/H_2O proportion of 1:10, using a reactor of 18.927 L, in batch mode. The hydro-chars were characterized by SEM, EDX, and XRD [6]. The hydro-char obtained at 250 °C, was activated chemically with a 2.0 M NaOH solution [6]. The influence of initial CH_3COOH concentrations on the adsorption kinetic was investigated [6]. The adsorption kinetics was investigated at 30, 60, 120, 240, 480, and 960 s [6]. The adsorption isotherms showed that chemically activated hydro-chars were able to recover acetic acid from aqueous solutions [6].

Sanette et al. [31], investigated the adsorption of AAEMs (alkaline and alkaline earth metals) (Ca^{2+}, K^+, Na^+, and Mg^{2+}), as well as phenolic compounds (guaiacol, phenol, vanillyl alcohol, and resorcinol) present in process water streams obtained by hydrothermal carbonization of a model (synthetic) mixture that mimics the organic fraction of municipal solid waste at 300 °C, 15 min, feedstock/H_2O proportion of 1:1, using a stainless steel reactor of 945 mL, on chemically activated hydro-char with KOH solutions of concentrations between 0.5 and 2.5 mol/L. For AAEM the order of adsorption was $Ca^{2+} > K^+ > Na^+ > Mg^{2+}$ [31]. The synthetic hydro-char was able to uptake 100% of guaiacol, phenol, and resorcinol and 61% of vanillyl alcohol present in process water streams [31]. The multilayer adsorption of phenolic compounds was correlated using the Dubinin-Radushkevich isotherm, showing equilibrium adsorbent-phase loadings of 68.7 mg/g and 50.3 mg/g for vanillyl alcohol and resorcinol, respectively [31]. On the other hand, Henry's law best correlated the adsorption of AAEMs [31].

In this work, the influence of process temperature on the textural, morphological, and crystalline characterization of hydro-chars produced by hydrothermal carbonization of corn stover at 175, 200, 225, and 250 °C, 240 min, heating rate of 2.0 °C/min, and biomass/H_2O proportion of 1:10, using a reactor of 18.927 L, was investigated systematically. In addition, this work also investigated the influence of alkali (NaOH) and acid (HCl) pre-treatment over the hydro-char produced at 250 °C, on the adsorption/sorption of acetic acid (CH_3COOH) on hydro-char by analyzing the effect of initial concentration of acetic acid (CH_3COOH), as well as acetic acid solution-to-adsorbent ratio on the adsorption kinetics, adsorption equilibrium, and sorption equilibrium capacity in laboratory scale, batch mode.

2. Materials and Methods

2.1. Methodology

Figure 1 outlines the methodology as a rational scheme of ideas, methods, and procedures to produce the bio-adsorbent. The chemically activated adsorbent was applied to uptake acetic acid from aqueous solutions. Initially, the corn stover was collected. Afterwards, it was submitted to pretreatments of drying, grinding and sieving. The hydrothermal carbonization was carried out in batch mode, closed system, and pilot scale, as described elsewhere [52]. The hydro-chars were characterized by TG/DTG/DTA, SEM/EDX, BET, and XRD. The hydro-char were obtained by processing of corn stover with hot compressed H_2O at 250 °C, chemically activated with NaOH and HCl. The adsorption of acetic acid (CH_3COOH) on hydro-char was investigated by analyzing the adsorption kinetics, adsorption equilibrium, and sorption equilibrium capacity in laboratory scale, batch mode.

2.2. Materials, Pre-Treatment, and Physicochemical Characterization of Corn Stover

The corn stover was supplied by ATB-Bornin [52]. The residues were pretreated by drying, grinding, and sieving, as described elsewhere [52]. Afterwards, the residues were physicochemically characterized for dry matter, organic matter, ash, and elemental analysis, while oxygen was computed by difference, as described elsewhere [52].

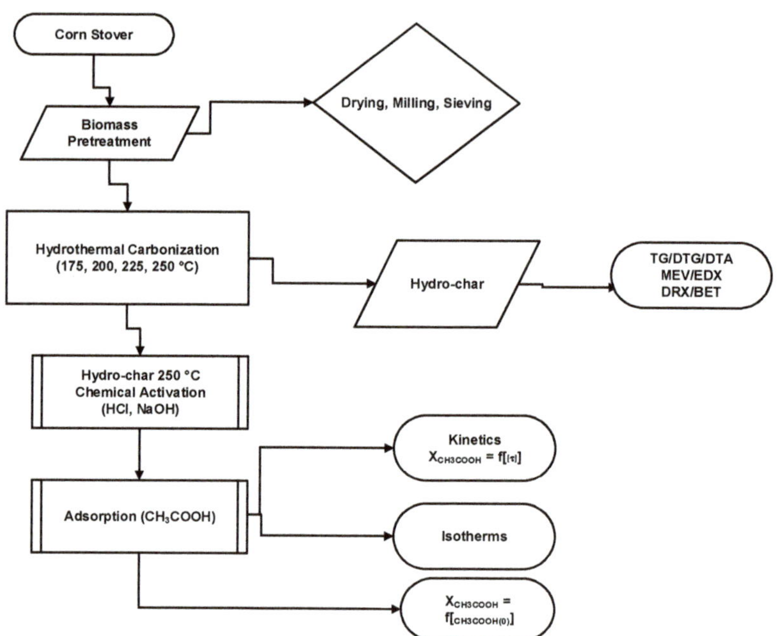

Figure 1. Process diagram of adsorption of acetic acid on hydro-char chemically activated with NaOH and HCl, obtained by HTC of corn stover at 250 °C, 240 min, and biomass/H_2O proportion of 1:10, batch mode, closed system, in pilot scale.

2.3. Experimental Apparatus and Procedures

Hydrothermal Carbonization

The experiments were performed at 175, 200, 225, and 250 °C, 240 min, and biomass/H_2O proportion of 1:10, using a reactor of 18.927 L, and the apparatus and procedures described elsewhere [52].

2.4. Adsorption of CH_3COOH

The adsorption kinetic of acetic acid into hydro-char produced by hydrothermal carbonization of corn stover at 250 °C, 240 min, biomass/H_2O proportion of 1:10, in pilot scale, chemically activated with NaOH and HCl, was investigated systematically. The uptake of CH_3COOH can be investigated by determining the acetic acid solution initial concentration $C^L_{CH3COOH}(0)$ and the concentrations of CH_3COOH in aqueous solutions at time (τ) $C^L_{CH3COOH}(\tau)$, until the time reaches its minimum value at (∞), that is, the acetic acid concentration in aqueous phase at equilibrium with hydro-char $C^L_{CH3COOH}(\infty)$ [6].

In cases, where only one organic acid specie is solvated in water (model solution), it is possible to determine the concentration of an organic acid in water by computing the acid value, as the proportion acidity/concentration of acetic acid ($C^L_{CH3COOH}$) is constant, given by Equation (1).

Applying a steady-state mole balanced, closed, and conservative system for the adsorption analysis (acetic acid aqueous solution (liquid phase) + hydro-char (solid phase)), the quantity of moles of acetic acid in water at the beginning n^L_0 is equal to quantity of moles of acetic acid in water at time (τ) n^L_τ and quantity of moles of acetic acid uptake into hydro-char at time (τ) n^S_τ defined by Equation (2). Dividing Equation (1) by solution volume V yields Equation (3), where $C^L_{CH3COOH}(0)$ is the initial acetic acid solution concentration, $C^L_{CH3COOH}(\tau)$ the acetic acid solution concentration at time (τ), and $C^S_{CH3COOH}(\tau)$ the acetic acid concentration in hydro-char at time (τ).

The acetic acid concentration in hydro-char at time (τ) is given by Equation (4). By substituting the relation given by Equation (1) in Equation (4), it is possible to compute, indirectly, the acetic acid concentration in hydro-char at time (τ), by the difference between the initial solution acid value $I^L_{CHeCOOH}(0)$ and the solution acid value at time (τ), $I^L_{CH3COOH}(\tau)$.

When acetic acid concentration in water reaches equilibrium, that is, $C^L_{CH3COOH}(\infty) = C^*_{CH3COOH}$, $I^L_{CH3COOH}(\infty) = I^*_{CH3COOH}$, the acetic acid concentration into hydro-char reaches its maximum at equilibrium $C^S_{CH3COOH}(\infty) = C^*_{CH3COOH}$, and the difference $I^L_{CH3COOH}(0) - I^L_{CH3COOH}(\infty)$, Equation (5), reaches its maximum.

$$I^L_{CH3COOH} \propto C^L_{CH3COOH}, \quad I^L_{CH3COOH} = K * C^L_{CH3COOH} \tag{1}$$

$$n^L_0 = n^L_\tau + n^S_\tau \tag{2}$$

$$C^L_{CH3COOH}(0) = C^L_{CH3COOH}(\tau) + C^S_{CH3COOH}(\tau) \tag{3}$$

$$C^S_{CH3COOH}(\tau) = C^L_{CH3COOH}(0) - C^L_{CH3COOH}(\tau), \quad I(\tau) = I^L_{CH3COOH}(0) - I^L_{CH3COOH}(\tau) \tag{4}$$

$$C^S_{CH3COOH}(\infty) = C^*_{CH3COOH}, \quad I^* = I^L_{CH3COOH}(0) - I^L_{CH3COOH}(\infty) \tag{5}$$

A first order kinetic was applied to describe the adsorption process, expressed as a dimensionless acidity, given by Equation (6).

$$I(\tau) = I^L_{CHeCOOH}(0) * [1 - \exp(-K * \tau)] \tag{6}$$

where K, is the adsorption kinetic constant.

2.4.1. Adsorption Isotherm of CH_3COOH

The Langmuir isotherm, given by Equation (7), was applied to analyze the adsorption equilibrium data of acetic acid in aqueous solutions within a porous solid matrix (adsorbent), where $\chi^*_{CH3COOH}$ is the equilibrium adsorbent-phase concentration of CH_3COOH, $C^*_{CH3COOH}$, is the equilibrium aqueous-phase concentration of acetic acid, and K_0, χ_{Max}, the adsorption equilibrium constant and the saturation adsorption loading, respectively.

$$\chi^*_{CH3COOH} = \frac{K_0 * \chi_{Max} * C^*_{CH3COOH}}{(1 + K_0 * C^*_{CH3COOH})} \tag{7}$$

2.4.2. Adsorption Apparatus and Procedures

The adsorption apparatus and procedures of CH_3COOH on hydro-char activated with NaOH (2.0 M) and HCl (2.0 M), was described in detail by Machado et al. [6].

2.5. Morphological, Crystalline, and Textural Characterization of Hydro-Chars

The morphological, crystalline, and textural characterization of hydro-chars obtained by hydrothermal processing of corn stover at 175, 200, 225, and 250 °C, 240 min, and biomass/H_2O proportion of 1:10, using a reactor of 18.927 L, was performed by SEM, EDX, and XRD (the equipment and procedures were described elsewhere [55,56]) as well as by thermo-gravimetric analysis (TG/DTG/DTA) and BET [56].

Thermo-Gravimetric Analysis (TG/DTG/DTA)

The weight loss of hydro-chars obtained by hydrothermal processing of corn stover at 175, 200, 225, and 250 °C, 240 min, and biomass/H_2O proportion of 1:10, using a reactor of 18.927 L, was analyzed by TG/DTG/DTA, and the equipment and procedures were described elsewhere [57].

3. Results

3.1. Morphological, Crystalline, and Textural Characterization of Hydro-Chars

3.1.1. Thermo-Gravimetric Analysis (TG/DTG/DTA)

Sittisun et al. [58], studied the thermal degradation of corn stover between 25 and 900 °C, heating rates of 10, 20, and 50 °C/min. The authors reported for heating rates of 10 °C/min, a mass loss of approximately 92 (wt.%), between 25 and 510 °C.

In the temperature region 510–900 °C, the mass loss was constant, and the remaining solid phase was composed of ash. The DTG curve shows three different thermal degradation steps. In the first one, between 25 and 167 °C, a mass loss of 8.5 (wt.%) was reported, representing the recovery of moisture within the solid phase. Afterwards, in the temperature region 167–368 °C, a mass loss of 56.59 (wt.%) occurred, mainly due to release of volatile compounds by the degradation reactions of cellulose and hemi-cellulose, described chemically by degradation of chemical functions containing oxygen, including hydroxyl, carbonyl, and carboxyl groups [58,59]. In the last step, between 368 and 514 °C, a mass loss of 26.50 (wt.%) occurred, associated with the thermal degradation and/or combustion of lignin, as well as residual char compounds, produced at the second step.

Figures 2–4 describe the TG, DTG, and DTA analysis of corn stover after hydrothermal processing at 175, 200, 225, and 250 °C, 240 min, and biomass/H_2O proportion of 1:10, respectively, using a reactor of 18.927 L, between 25 and 800 °C, 10 °C/min, under N_2 atmosphere.

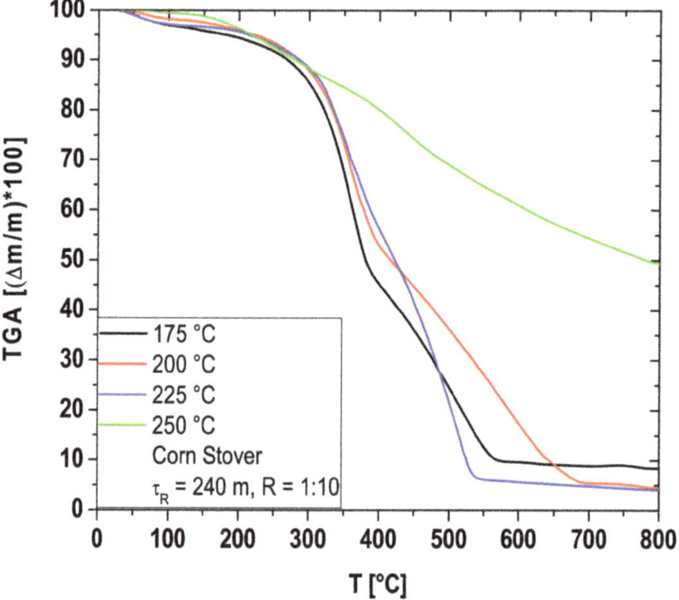

Figure 2. TG of hydro-chars obtained by hydrothermal carbonization of corn stover at 175, 200, 225, and 250 °C, 240 min, and biomass/H_2O proportion of 1:10, using a reactor of 18.927 L.

One observes, for the thermal degradation of corn stover after hydrothermal processing at 175 °C, the occurrence of 03 reaction steps. In the first one, between 25 and 150 °C, a mass loss of approximately 3.0–4.0 (wt.%) occurred, due to the release of moisture. It is also probable that degradation of volatile compounds selectively adsorbed into the pores of solid phase products has taken place, including alcohols (methanol, ethanol), carboxylic acids (acetic acid, propanoic acid) [6], and aldehydes (HMF). In the second degradation step, between 150 and 530 °C, a mass loss of approximately 79–80 (wt.%) occurred.

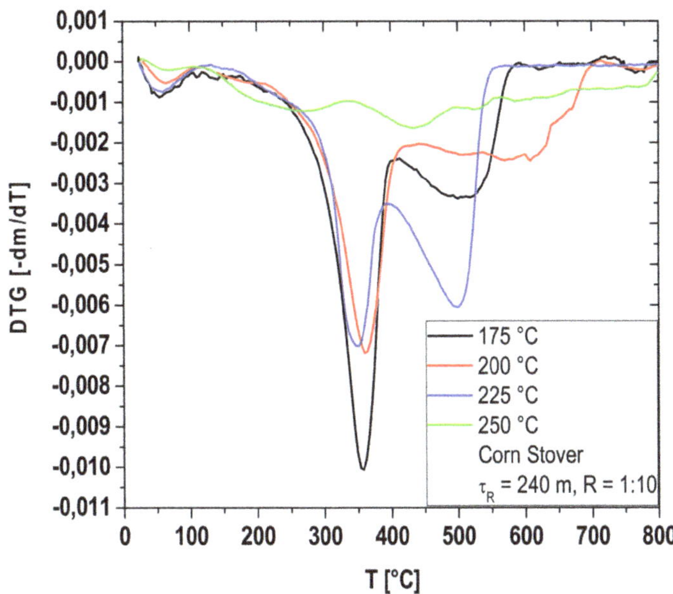

Figure 3. DTG of hydro-chars obtained by hydrothermal carbonization of corn stover at 175, 200, 225, and 250 °C, 240 min, and biomass/H_2O proportion of 1:10, using a reactor of 18.927 L.

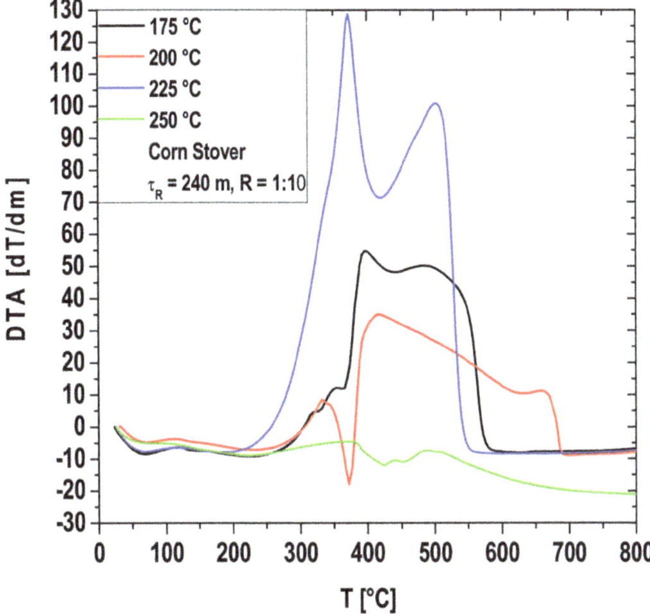

Figure 4. DTA of hydro-chars obtained by hydrothermal carbonization of corn stover at 175, 200, 225, and 250 °C, 240 min, and biomass/H_2O proportion of 1:10, using a reactor of 18.927 L.

The mass balance by hydrothermal carbonization of corn stover at 175 °C shows that approximately 63.0 (wt.%) of initial biomass still remains as solid phase reaction products [52]. In addition, the low concentration of lignin-derived reaction compounds such as phenols and guajacol in the aqueous phase indicates that only a small portion of lignin has been thermally degraded [52]. Contrarily, the high concentration of cellulose/hemi-cellulose-derived reaction compounds in the aqueous phase including furfural, HMF, and acetic acids is a sign that higher amounts of cellulose/hemi-cellulose compared to lignin have been thermally degraded [52].

Since corn stover consist basically by cellulose + hemi-cellulose (69.0%), lignin (20.0%), ash (8.0%) [60–62], and small amounts of soluble substances [61], the TG/DTG analysis shows that almost all cellulose + hemi-cellulose and soluble substances in the aqueous phase have been thermally degraded, as well as approximately the haft of lignin, thus remaining within the solid phase reaction products the inorganic matter and lignin. In the last step, between 530 and 800 °C, a mass loss of approximately 11.0–12.0 (wt.%) occurred, associated with the thermal degradation and/or combustion of lignin. The ash content of 8.0 (wt.%) is according to that reported in the literature [60–62].

The thermal degradation of corn stover after hydrothermal processing at 200 °C shows the occurrence of 03 reaction steps. In the first one, between 25 and 110 °C, a mass loss of approximately 2.0 (wt.%) occurred, because of humidity removal, as well as the degradation of low molecular weight compounds (volatile compounds) adsorbed into the pores of solid phase products such as methanol. [6,52]. In the second degradation step, between 110 and 520 °C, a mass loss of approximately 62.0 (wt.%) was observed. The mass balance by hydrothermal carbonization of corn stover at 200 °C shows that approximately 57.4 (wt.%) of initial biomass still remains as solid phase reaction products [52]. Although, the concentrations of lignin-derived reaction compounds in process water such as phenols and guajacol has increased between 175 and 200 °C [52], still a small portion of lignin has been thermally degraded. The high concentration of cellulose/hemi-cellulose-derived reaction compounds in the aqueous phase including furfural, HMF, and acetic acids show that high amount of cellulose/hemi-cellulose has been thermally degraded [52], remaining within the solid phase reaction products the inorganic matter and lignin. In the third step, between 520 and 800 °C, a mass loss of approximately 32.0 (wt.%) occurred, associated with the thermal degradation and/or combustion of lignin. The ash content of 4.0 (wt.%) is according to that reported in the literature [60–62].

One observes, for the thermal degradation of corn stover after hydrothermal processing at 225 °C, the occurrence of 03 reaction steps. In the first one, between 25 and 130 °C, a mass loss of approximately 2.0 (wt.%) occurred, because of humidity removal, as well as the degradation of low molecular weight compounds (volatile compounds) adsorbed into the pores of solid phase products including alcohols (methanol, ethanol), volatile carboxylic acids (acetic acid, propanoic acid) [6], and aldehydes (HMF). In the second degradation step, between 130 and 530 °C, a mass loss of approximately 92.0 (wt.%) occurred. The mass balance by hydrothermal carbonization of corn stover at 225 °C shows that approximately 41.0 (wt.%) of initial biomass still remains as solid phase reaction products [52]. The high concentration of lignin-derived reaction compounds in process water such as phenols and guajacol, as well as cellulose/hemi-cellulose-derived reaction chemicals dissolved in process water such as carboxylic acid, show that most of lignin and cellulose/hemi-cellulose have been thermally degraded [52], remaining within the solid phase reaction products the inorganic matter. In the third step, between 530 and 800 °C, a mass loss of approximately 3.0 (wt.%) occurred, associated with the thermal degradation and/or combustion of lignin. The ash content of 3.0 (wt.%) is according to that reported in the literature [60–62].

The thermal degradation of corn stover after hydrothermal processing at 250 °C shows the occurrence of 04 reaction steps. In the first one, between 25 and 110 °C, a mass loss of approximately 1.0 (wt.%) occurred, due to the release of moisture. In the second degradation step, between 110 and 270 °C, a mass loss of approximately 9.0 (wt.%) has been observed. In the third degradation step, between 270 and 430 °C, a mass loss of approximately 15.0 (wt.%) occurred. In the last degradation step, between 430 and 800 °C, a mass loss of approximately 25.0 (wt.%) occurred. The mass balance by hydrothermal carbonization of corn stover at 250 °C shows that approximately 35.8 (wt.%) of initial biomass still remains as solid phase reaction products [52]. The high concentration of cellulose/hemi-cellulose-derived reaction chemicals in process water such as acetic acid, as well as lignin-derived reaction compounds solvated in process water such as guajacol and phenols, show that high amounts of cellulose/hemi-cellulose and lignin have been thermally degraded [52], being the solid phase reaction products a carbon rich material. Finally, the results illustrated in Figures 2 and 3 are in accord to a similar study described in the literature, for the thermal analysis of corn stover, by Mohammed et al. [63].

Finally, one observes in Figure 3, for the hydro-chars obtained at 175 and 200 °C, that highest mass loss are associated with the occurrence of endothermic peaks at 355.2 and 362.9 °C, respectively. This behavior, was also observed in Figure 4 by the presence of exothermic peaks around 360 °C. For the hydro-chars obtained at 225 and 250 °C, the endothermic peaks occurred at 397 and 434 °C, respectively, showing that the higher the temperature the higher the thermal stability of solid reaction products. In addition, the hydro-char obtained at 250 °C presented the highest thermal stability compared to the hydro-chars obtained at 175, 200, and 225 °C, as it presented the lowest mass loss of 50.51% (wt.) between the temperatures of 144 and 768.5 °C.

3.1.2. SEM Analysis

The microscopic analyses of hydro-chars obtained at 175, 200, 225, and 250 °C, 240 min, and biomass/H_2O proportion of 1:10, using a reactor of 18.927 L, are illustrated in Figure 5a,b and Figure 6a,b, respectively. The SEM image in Figure 5a indicates a rigid and well-organized fiber structure, demonstrating that temperature had almost no effect on the vegetal structure, as it largely retained the original morphological microscopic characteristics. The results are in agreement to analogous investigations on the effect of process temperature over the morphology of hydro-chars produced by corn stover [63], and corn straw [64].

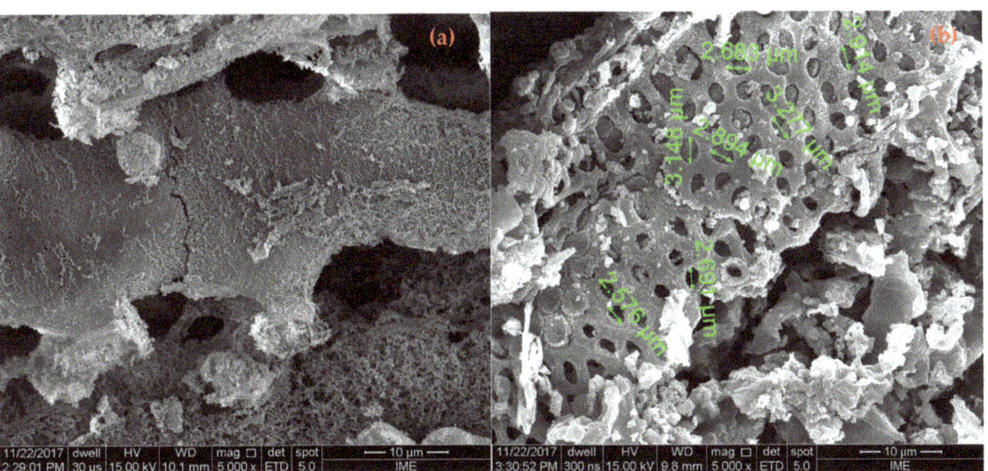

Figure 5. SEM of corn stover after hydrothermal processing at 175 °C (**a**) and 200 °C (**b**), 240 m, and biomass/H_2O proportion of 1:10, using a reactor of 18.927 L (Mag: 5000×).

The SEM image illustrated in Figure 5b shows that disaggregate, amorphous, and heterogeneous structures with nonuniform geometry dominated, demonstrating that temperature had caused the appearance of micropores with average diameter of 3.00 μm, indicating that the lignocellulose material started to decompose. However, process temperature still had little effect on the vegetal structure, as it largely retained the original microscopic characteristics. The results are in agreement to similar investigations described by Xing et al. [64], and Mohammed et al. [63].

Figure 6. SEM of corn stover after hydrothermal processing at 225 °C (**a**) and 250 °C (**b**), 240 m, and biomass/H_2O proportion of 1:10, using a reactor of 18.927 L (Mag: 5000×).

The SEM image in Figure 6a indicates that hydro-char structure consists of agglomerated micro-spheres and heterogeneous structures with irregular shapes (fragmentation), showing particle size averaging 0.5 µm as depicted in detail by Figures S1 and S2 in Supplementary Materials. The agglomerated micro-spheres and fragmentation indicates that cellulose and hemi-cellulose were decomposed, as reported by Xing et al. [64], demonstrating that temperature had generated significant alterations on the morphological structure of corn stover by destructing the plant cell walls [6]. Xing et al. [64] reported the occurrence of micro-spheres by the hydrothermal carbonization of corn straw at 230 °C, 30 min, and biomass/H_2O proportion of 1:8.

One observes in Figure 6b an aggregate amorphous solid phase consisting of micro-spheres and heterogeneous structures with nonuniform geometry (fragmentation), showing that original vegetal surface structure had been drastic changed, and was the carbonization grade higher compared to the SEM image at 225 °C. In fact, process temperature had generated significant alterations on the morphological structure of corn stover by destructing the plant cell structure. The agglomerated micro-spheres and fragmentation indicates that cellulose and hemi-cellulose were decomposed. Xing et al. [64] reported the occurrence of a similar structure by the hydrothermal carbonization of corn straw at 260 °C, 30 min, and biomass/H2O proportion of 1:8.

3.1.3. EDX Analysis

Table 1 shows the EDX (energy dispersive X-ray spectroscopy) analysis of hydro-chars. The samples were analyzed at five different points. By increasing the temperature, the carbon content increased and that of oxygen diminished. By comparing the results for the carbon content in Table 1 and those described in Supplementary Table S1, one observes that the EDX technique exhibited higher carbon contents but similar oxygen contents. This was probably due to limitations of the technique, as it was not adequate to identify/recognize elements such as nitrogen and hydrogen. In addition, calculation of oxygen and carbon contents, described in Table S1, was computed by subtracting the ash content [52]. In addition, Ca, Zn, Cu, Mo, Na, and P, as well as Si were identified in almost all the points marked by EDX. The inorganics identified in hydro-chars by EDX in hydro-chars are in agreement to those identified in corn stover after drying at 105 °C [65]. Finally, Table 1 shows an increase on carbon content (carbonization) with increasing process temperature.

Table 1. Percentages in mass and atomic mass of hydro-chars, obtained by hydrothermal carbonization of corn stover at 175, 200, 225, and 250 °C, 240 min, biomass/H2O proportion of 1:10, using a reactor of 18.927 L, at the point marked by EDX technique [6].

	Hydro-Chars											
	175 °C			**200 °C**			**225 °C**			**250 °C**		
Chemical Elements	**Mass (wt.%)**	**Atomic Mass (wt.%)**	**SD**	**Mass (wt.%)**	**Atomic Mass (wt.%)**	**SD**	**Mass (wt.%)**	**Atomic Mass (wt.%)**	**SD**	**Mass (wt.%)**	**Atomic Mass (wt.%)**	**SD**
C	60.69	67.56	0.503	71.98	77.43	0.818	73.89	79.17	0.654	76.04	81.05	0.631
O	38.57	32.24	0.504	27.83	22.47	0.819	25.73	20.70	0.656	23.59	18.87	0.632
Si	0.136	0.065	0.020	0.194	0.089	0.043	-	-	-	-	-	-
Ca	0.181	0.061	0.018	-	-	-	0.374	0.120	0.032	-	-	-
Zn	-	-	-	-	-	-	-	-	-	0.366	0.072	0.065
Cu	0.210	0.044	0.042	-	-	-	-	-	-	-	-	-

SD = standard deviation.

3.1.4. XRD Analysis

The XRD (X-ray diffraction) data of hydro-chars obtained by hydrothermal processing of corn stover at 175, 200, 225, and 250 °C, 240 min, and biomass/H_2O proportion of 1:10, using a reactor of 18.927 L, is illustrated in Figure 7a,b and Figure 8a,b, respectively.

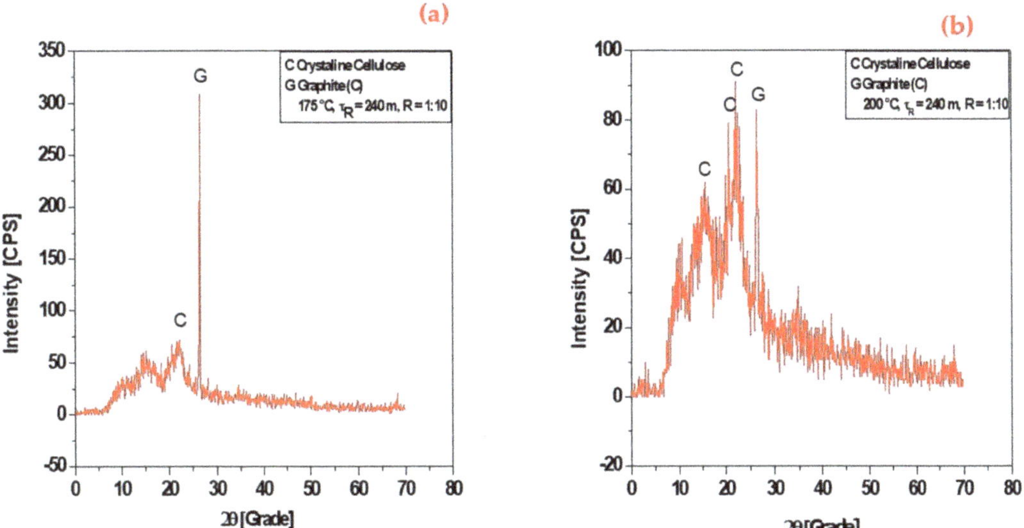

Figure 7. XRD of hydro-chars obtained by hydrothermal carbonization of corn stover at 175 °C (**a**) and 200 °C (**b**), 240 min, and biomass/H_2O proportion of 1:10, using a reactor of 18.927 L.

The diffractogram of corn stover obtained by hydrothermal carbonization at 175 °C, is illustrated in Figure 7a. It shows the presence of two crystalline phases—graphite (C) with a peak of higher intensity (100%) on the position 2θ: 26.47, and crystalline cellulose with a peak of higher intensity (12.95%) on the position 2θ: 22.20. The peak on the position 2θ: 22.20 is characteristics of crystalline cellulose structure, as described by Regmi et al. [5] and Kang et al. [66], who identified diffraction peaks on the positions 2θ: 15.3 and 22.3 [5,66].

At 200 °C, X-ray-diffraction data identified two crystalline phases—crystalline cellulose with a peak of higher intensity (100%) on the positions 2θ: 20.68, 15.53 (79%), and 21.96 (90.07%), because of amorphous cellulose matrix degradation and the exposure of the crystalline cellulose [57], and graphite (C) with a peak of medium intensity (59.38%) on the position 2θ: 26.45, as shown in Figure 7b.

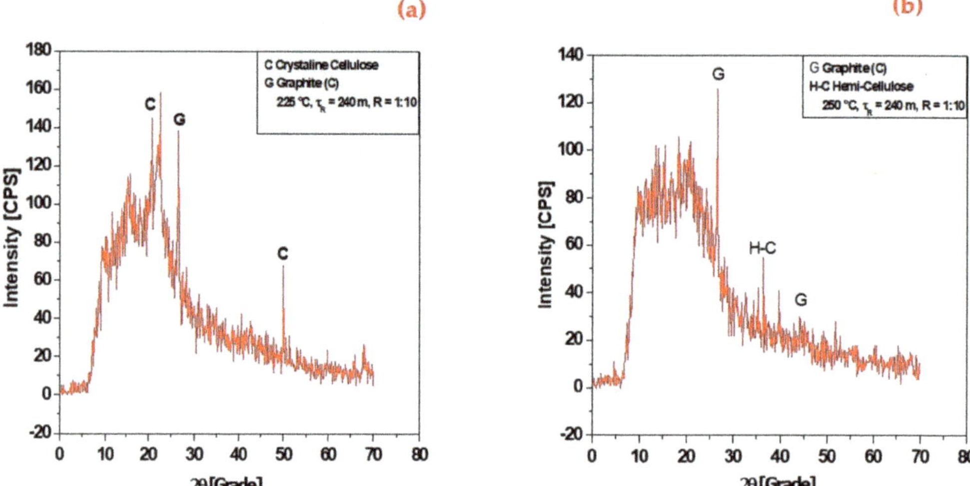

Figure 8. XRD of hydro-chars obtained by hydrothermal carbonization of corn stover at 225 °C (**a**) and 250 °C (**b**), 240 min, and biomass/H_2O proportion of 1:10, using a reactor of 18.927 L.

The XRD at 225 °C identified two crystalline phases—graphite (C) with a peak of medium intensity (53.75%) on the position 2θ: 26.57, and crystalline cellulose with peaks of high and low intensity on the positions 2θ: 22.47 (100%) and 49.94 (49.86%), as well as three peaks of medium and low intensity on the positions 2θ: 20.73 (65.99%) and 49.94 (49.86%), as shown in Figure 8a.

The diffractograms (XRD) identified the occurrence of peaks of higher intensity of graphite (C) as the temperature increased, as well as a decrease of peaks intensity for crystalline cellulose, demonstrating that higher temperatures favor the formation of crystalline-phase graphite (C), being in agreement to the results illustrated in Table 1.

3.1.5. BET

Figure 9 shows the BET analysis of hydro-char obtained by HTC of corn stover at 250 °C, 240 min, and biomass/H_2O of 1:10, using a reactor of 18.927 L. The N_2 capacity increased as the relation (P/P$_0$) increased, showing a maximum capacity of approximately 12 cm^2/g as the relative pressure approached 1.0. The density of hydro-char was 2.10 g/cm^3, and the surface area measured by relative pressure (P/P$_0$ = 0.201) was 4.02 m^2/g, while the surface area was 4.3498 m^2/g. The pore volume measured by reduced pressure (P/P$_0$ = 0.988) was 0.01857 cm^3/g. The average pore width was 17.079 μm.

3.2. Adsorption Kinetics of Acetic Acid (CH_3COOH) on Hydro-Char

One of the great disadvantages of hydrothermal processing of lignocellulose-rich materials, such as biomass, is the occurrence of hazardous lignin and cellulose-derived reaction products including phenols, furfural, and hydroxymethylfurfural in the aqueous phase [52–55,67,68]. In addition, the high concentration of carboxylic acids (acetic acid, propionic acid, etc.), in process water confers its high acidity [52–55,68,69]. From this perspective, the hydro-char obtained by HTC at 250 °C, 240 min, biomass/H_2O proportion of 1:10, was chemically activated with alkali (2.0 M NaOH) and acid (2.0 M HCl) solutions in order to investigate the its capacity to selectively uptake (adsorb) acetic acid from model aqueous solutions (1.0, 2.0, 3.0, and 4.0 mg/mL), the major carboxylic acid within hydrothermal carbonization process water streams.

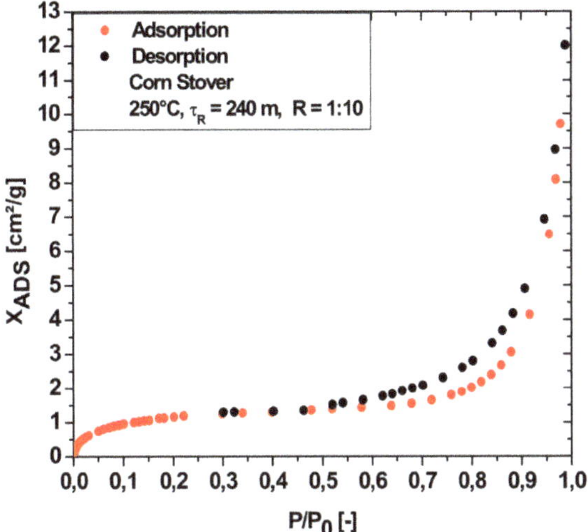

Figure 9. BET of hydro-char obtained by hydrothermal carbonization of corn stover at 250 °C, 240 min, and biomass/H_2O of 1:10, using a reactor of 18.927 L.

3.2.1. Bio-Adsorbent Activation with NaOH

Influence of Acetic Acid Concentration

Figure 10 illustrates the influence of acetic acid initial solution concentration (2.0, 3.0, and 4.0 mg mL^{-1}) on the adsorption kinetics within hydro-char produced by HTC at 250 °C, 240 min, biomass/H_2O proportion of 1:10, alkali-activated (2.0 M NaOH), 0.1 $g_{Adsorbent}$/10 mL$_{CH3COOH}$. The adsorption kinetic is rapid and equilibrium is reached between 240 and 480 s. The activation of hydro-char with a strong alkali such as NaOH causes significant changes in the surface and porous (micropores, mesopores, and macropores) structure [8,9]. The acetic acid molecules solvated in H_2O are selectively caught by the negative-charged active sites, which is according to Liang et al. [69], who reported that adsorption process of CH_3COOH on carbon microspheres, prepared by hydrothermal carbonization of starch, was mainly due to the porous (micropores, mesopores, and macropores) structure transformations caused by destruction of hydrophilic oxygen functional groups. In addition, Liang et al. [69] proved that pore structure determines the adsorption capacity and diffusion rate and the process is physical.

The absorption of acetic acid molecules occurs because of a concentration difference between the solution and hydro-char surface and internal porous, micro-porous, and macroporous structure, as well as the appearance of dipole–dipole electrostatic attraction forces between the negative charged sites within the porous and the dissociated H$^+$ of R-COOH (H$^+$ + R-COO$^-$). The adsorption kinetic data of acetic acid on hydro-char was correlated with a first order model, exhibiting *root-mean-square error* (r^2) between 0.969 and 0.999, as shown in Table 2, which is in agreement with the results described by Machado et al. [6]. It can be also observed that dimensionless acid value I(τ), increases with decreasing acetic acid solutions concentrations (2.0, 3.0, and 4.0 mg mL^{-1}), that is, the lower the acetic acid solution concentration, the higher the adsorption efficiency.

The influence of acetic acid initial solution concentration (1.0, and 2.0 mg mL^{-1}) on the adsorption kinetics within hydro-char obtained by hydrothermal processing of corn stover at 250 °C, 240 min, biomass/H_2O proportion of 1:10, chemically activated with

NaOH (2.0 M), 0.2 $g_{Adsorbent}$/10 mL$_{CH3COOH}$, is illustrated in Figure 11. One can observe that dimensionless acid value I(τ), increases with decreasing CH_3COOH concentrations (1.0, and 2.0 mg/mL), that is, the lower the acetic acid solution concentration, the higher the adsorption efficiency. In addition, by comparing Figures 10 and 11, it is easy to observe that higher adsorbent-to-solution ratios favors the adsorption efficiency, due to an increase on the surface area. The adsorption kinetic data of acetic acid on hydro-char was correlated with a first order model, exhibiting root-mean-square error (r^2) between 0.969 and 0.982, as shown in Table 3.

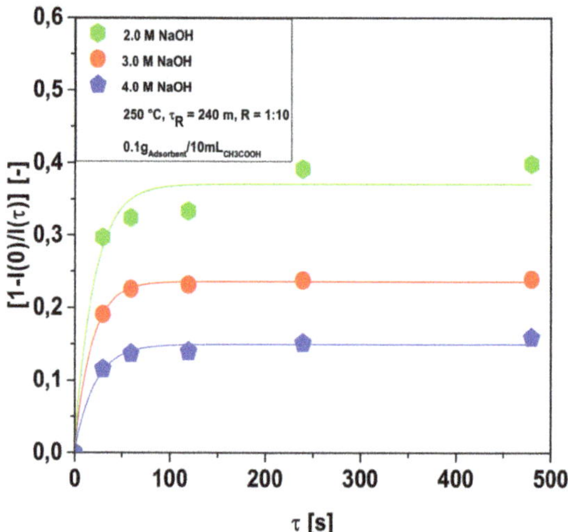

Figure 10. Adsorption kinetic of CH_3COOH solutions, expressed as dimensionless acid value I(τ), on NaOH-activated hydro-char obtained at 250 °C.

Table 2. Regression parameters by adsorption kinetic of CH_3COOH solutions (2.0, 3.0, and 4.0 mg mL^{-1}), expressed as dimensionless acid value, on NaOH (2.0 M)-activated hydro-char produced by hydrothermal processing of corn stover at 250 °C, 240 min, and biomass/H_2O proportion of 1:10, 0.1 $g_{Adsorbent}$/10 mL$_{CH3COOH}$.

Parameters	$C_{CH3COOH}$ (mg/mL)		
	2	3	4
Pseudo-first-order			
$I^L_{CHeCOOH}(0)$ (-)	0.3694	0.2361	0.1497
κ (s^{-1})	0.0479	0.0545	0.0468
r^2	0.969	0.999	0.988

Influence of Adsorbent-to-Solution Ratio

The effect of hydro-char/solution ratio on the adsorption kinetics of acetic acid solution (2.0 mg mL^{-1}) within NaOH (2.0 M)-activated hydro-char obtained by hydrothermal processing of corn stover at 250 °C, 240 min, biomass/H_2O proportion of 1:10, is illustrated in Figure 12. The adsorption performance increased with increasing hydro-char/solution ratio increase as the number of active sites on surface and porous (micropores, mesopores, and macropores) structure increased, that is, by increasing the hydro-char/solution ratio, the surface area increased. The adsorption kinetic data of acetic acid on hydro-char was correlated with a first order model, exhibiting root-mean-square error (r^2) between 0.973 and 0.982, as shown in Table 4.

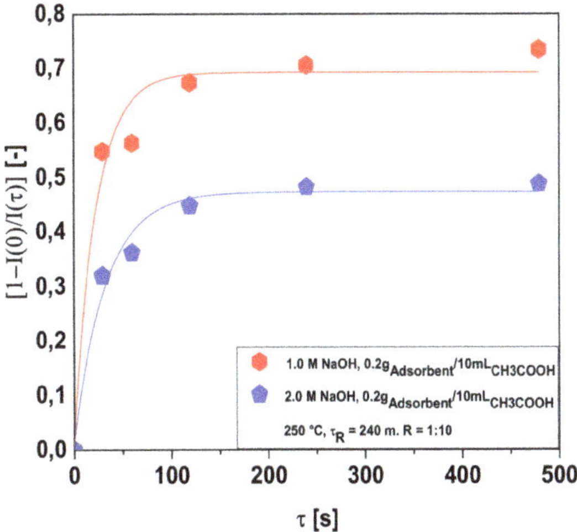

Figure 11. Adsorption kinetic of CH_3COOH solutions (1.0 and 2.0 mg mL^{-1}), expressed as dimensionless acid value I(τ), on NaOH (2.0 M)-activated hydro-char produced by hydrothermal processing of corn stover at 250 °C, 240 min, and biomass/H_2O proportion of 1:10, 0.2 g$_{Adsorbent}$/10 mL$_{CH3COOH}$.

Table 3. Regression parameters by adsorption kinetic of CH_3COOH solutions (1.0 and 2.0 mg mL^{-1}), expressed as dimensionless acid value, on NaOH (2.0 M)-activated hydro-char produced by hydrothermal processing of corn stover at 250 °C, 240 min, and biomass/H_2O proportion of 1:10, 0.2 g$_{Adsorbent}$/10 mL$_{CH3COOH}$.

Parameters	$C_{CH3COOH}$ (mg/mL)	
	1	2
Pseudo-first-order		
$I^L_{CHeCOOH}(0)$ (-)	0.4689	0.3694
κ (s^{-1})	0.0305	0.0479
r^2	0.982	0.969

Table 4. Regression parameters by the effect of hydro-char/solution ratio (0.1 g$_{Adsorbent}$/10 mL$_{CH3COOH}$, 0.2 g$_{Adsorbent}$/10 mL$_{CH3COOH}$) on adsorption kinetic of CH_3COOH, expressed as dimensionless acid value I(τ), on NaOH (2.0 M)-activated hydro-char produced by hydrothermal processing of corn stover at 250 °C, 240 min, and biomass/H_2O proportion of 1:10.

Parameters	$C_{CH3COOH}$ (mg/mL)	
	2	2
Pseudo-first-order	0.1 g$_{Adsorbent}$/10 mL$_{CH3COOH}$	0.2 g$_{Adsorbent}$/10 mL$_{CH3COOH}$
$I^L_{CHeCOOH}(0)$ (-)	0.4689	0.6889
κ (s^{-1})	0.0305	0.0430
r^2	0.982	0.973

3.2.2. Bio-Adsorbent Activation with HCl

Influence of Hydrochloric Acid Concentration

The influence of acetic acid initial solution concentration (1.0, 2.0, 3.0, and 4.0 mg mL^{-1}) on the adsorption kinetics within HCl (2.0 M)-activated hydro-char obtained by on hydro-char obtained by hydrothermal processing of corn stover at 250 °C, 240 min, biomass/H_2O proportion of 1:10, 0.2 g$_{Adsorbent}$/10 mL$_{CH3COOH}$, is shown in Figure 13.

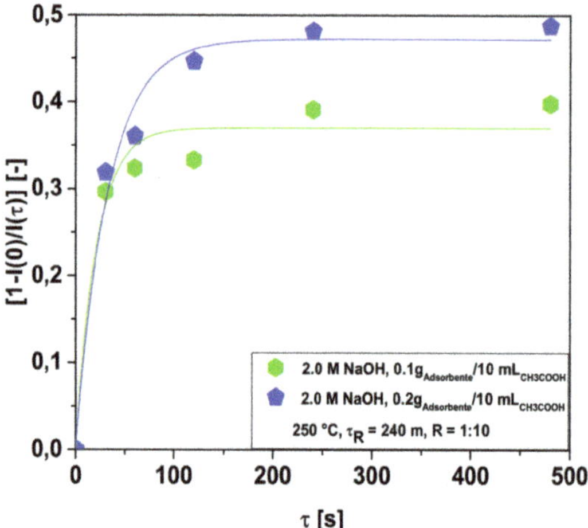

Figure 12. Effect of hydro-char/solution ratio (0.1 $g_{Adsorbent}$/10 $mL_{CH3COOH}$, 0.2 $g_{Adsorbent}$/10 $mL_{CH3COOH}$) on adsorption kinetic of CH_3COOH, expressed as dimensionless acid value $I(\tau)$, on NaOH (2.0 M)-activated hydro-char produced by hydrothermal processing of corn stover at 250 °C, 240 min, and biomass/H_2O proportion of 1:10.

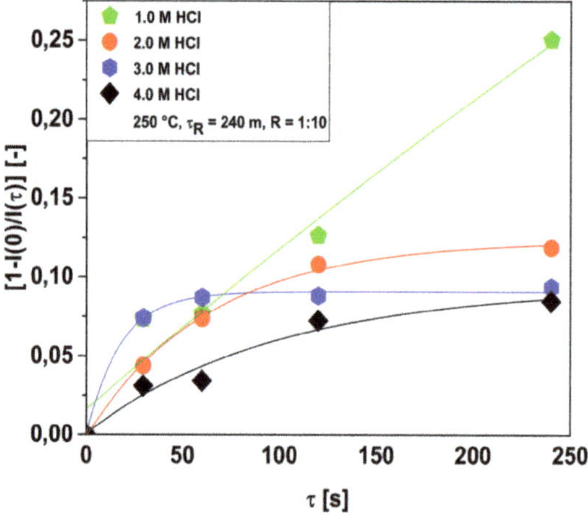

Figure 13. Adsorption kinetic of CH_3COOH solutions (1.0, 2.0, 3.0, and 2.0 mg mL^{-1}), expressed as dimensionless acid value $I(\tau)$, on HCl (2.0 M)-activated hydro-char produced by hydrothermal processing of corn stover at 250 °C, 240 min, and biomass/H_2O proportion of 1:10, 0.1 $g_{Adsorbent}$/10 $mL_{CH3COOH}$.

The adsorption kinetic was rapid and equilibrium was reached around 240 s. The activation of hydro-char with a strong acid such as HCl causes significant changes and/or damages on the surface and porous (micropores, mesopores, and macropores) structure [51]. The acetic acid molecules solvated/dissociated in water are not efficiently captured by the positive charged active sites. The absorption of acetic acid molecules is due basically to a concentration difference between bulk solution and hydro-char surface and porous (micropores, mesopores, and macropores) structure, as the positive-charged sites on the surface and within the internal porous structure contribute to the appearance of electrostatic repulsion forces between the dissociated H^+ of R-COOH (H^+ + R-COO$^-$) and the positive charged sites. This is according to Liang et al. [69], who proved that pore structure determines the adsorbent capacity and that process is diffusion-rate controlled. The adsorption kinetic data of acetic acid on hydro-char was correlated with a first order model, exhibiting root-mean-square error (r^2) between 0.967 and 0.997, as shown in Table 5.

Table 5. Regression parameters by adsorption kinetic of CH_3COOH solutions (1.0, 2.0, 3.0, and 4.0 mg mL^{-1}), expressed as dimensionless acid value I(t), on HCl (2.0 M) hydro-char produced by HTC of corn stover at 250 °C, 240 min, and biomass/H_2O proportion of 1:10, 0.1 $g_{Adsorbent}$/ 10 $mL_{CH3COOH}$.

Parameters	$C_{CH3COOH}$ (mg/mL)			
	1	2	3	4
Pseudo-first-order				
$I^L_{CHeCOOH}(0)$ (-)	-	0.1241	0.0908	0.0941
κ (s^{-1})	-	0.0156	0.0563	0.0101
r^2	0.969	0.998	0.978	0.967

The results presented in Figures 10 and 13 show that NaOH (2.0 M) chemically activated hydro-char is not only more efficient than that activated with HCl (2.0 M), but also presents the highest adsorption performance and capacity.

3.2.3. Adsorption Equilibrium Isotherms of Acetic Acid (CH_3COOH) on Hydro-Char

The Langmuir isotherm was applied to correlate the equilibrium adsorption data of CH_3COOH on alkali (2.0 M NaOH)- and acid (2.0 M HCl)-activated hydro-char produced by hydrothermal processing of corn stover at 250 °C, 240 min, biomass/H2O proportion of 1:10, 0.1 $g_{Adsorbent}$/10 $mL_{CH3COOH}$, as illustrated by Figure 14.

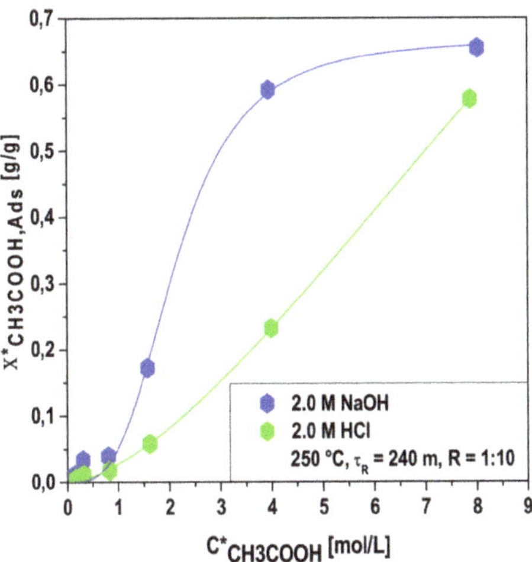

Figure 14. Langmuir adsorption isotherm of CH_3COOH solutions (0.5, 0.25, 0.1, 0.05, 0.02, and 0.01 M) within NaOH (2.0 M)- and HCl (2.0 M)-activated hydro-chars produced by hydrothermal carbonization of corn stover at 250 °C, 240 min, and biomass/H_2O proportion of 1:10.

The adsorption isotherm of acetic acid on NaOH- and HCl-activated hydro-char was correlated with the Langmuir model, exhibiting root-mean-square errors (r^2) of 0.994 and 0.989, respectively. The equilibrium adsorbent-phase concentration of CH_3COOH for NaOH- and HCl-activated hydro-chars were approximately 650 and 575 mg/g, respectively. The bio-adsorbent (hydro-chars) equilibrium loadings were in agreement with the adsorption of MB (methylene blue) from aqueous solution on NaOH-activated hydro-chars produced by HTC of factory-rejected tea and palm date seeds, as reported by Islam et al. [8,9], correlated using a pseudo-second-order model and the adsorption isotherms by the Langmuir [8], and the Freundlich models [9], respectively. For the adsorption kinetic of MB (methylene blue) on NaOH-activated hydro-chars produced by HTC of from factory-rejected tea and palm date seeds, maximum adsorption loadings of 487.4 mg/g at 30 °C [8], and 612.1, 464.3, and 410.0 mg/g at 30, 40, and 50 °C [9], were reported, respectively. Finally, the adsorption of CH_3COOH on carbon microspheres synthesized by hydrothermal carbonization was investigated by Liang et al. [69], reporting equilibrium adsorbent-phase concentration of CH_3COOH of 260 mg/g at 25 °C.

4. Conclusions

TG/DTG analysis showed, for the hydro-chars obtained at 175 and 200 °C, that highest mass loss was associated with the occurrence of endothermic peaks at 355.2 and 362.9 °C, respectively. This was also confirmed by DTA analysis by the presence of exothermic peaks around 360 °C. For the hydro-chars obtained at 225 and 250 °C, the endothermic peaks occurred at 397 and 434 °C, respectively, showing that the higher the temperature the higher the thermal stability of solid reaction products. The hydro-char obtained at 250 °C presented the highest thermal stability compared to the hydro-chars obtained at 175, 200, and 225 °C, as it presented the lowest mass loss of 50.51% (wt.) between the temperatures 144 and 768.5 °C.

SEM images indicate a rigid and well-organized fiber structure by hydrothermal processing of corn stover at 175 and 200 °C, demonstrating that temperature had almost no effect on the vegetal structure, as it largely retained the original morphological microscopic characteristics. On the other hand, SEM images of hydro-char produced by hydrothermal carbonization of corn stover at 225 and 250 °C indicate that hydro-char structure consisted of agglomerated micro-spheres and heterogeneous structures with irregular shapes (fragmentation). The agglomerated micro-spheres and fragmentation indicates that cellulose and hemi-cellulose were decomposed, as reported by Xing et al. [62], demonstrating that temperature had generated significant alterations on the morphological structure of corn stover by destructing the plant cell structure/walls [6].

As observed by EDX analysis, by increasing the temperature, the carbon content increased and that of oxygen diminished. The diffractograms (XRD) identified the occurrence of peaks of higher intensity of graphite (C) as the temperature increased, demonstrating that higher temperatures favors the formation of crystalline-phase graphite (C).

The analysis of acetic acid adsorption kinetics data, the main volatile carboxylic acid identified in the hydrothermal carbonization liquid phase, showed that NaOH (2.0 M)-activated hydro-char obtained by hydrothermal processing of corn stover at 250 °C, 240 min, biomass/H_2O proportion of 1:10, presented the highest adsorption performance and capacity. The chemically activated hydro-chars were selective to uptake of CH_3COOH, demonstrating that enriching/recovery of CH_3COOH from HTC process water streams is possible.

Supplementary Materials: The following are available online at https://www.mdpi.com/article/10.3390/en14238154/s1, Figure S1: SEM of corn stover after hydrothermal processing at 225 °C, 240 min, and biomass/H_2O ratio of 1:10, using a reactor of 18,927 L (Mag: 30,000×). Figure S2: SEM of corn stover after hydrothermal processing at 225 °C, 240 min, and biomass/H_2O ratio of 1:10, using a reactor of 18.927 L (Mag: 30,000×). Table S1: Physical chemistry and elemental analysis of corn stover after hydrothermal processing at 175, 200, 225, and 250 °C, 240 min, and biomass/H_2O ratio of 1:10, using a reactor of 18.927 l. (MM = moist matter, TS = total solids.) [52].

Author Contributions: The individual contributions of all the co-authors are provided as follows: M.E.G.C. contributed to the methodology, formal analysis, and writing of the original draft paper; M.C.S. contributed to the formal analysis, methodology, and software; L.M.P. contributed to the formal analysis and software; F.P.d.C.A. contributed to the formal analysis and software; T.T. contributed to the formal analysis and software; C.E.F.d.C. contributed to the chemical analysis;

M.S. contributed to the investigation, methodology, and chemical analysis; T.H. contributed to the resources and infrastructure; D.A.R.d.C. to the methodology, formal analysis, and co-supervision; and N.T.M. contributed to the supervision, conceptualization, and data curation. All authors have read and agreed to the published version of the manuscript.

Funding: This research was partially funded by CNPq-Brazil, grant number: 207325/2014-6.

Institutional Review Board Statement: Not applicable.

Informed Consent Statement: Not applicable.

Acknowledgments: I would like to acknowledge and dedicate this research in memory to Hélio da Silva Almeida, Professor at the Faculty of Sanitary and Environmental Engineering/UFPa, who passed away in 13 March 2021. His contagious joy, dedication, intelligence, honesty, seriousness, and kindness will always be remembered in our hearts.

Conflicts of Interest: The authors declare no conflict of interest.

Abbreviations

HTC	Hydrothermal carbonization
SEM	Scanning electron microscopy
TG/DTG/DTA	Thermogravimetric analysis
EDX	Energy dispersive X-ray spectroscopy
XRD	X-ray diffraction
BET	Surface area and pore size distribution analysis
AAEMs	Alkaline and alkaline earth metals
HMF	Hydroxymethylfurfural
CH_3COOH	Acetic acid

References

1. Kumar, S.; A Loganathan, V.; Gupta, R.B.; Barnett, M.O. An Assessment of U(VI) removal from groundwater using biochar produced from hydrothermal carbonization. *J. Environ. Manag.* **2011**, *92*, 2504–2512. [CrossRef]
2. Elaigwu, S.E.; Rocher, V.; Kyriakou, G.; Greenway, G.M. Removal of Pb2+ and Cd2+ from aqueous solution using chars from pyrolysis and microwave-assisted hydrothermal carbonization of Prosopis Africana shell. *J. Ind. Eng. Chem.* **2014**, *20*, 3467–3473. [CrossRef]
3. Liu, Z.; Zhang, F.-S. Removal of lead from water using biochars prepared from hydrothermal liquefaction of biomass. *J. Hazard. Mater.* **2009**, *167*, 933–939. [CrossRef]
4. Xue, Y.; Gao, B.; Yao, Y.; Inyang, M.; Zhang, M.; Zimmerman, A.; Ro, K.S. Hydrogen peroxide modification enhances the ability of biochar (hydrochar) produced from hydrothermal carbonization of peanut hull to remove aqueous heavy metals: Batch and column tests. *Chem. Eng. J.* **2012**, *200–202*, 673–680. [CrossRef]
5. Regmi, P.; Moscoso, J.L.G.; Kumar, S.; Cao, X.; Mao, J.; Schafran, G. Removal of copper and cadmium from aqueous solution using switchgrass biochar produced via hydrothermal carbonization process. *J. Environ. Manag.* **2012**, *109*, 61–69. [CrossRef]
6. Machado, N.; Castro, D.; Queiroz, L.; Santos, M.; Costa, C. Production and Characterization of Energy Materials with Adsorbent Properties by Hydrothermal Processing of Corn Stover with Subcritical H_2O. *J. Appl. Solut. Chem. Model.* **2016**, *5*, 117–130. [CrossRef]
7. Libra, J.; Ro, K.; Kammann, C.; Funke, A.; Berge, N.D.; Neubauer, Y.; Titirici, M.; Fühner, C.; Bens, O.; Kern, J.; et al. Hydrothermal carbonization of biomass residuals: A comparative review of the chemistry, processes and applications of wet and dry pyrolysis. *Biofuels* **2011**, *2*, 71–106. [CrossRef]
8. Islam, A.; Benhouria, A.; Asif, M.; Hameed, B.H. Methylene blue adsorption on factory-rejected tea activated carbon prepared by conjunction of hydrothermal carbonization and sodium hydroxide activation processes. *J. Taiwan Inst. Chem. Eng.* **2015**, *52*, 57–64. [CrossRef]
9. Islam, A.; Tan, I.A.W.; Benhouria, A.; Asif, M.; Hameed, B.H. Mesoporous and adsorptive properties of palm date seed activated carbon prepared via sequential hydrothermal carbonization and sodium hydroxide activation. *Chem. Eng. J.* **2015**, *270*, 187–195. [CrossRef]
10. Sun, K.; Tang, J.; Gong, Y.; Zhang, H. Characterization of potassium hydroxide (KOH) modified hydrochars from different feedstocks for enhanced removal of heavy metals from water. *Environ. Sci. Pollut. Res.* **2015**, *22*, 16640–16651. [CrossRef]
11. Li, Y.; Meas, A.; Shan, S.; Yang, R.; Gai, X. Production and optimization of bamboo hydrochars for adsorption of Congo red and 2-naphthol. *Bioresour. Technol.* **2016**, *207*, 379–386. [CrossRef]
12. Han, L.; Ro, K.S.; Sun, K.; Sun, H.; Wang, Z.; Libra, J.A.; Xing, B. New evidence for high sorption capacity of hydrochar for hydrophobic organic pollutants. *Environ. Sci. Technol.* **2016**, *50*, 13274–13282. [CrossRef]

13. Han, L.; Sun, H.; Ro, K.; Sun, K.; Libra, J.; Xing, B. Removal of antimony (III) and cadmium (II) from aqueous solution using animal manure-derived hydrochars and pyrochars. *Bioresour. Technol.* **2017**, *234*, 77–85. [CrossRef]

14. Pellera, F.-M.; Giannis, A.; Kalderis, D.; Anastasiadou, K.; Stegmann, R.; Wang, J.-Y.; Gidarakos, E. Adsorption of Cu(II) ions from aqueous solutions on biochars prepared from agricultural by-products. *J. Environ. Manag.* **2012**, *96*, 35–42. [CrossRef]

15. Zhou, N.; Chen, H.; Xi, J.; Yao, D.; Zhou, Z.; Tian, Y.; Lu, X. Biochars with excellent Pb(II) adsorption property produced from fresh and dehydrated banana peels via hydrothermal carbonization. *Bioresour. Technol.* **2017**, *232*, 204–210. [CrossRef]

16. Kirschhöfer, F.; Sahin, O.; Becker, G.C.; Meffert, F.; Nusser, M.; Anderer, G.; Kusche, S.; Klaeusli, T.; Kruse, A.; Brenner-Weiss, G. Wastewater treatment-adsorption of organic micropollutants on activated HTC-carbon derived from sewage sludge. *Water Sci. Technol.* **2015**, *73*, 607–616. [CrossRef]

17. Koottatep, T.; Fakkaew, K.; Tajai, N.; Polprasert, C. Isotherm models and kinetics of copper adsorption by using hydrochar produced from hydrothermal carbonization of faecal sludge. *J. Water, Sanit. Hyg. Dev.* **2017**, *7*, 102–110. [CrossRef]

18. Sun, K.; Gao, B.; Ro, K.; Novak, J.M.; Wang, Z.; Herbert, S.; Xing, B. Assessment of herbicide sorption by biochars and organic matter associated with soil and sediment. *Environ. Pollut.* **2012**, *163*, 167–173. [CrossRef]

19. Eibisch, N.; Schroll, R.; Fuß, R.; Mikutta, R.; Helfrich, M.; Flessa, H. Pyrochars and hydrochars differently alter the sorption of the herbicide isoproturon in an agricultural soil. *Chemosphere* **2015**, *119*, 155–162. [CrossRef]

20. Zhu, X.; Liu, Y.; Zhou, C.; Luo, G.; Zhang, S.; Chen, J. A novel porous carbon derived from hydrothermal carbon for efficient adsorption of tetracycline. *Carbon* **2014**, *77*, 627–636. [CrossRef]

21. Fernandez, M.E.; Ledesma, B.; Román, S.; Bonelli, P.R.; Cukierman, A.L. Development and characterization of activated hydrochars from orange peels as potential adsorbents for emerging organic contaminants. *Bioresour. Technol.* **2015**, *183*, 221–228. [CrossRef]

22. Sun, K.; Ro, K.; Guo, M.; Novak, J.; Mashayekhi, H.; Xing, B. Sorption of bisphenol A, 17α-ethinyl estradiol and phenanthrene on thermally and hydrothermally produced biochars. *Bioresour. Technol.* **2011**, *102*, 5757–5763. [CrossRef]

23. Fang, J.; Gao, B.; Chen, J.; Zimmerman, A.R. Hydrochars derived from plant biomass under various conditions: Characteri-zation and potential applications and impacts. *Chem. Eng. J.* **2015**, *267*, 253–259. [CrossRef]

24. Fang, J.; Gao, B.; Zimmerman, A.R.; Ro, K.S.; Chen, J. Physically (CO2) activated hydrochars from hickory and peanut hull: Preparation, characterization, and sorption of methylene blue, lead, copper, and cadmium. *RSC Adv.* **2016**, *6*, 24906–24911. [CrossRef]

25. Xia, Y.; Yang, T.; Zhu, N.; Li, D.; Chen, Z.; Lang, Q.; Liu, Z.; Jiao, W. Enhanced adsorption of Pb(II) onto modified hydrochar: Modeling and mechanism analysis. *Bioresour. Technol.* **2019**, *288*, 121593. [CrossRef]

26. Islam, M.A.; Ahmed, M.J.; Khanday, W.A.; Asif, M.; Hameed, B.H. Mesoporous activated coconut shell-derived hydrochar prepared via hydrothermal carbonization-NaOH activation for methylene blue adsorption. *J. Environ. Manag.* **2017**, *203*, 237–244. [CrossRef]

27. Islam, A.; Ahmed, M.; Khanday, W.A.; Asif, M.; Hameed, B. Mesoporous activated carbon prepared from NaOH activation of rattan (*Lacosperma secundiflorum*) hydrochar for methylene blue removal. *Ecotoxicol. Environ. Saf.* **2017**, *138*, 279–285. [CrossRef]

28. Petrović, J.; Stojanović, M.; Milojković, J.V.; Petrović, M.; Šoštarić, T.; Laušević, M.D.; Mihajlović, M.L. Alkali modified hydrochar of grape pomace as a perspective adsorbent of Pb 2+ from aqueous solution. *J. Environ. Manag.* **2016**, *182*, 292–300. [CrossRef]

29. Lv, B.W.; Xu, H.; Guo, J.Z.; Bai, L.Q.; Li, B. Efficient adsorption of methylene blue on carboxylate-rich hydrochar prepared by one-step hydrothermal carbonization of bamboo and acrylic acid with ammonium persulphate. *J. Hazard. Mater.* **2022**, *421*, 126741.

30. Zhou, N.; Chen, H.; Feng, Q.; Yao, D.; Chen, H.; Wang, H.; Zhou, Z.; Li, H.; Tian, Y.; Lu, X. Effect of phosphoric acid on the surface properties and Pb(II) adsorption mechanisms of hydrochars prepared from fresh banana peels. *J. Clean. Prod.* **2017**, *165*, 221–230. [CrossRef]

31. Marx, S.; Venter, R.J.; Louw, A.; Dewah, C.T. Upgrading of the aqueous product stream from hydrothermal liquefaction: Simultaneous removal of minerals and phenolic components using waste-derived hydrochar. *Biomass-Bioenergy* **2021**, *151*, 106170. [CrossRef]

32. Demir-Cakan, R.; Baccile, N.; Antonietti, M.; Titirici, M.M. Carboxylate-rich carbonaceous materials via one-step hy-drothermal carbonization of glucose in the presence of acrylic acid. *Chem. Mater.* **2009**, *21*, 484–490. [CrossRef]

33. Ghadikolaei, N.F.; Kowsari, E.; Balou, S.; Moradi, A.; Taromi, F.A. Preparation of porous biomass-derived hydrothermal carbon modified with terminal amino hyperbranched polymer for prominent Cr(VI) removal from water. *Bioresour. Technol.* **2019**, *288*, 121545. [CrossRef]

34. Takaya, C.; Fletcher, L.; Singh, S.; Anyikude, K.; Ross, A. Phosphate and ammonium sorption capacity of biochar and hydrochar from different wastes. *Chemosphere* **2016**, *145*, 518–527. [CrossRef]

35. Han, B.; Zhang, E.; Cheng, G.; Zhang, L.; Wang, D.; Wang, X. Hydrothermal carbon superstructures enriched with carboxyl groups for highly efficient uranium removal. *Chem. Eng. J.* **2018**, *338*, 734–744. [CrossRef]

36. Liu, J.-L.; Qian, W.-C.; Guo, J.-Z.; Shen, Y.; Li, B. Selective removal of anionic and cationic dyes by magnetic Fe3O4-loaded amine-modified hydrochar. *Bioresour. Technol.* **2021**, *320*, 124374. [CrossRef]

37. Liu, Z.; Wang, Z.; Chen, H.; Cai, T.; Liu, Z. Hydrochar and pyrochar for sorption of pollutants in wastewater and exhaust gas: A critical review. *Environ. Pollut.* **2021**, *268*, 115910. [CrossRef]

38. Li, B.; Guo, J.; Lv, K.; Fan, J. Adsorption of methylene blue and Cd(II) onto maleylated modified hydrochar from water. *Environ. Pollut.* **2019**, *254*, 113014. [CrossRef]
39. Li, B.; Lv, J.-Q.; Guo, J.-Z.; Fu, S.-Y.; Guo, M.; Yang, P. The polyaminocarboxylated modified hydrochar for efficient capturing methylene blue and Cu(II) from water. *Bioresour. Technol.* **2019**, *275*, 360–367. [CrossRef]
40. Li, H.Z.; Zhang, Y.N.; Guo, J.Z.; Lv, J.Q.; Huan, W.W.; Li, B. Preparation of hydrochar with high adsorption perfor-mance for methylene blue by cohydrothermal carbonization of polyvinyl chloride and bamboo. *Bioresour. Technol.* **2021**, *337*, 125442. [CrossRef]
41. Li, F.; Zimmerman, A.R.; Hu, X.; Yu, Z.; Huang, J.; Gao, B. One-pot synthesis and characterization of engineered hy-drochar by hydrothermal carbonization of biomass with ZnCl2. *Chemosphere* **2020**, *254*, 126866. [CrossRef]
42. Madduri, S.; Elsayed, I.; Hassan, E.B. Novel oxone treated hydrochar for the removal of Pb(II) and methylene blue (MB) dye from aqueous solutions. *Chemosphere* **2020**, *260*, 127683. [CrossRef]
43. Nguyen, D.H.; Tran, H.N.; Chao, H.-P.; Lin, C.-C. Effect of nitric acid oxidation on the surface of hydrochars to sorb methylene blue: An adsorption mechanism comparison. *Adsorpt. Sci. Technol.* **2019**, *37*, 607–622. [CrossRef]
44. Parra-Marfíl, A.; Ocampo-Pérez, R.; Collins-Martínez, V.H.; Flores-Vélez, L.M.; Gonzalez-Garcia, R.; Medellín-Castillo, N.A.; Labrada-Delgado, G.J. Synthesis and characterization of hydrochar from industrial Capsicum annuum seeds and its application for the adsorptive removal of methylene blue from water. *Environ. Res.* **2020**, *184*, 109334. [CrossRef]
45. Qian, W.-C.; Luo, X.-P.; Wang, X.; Guo, M.; Li, B. Removal of methylene blue from aqueous solution by modified bamboo hydrochar. *Ecotoxicol. Environ. Saf.* **2018**, *157*, 300–306. [CrossRef]
46. Ronix, A.; Pezoti, O.; de Souza, L.S.; Souza, I.P.A.F.; Bedin, K.C.; Souza, P.S.C.; Silva, T.L.; Melo, S.A.R.; Cazetta, A.L.; Almeida, V.C. Hydrothermal carbonization of coffee husk: Optimization of experimental parameters and adsorption of methylene blue dye. *J. Environ. Chem. Eng.* **2017**, *5*, 4841–4849. [CrossRef]
47. Tran, T.H.; Le, A.H.; Pham, T.H.; Nguyen, D.T.; Chang, S.W.; Chung, W.J.; Nguyen, D.D. Adsorption isotherms and kineticmodel-ing of methylene blue dye onto a carbonaceous hydrochar adsorbent derived from coffee husk waste. *Sci. Total Environ.* **2020**, *725*, 138325. [CrossRef]
48. Tran, H.N.; You, S.-J.; Chao, H.-P. Insight into adsorption mechanism of cationic dye onto agricultural residues-derived hydrochars: Negligible role of π-π interaction. *Korean J. Chem. Eng.* **2017**, *34*, 1708–1720. [CrossRef]
49. Li, B.; Wang, Q.; Guo, J.-Z.; Huan, W.-W.; Liu, L. Sorption of methyl orange from aqueous solution by protonated amine modified hydrochar. *Bioresour. Technol.* **2018**, *268*, 454–459. [CrossRef]
50. Spataru, A.; Jain, R.; Chung, J.W.; Gerner, G.; Krebs, R.; Lens, P.N.L. Enhanced adsorption of orthophosphate and copper onto hydrochar derived from sewage sludge by KOH activation. *RSC Adv.* **2016**, *6*, 101827–101834. [CrossRef]
51. Flora, J.F.; Lu, X.; Li, L.; Flora, J.R.; Berge, N.D. The effects of alkalinity and acidity of process water and hydrochar washing on the adsorption of atrazine on hydrothermally producedhydrochar. *Chemosphere* **2013**, *93*, 1989–1996. [CrossRef]
52. Machado, N.; de Castro, D.; Santos, M.; Araújo, M.; Lüder, U.; Herklotz, L.; Werner, M.; Mumme, J.; Hoffmann, T. Process analysis of hydrothermal carbonization of corn Stover with subcritical H$_2$O. *J. Supercrit. Fluids* **2018**, *136*, 110–122. [CrossRef]
53. Hoekman, S.K.; Broch, A.; Robbins, C.; Zielinska, B.; Felix, L. Hydrothermal carbonization (HTC) of selected woody and herbaceous biomass feedstocks. *Biomass-Convers. Biorefinery* **2012**, *3*, 113–126. [CrossRef]
54. Silva, C.D.M.S.D.; de Castro, D.A.R.; Santos, M.C.; Almeida, H.D.S.; Schultze, M.; Lüder, U.; Hoffmann, T.; Machado, N.T. Process analysis of main organic compounds dissolved in aqueous phase by hydrothermal processing of Açaí (Euterpe oleraceae, Mart.) Seeds: Influence of process temperature, biomass-to-water ratio, and production scales. *Energies* **2021**, *14*, 5608. [CrossRef]
55. Mota, S.; Mancio, A.D.A.; Lhamas, D.; de Abreu, D.; da Silva, M.; dos Santos, W.; de Castro, D.; de Oliveira, R.; Araújo, M.; Borges, L.E.; et al. Production of green diesel by thermal catalytic cracking of crude palm oil (Elaeis guineensis Jacq) in a pilot plant. *J. Anal. Appl. Pyrolysis* **2014**, *110*, 1–11. [CrossRef]
56. Almeida, H.D.S.; Corrêa, O.; Eid, J.; Ribeiro, H.; de Castro, D.; Pereira, M.; Pereira, L.; Mâncio, A.D.A.; Santos, M.; Souza, J.D.S.; et al. Production of biofuels by thermal catalytic cracking of scum from grease traps in pilot scale. *J. Anal. Appl. Pyrolysis* **2016**, *118*, 20–33. [CrossRef]
57. Serrão, A.C.M.; Silva, C.M.S.; Assunção, F.P.D.C.; Ribeiro, H.J.D.S.; Santos, M.C.; Almeida, H.D.S.; Junior, S.D.; Borges, L.E.P.; de Castro, D.A.R.; Machado, N.T. Análise do processo de pirólise de sementes de açaí (euterpe oleracea, mart): Influência da temperatura no rendimento dos produtos de reação e nas propriedades físico-químicas do bio-óleo/process analysis of pyrolise of açaí (euterpe oleracea, mart) seeds: Influence of temperature on the yield of reaction products and physico-chemical properties of bio-oil. *Braz. J. Dev.* **2021**, *7*, 18200–18220. [CrossRef]
58. Sittisun, P.; Tippayawong, N.; Wattanasiriwech, D. Thermal Degradation characteristics and kinetics of oxy combustion of corn residues. *Adv. Mater. Sci. Eng.* **2015**, *2015*, 1–8. [CrossRef]
59. Zhao, M.; Li, B.; Cai, J.-X.; Liu, C.; McAdam, K.; Zhang, K. Thermal & chemical analyses of hydrothermally derived carbon materials from corn starch. *Fuel Process. Technol.* **2016**, *153*, 43–49.
60. Kumar, S.; Kothari, U.; Kong, L.; Lee, Y.; Gupta, R.B. Hydrothermal pretreatment of switchgrass and corn stover for production of ethanol and carbon microspheres. *Biomass-Bioenergy* **2011**, *35*, 956–968. [CrossRef]
61. Yang, B.; Wyman, C.E. Effect of xylan and lignin removal by batch and flowthrough pretreatment on the enzymatic digestibility of corn stover cellulose. *Biotechnol. Bioeng.* **2004**, *86*, 88–98. [CrossRef]

62. Varga, E.; Szengyel, Z.; Réczey, K. Chemical pretreatments of corn stover for enhancing enzymatic digestibility. *Appl. Biochem. Biotechnol.* **2002**, *98*, 73–87. [CrossRef]
63. Mohammed, I.; Na, R.; Kushima, K.; Shimizu, N. Investigating the effect of processing parameters on the products of hydrothermal carbonization of corn stover. *Sustainability* **2020**, *12*, 5100. [CrossRef]
64. Xing, X.; Fan, F.; Shi, S.; Li, Y.; Zhang, X.; Yang, J. Fuel properties and combustion kinetics of hydrochar prepared by hydrothermal carbonization of corn straw. *BioResources* **2016**, *11*, 9190–9204. [CrossRef]
65. Hoskinson, R.L.; Karlen, D.L.; Birrell, S.J.; Radtke, C.W.; Wilhelm, W.W. Engineering, nutrient removal, and feedstock conversion evaluations of four corn stover harvest scenarios. *Biomass Bioenergy* **2007**, *31*, 126–136. [CrossRef]
66. Kang, K.; Sonil, N.; Guotao, S.; Ling, Q.; Yongqing, G.; Tianle, Z.; Mingqiang, Z.; Runcang, S. Microwaveassisted hydrothermal carbonizatijon of corn stalk for solid biofuel production: Optimization of process parameters and characterization of hydrochar. *Energy* **2019**, *186*, 115–125. [CrossRef]
67. Becker, R.; Dorgerloh, U.; Helmis, M.; Mumme, J.; Diakité, M.; Nehls, I. Hydrothermally carbonized plant materials: Patterns of volatile organic compounds detected by gas chromatography. *Bioresour. Technol.* **2013**, *130*, 621–628. [CrossRef] [PubMed]
68. Reza, M.T.; Wirth, B.; Lueder, U.; Werner, M. Behavior of selected hydrolyzed and dehydrated products during hydrothermal carbonization of biomass. *Bioresour. Technol.* **2014**, *169*, 352–361. [CrossRef] [PubMed]
69. Liang, J.; Liu, Y.; Zhang, J. Effect of solution pH on the carbon microsphere synthesized by hydrothermal carbonization. *Procedia Environ. Sci.* **2011**, *11*, 1322–1327. [CrossRef]

Article

Study of Energy Valorization of Disposable Masks via Thermochemical Processes: Devolatilization Tests and Simulation Approach

Rachele Foffi [1], Elisa Savuto [1], Matteo Stante [2], Roberta Mancini [2] and Katia Gallucci [1,*]

1   Department of Industrial and Information Engineering and Economics, University of L'Aquila, Monteluco di Roio, 67100 L'Aquila, Italy; rachele.foffi@student.univaq.it (R.F.); elisa.savuto@univaq.it (E.S.)
2   ALMA C.I.S., 66013 Chieti, Italy; matteo.stante@almacis.it (M.S.); roberta.mancini@almacis.it (R.M.)
*   Correspondence: katia.gallucci@univaq.it; Tel.: +39-0862434213

Abstract: The COVID-19 pandemic exacerbated the use of medical protective equipment, including face masks, to protect the individual from the virus. This work studies the feasibility of using these materials as fuel for thermochemical processes for the production of syngas. A preliminary physic-chemical characterization was made by means of moisture and ash determination, thermogravimetric analysis, X-ray fluorescence. Afterward, pyrolysis and gasification tests were executed in a laboratory-scale fluidized bed reactor with chirurgical and FFP2 masks investigating four temperature levels and three different operating conditions (fluidizing agents and dry/wet sample). A qualitative and quantitative analysis of condensable aromatic hydrocarbons in the produced gas, collected during the test campaign, was performed employing a gas chromatograph-mass spectrometer. The experimental data from the tests were used to propose a hybrid approach to simulate the gasification process, based on experimental laws for the devolatilization step and a thermodynamic equilibrium approach for char gasification. The resulting data were compared with a thermodynamic equilibrium model, showing that the new approach captures non-equilibrium effects always present in real gasifiers operation.

Keywords: disposable masks; devolatilization tests; Aspen Plus ®; tar analysis

Citation: Foffi, R.; Savuto, E.; Stante, M.; Mancini, R.; Gallucci, K. Study of Energy Valorization of Disposable Masks via Thermochemical Processes: Devolatilization Tests and Simulation Approach. *Energies* **2022**, *15*, 2103. https://doi.org/10.3390/en15062103

Academic Editor: Mark Laser

Received: 14 January 2022
Accepted: 11 March 2022
Published: 13 March 2022

Publisher's Note: MDPI stays neutral with regard to jurisdictional claims in published maps and institutional affiliations.

1. Introduction

In modern society, the demands of higher consumption of goods, waste management, and energy supply are among the most significant challenges humans must deal with. Sustainable development has become part of EU legislation and policies: in 2015, the Commission launched the ambitious "Closing the Loop–An EU Action Plan for the Circular Economy". According to the Agenda 2030 goals, the EU requires the transformation of waste management into sustainable material management. This one embeds the principles of the circular economy, enhances the diffusion of renewable energy and provides economic opportunities and moreover reduces the dependence of UE on imported resources [1].

In this framework, Waste-to-Energy (WtE) technologies play a significant role: thanks to WtE, waste sources produce energy in the form of power, heat, high-value chemicals, or transport fuels. Incineration (combustion) is the most established and widespread WtE technology: it is used to remedy a broad variety of trashes [2]. However, in the 1980s dissent emerged among the population due to its contribution to pollutant emissions, especially the toxic furans/dioxins. This shifted the interest towards advanced thermal conversion processes like gasification and pyrolysis, which are perceived to have the potential of being more efficient in the energy recovery from solid waste and in reducing pollutant emissions, especially the toxic ones.

Regarding the latter issue, the situation has worsened since 2020 because of the magnitude of the COVID-19 emergency. In this context, the demand for Personal Protect

Equipment (PPEs), especially face masks, face shields, and gloves, increased between medical care workers and the general public; however, no clear instructions of their disposing mechanisms were provided [3]. It was estimated that about 129 billion face masks are being used globally every month (3 million/min) [4]. Every day, China manufactures around 240 tons of pharmaceutical trash. Similarly, cities like Manila, Bangkok, Kuala Lumpur and Hanoi manufacture about 154–280 tons/day of additional medical rubbish as the outbreak compared to the previous period. In another town, throughout the lockdown, medical waste generation increased from 550 to 600 kg/day to around 1000 kg/day [2,5].

The use of personal protective equipment and single-use plastics has increased the amount of contagious medical waste that needs additional careful handling. There is an immediate threat that the unsafe disposal of medical waste will add to the environmental pollution crisis [6]. As stated by the WHO's health guidelines, used face masks and soiled tissues must be thrown only into lidded litter bins; on the other hand, any medical gear used by affected patients and hospital staff must be burnt in dedicated incinerators. Like common biomedical wastes, the COVID refuse is incinerated at a temperature above 1100 °C. However, as already mentioned, incineration is famous for being a dirty disposal method as it creates harmful dioxins and other emissions [7].

Despite different approaches to pyrolytic studies of polypropylene and polyethylene recovery into liquid products and condensable products [8,9] and catalytic degradation of waste plastic [10], gasification and high-temperature pyrolysis are mature processes that convert solid fuels (coal, waste materials, petroleum coke, or biomasses) to usable synthesis gas, or syngas (i.e., a mix of CO, H_2, CH_4, and CO_2 in variable proportions) [11,12]. The gasification process is fundamentally different from combustion: instead of producing only heat and electricity, the syngas produced can be converted into higher-value commercial products such as chemicals, transportation fuels, fertilizers, and substitute natural gas [13]. The gasification process takes place in reactors that may utilize oxygen, steam, CO_2, or a blend of these being gasification agents [2].

Due to the high complexity of the residual biomass/waste gasifier, mathematical models can be a useful tool to provide a better understanding of the thermochemical processes. Moreover, gasification models enable the study of the influences of several input variables (biomass moisture content, air/biomass ratio, etc.) on the key output variables (syngas composition, calorific value, etc.) [14,15]. Simulating accurately the performance of a biomass gasifier additionally allows optimizing the gasifier integration in complex energy systems [16,17].

In general, mathematical models of gasification are divided into thermodynamic-equilibrium and kinetic models. The former predicts the thermodynamic limits of the gasification reaction without accounting for the reactor design. This approach assumes that the residence time in the reactor is long enough to reach thermodynamic equilibrium, namely the state minimizing the Gibbs free energy. In non-stoichiometric equilibrium models, the biomass elemental composition, readily available from ultimate analysis, is the only input needed for the calculation [18–22]. Despite the simplicity of this approach, equilibrium is never reached under normal gasifier operations [23], and phenomena under a kinetic regime, such as tar formation and char residual, cannot be considered [24]. The advantage of this approach is the ease of integration into complex process models and optimization routines [25]. Kinetic models [26,27], on the other hand, take detailed information on the reaction rates, diffusivity phenomena, and reactor geometry into account. The rate of reactions is usually estimated by experimental data specific for a certain feedstock and reactor design, making the results not easily extrapolated to different conditions. Moreover, kinetic models require higher efforts in modeling as well as higher computational time.

Today, syngas produced by coal gasification stands at around 70% of the production, followed by a 15% of syngas production from natural gas and petroleum. On the other hand, syngas deriving from biomass or waste gasification constitutes only a few percentages of the worldwide syngas production [4,24]. However, four billion tons of waste are produced yearly worldwide [27].

In this scenario, the present work aims at evaluating the possibility of recovering these waste materials through thermochemical processes to convert them into useful and value-added fuel (syngas), overcoming one of the main bottlenecks of thermochemical processes that consist of the lack of information for the proper design of "unconventional" materials through process simulation tools. It is often necessary to resort to extensive and costly experimentation to obtain process data. Here is a quick approach to gather this information and translate it into a simulation model useful as a design tool. Furthermore, this approach allows investigating the ashes behavior in the bed, one of the major drawbacks that occur in real waste gasification processes.

The activities carried out to study the potential use of this kind of waste for gasification purposes are presented. The activities included:

1. Physicochemical characterization of selected materials.
2. Devolatilization tests of the characterized materials at semi-batch lab-scale apparatus that allows a fast determination of gas-phase composition in terms of permanent gas (H_2, CO, CO_2, CH_4).

Implementation of a novel approach to simulate the gasification process, based on experimental results, developed using the commercial software Aspen Plus ®12.0 [28]. The novel approach is based on deriving experimental correlations using the devolatilization data obtained during the experimental campaign to simulate the devolatilization step. The objective of the work is to demonstrate a simplified methodology to obtain an improved thermodynamic model accounting for experimental data of devolatilization tests, and thus a useful tool to simulate fluidized bed gasifiers.

2. Materials and Methods

2.1. Materials for Devolatilization Tests

The materials selected and characterized in this work consist of chirurgical and FFP2 masks commonly used to help slow the diffusion of COVID-19. Typically, common facial masks are composed of three layers: the outer layer is a non-absorbent material (e.g., polyester), the middle layer is non-woven fabrics (e.g., polystyrene and polypropylene) and the inner layer is usually realized with an absorbent material, like cotton [29]. This composition varies according to the type of mask (FFP2 masks usually have four filter layers, while chirurgical masks only three). Overall, the major constitutes are PP (73.33 wt%) and PE (13.77 wt%), which are used to produce mask filters [30].

It is therefore essential to carry out a preliminary sterilization operation to avoid contagion. Exposure to high temperatures can kill Coronaviruses: to estimate the temperature that could kill SARS-CoV-2, a group of researchers investigated the effect of temperature on other coronaviruses [31]. One of these viruses was SARS-CoV, which causes SARS and is closely related to SARS-CoV-2. Based on the data, the researchers state that most coronaviruses would be almost completely killed after exposure to temperatures of 65 °C (149 °F) or higher for longer than 3 min.

In this work, the sterilization operation takes place simultaneously with the material characterization during the moisture determination due to the temperature reached in the oven.

To load the reactor with the samples, a preliminary pelletizing operation is necessary. This facilitates the reactor loading with the pellets and allows compacting a large quantity of materials in a small volume. For this purpose, masks are deprived of their nose metallic wire and are shredded with the help of a blender and then put into 13-mm pellet dies. The cylindrical dies are then placed in a Specac manual hydraulic press, producing pellets.

2.2. Materials Characterization

Physicochemical characterization of materials selected was conducted by means of appropriate instrumental techniques such as tests in the oven and muffle furnace, X-ray fluorescence (XRF), thermo-gravimetric analyzer (TGA), and CHNS elemental analyzer.

Moisture fractions of the coarsely ground disposable masks were quantified by measuring the samples' mass variation through a drying treatment using an oven (model MPIM factory Oven SL, MPIM srl, Sambuceto, Italy), keeping the temperature at 105 °C overnight according to EN 15414-3:2011 [32] standard indications, as shown in Figure 1. Due to the temperature reached in the oven, the sterilization operation took place simultaneously with this first analysis.

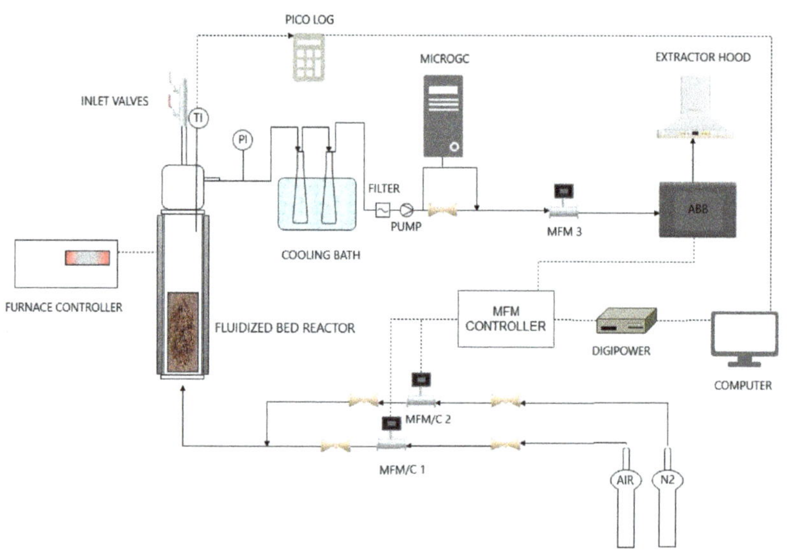

Figure 1. Schematic view of the laboratory-scale experimental apparatus for devolatilization tests.

Ash content was determined using a LENTON type ECF 12/10 equipment, setting the thermal program defined in ISO 1171-2010 [33] standard: (i) 8 °C min^{-1} heating ramp- up from room temperature to 500 °C, followed by 1 h dwell at 500 °C; (ii) 8 °C min^{-1} heating ramp-up from 500 °C to 815 °C and 2 h dwell at 500 °C. To carry out the measurements, disposable masks were coarsely ground into porcelain crucibles.

Thermophysical properties of samples were determined through TGA; the device used for this purpose is a Linseis STA PT 1000; the operating conditions used are defined by the proximate and compositional analysis. Since small quantities of samples are required for TGA analysis, fragments of masks were placed in the crucibles with an average weight of 0.15 g of material. Three tests were performed for each kind of disposable mask.

Proximate analysis was adapted from the standard test method, ASTM E1131-20 (used for coal, lubricant, polymers, etc.) [34]. It led to the determination of medium volatile, highly volatile, combustible material and ash and contents setting the first treatment in nitrogen atmosphere (3 Nl h^{-1}): (i) 10 °C min^{-1} heating ramp-up from room temperature to 150 °C; (ii) 10 °C min^{-1} heating ramp-up from 150 °C to 600 °C (iii) 10 °C min^{-1} heating ramp-up from 600 °C to 750 °C; followed by treatment in air flux (3 Nl h^{-1}): (iv) 10 °C min^{-1} heating ramp-up from 750 °C to 800 °C, respectively.

The elemental analyzer CHNS (PerkinElmer 2400 Series-II) operated to perform the ultimate analysis of FPP2 and chirurgical masks, measuring their carbon (C), hydrogen (H), nitrogen (N), and sulfur (S) mass percentages on a dry basis. On average, 10 mg of fragmented masks was used and two repetitions were carried out for each type of mask. The instrument error is the order of 0.3%.

Ash from the above-mentioned muffle treatment at 815 °C was analyzed by XRF to determine FFP2 and chirurgical masks' semi-quantitative elemental composition. The device used was a Spectro Xepos I instrument, with wavelength dispersion method, equipped with a detector. This one does not allow to detect elements with an atomic mass lower than that of magnesium (Mg); however, it can lead to the determination of the presence of mineral elements (e.g., Si, K, Ca, Al, Mg), which are responsible for the ash-melting behavior [35].

2.3. Devolatilization Tests: Experimental Plan and Layout

Devolatilization tests were carried out on dry and wet samples of chirurgical and FFP2 masks by means of semi-batch tests, in a quartz fluidized bed reactor, that investigated four temperature levels (from 500 °C to 800 °C, with 100 °C steps), using two fluidizing agents (N_2, or air diluted by N_2 to have 1.3–2.3 vol% of O_2 at the bed inlet).

Tests in an inert environment simulated the devolatilization process, whereas those with diluted air have the same role regarding pyrolysis; by wetting the samples and blowing nitrogen and air in the reactor, a pseudo-steam gasification process was emulated.

The layout of the lab-scale plant is depicted in Figure 1. There was a quartz fluidized bed reactor, placed in a cylindrical electric furnace, fed upward by nitrogen or a mixture of nitrogen and air. Inlet gas flowrate is set manually or by PC, using the Bronkorst El-Flow control unit.

As the reaction took place, gaseous products left the reactor from the top and passed through two impinger bottles in a cooling bath to collect condensable gases: 25 mL of isopropyl alcohol (most ideal no-toxic solvent as ascribed by the Standard UNI CEN/TS 15439 [36]) was poured the first of two flasks in series, settled in an ice-water mixture bath. Isopropyl alcohol sampling was performed at the end of the three tests conducted for each operating condition.

The dry gas was sent to the ABB online analysis system that allowed to measure the volume concentration of CO, CO_2, CH_4, and H_2; and a micro gas-chromatograph (μGC AGILENT 490, Agilent Technologies Italia S.p.A., Milan, Italy).

To load the reactor with the samples, a preliminary pelletizing operation facilitated the reactor loading and allowed compacting a large amount of material in a small volume. Masks were deprived of their nose metallic wire, shredded with the help of a blender and then put into 13 mm pellet dies.

The pellets were fed in the reactor individually by hand, thanks to a vertical system composed of two consecutive valves at the top of the reactor, to minimize the air entering the reactor while feeding the pellet.

For each set "material/fluidizing agent/bed temperature", three pellets were devolatilized to obtain three repetitions of a specific test condition; the process is in an unsteady state as tests conducted on each sample were carried out until completion, before loading the following sample.

The collected samples of condensable gasses, composed of aromatic hydrocarbons (HCs), were analyzed in a gas chromatograph with a mass spectrometer (GC–MS), Agilent GC7890A with an MSD-Triple Axis Detector 5975C, for the identification and quantification of the compounds produced during the devolatilization tests.

2.4. Devolatilization Tests: Reaction Section

The reactor used was made of quartz, able to withstand temperatures up to 1000 °C. It was one meter high and had a 5-cm inner diameter. A packed bed made of silicon carbide (SiC) particles was placed in the windbox of the reactor to work as a preheater for the inlet gas flow. The sand was loaded into the reactor to have a bed height equal to 1.5 times its diameter.

To determine the minimum gas flow rate to achieve bed fluidization, the minimum fluidization velocity must be determined. The reactor bed consists of sand particles having the following characteristics, as summarized in Table 1.

Table 1. Bed material physical properties of fluid-dynamic interest and minimum fluidization velocity (u_{mf}) as a function of temperature at devolatilization conditions are described in Section 2.3.

Bed Material	Sand
d_p (µm)	212–250
ρ_p (kg m^{-3})	2587
T (°C)	u_{mf} (cm s^{-1})
500	2.9
600	2.7
700	2.5
800	2.3

In the light of the sand bed characteristics reported in Table 1, the sand group according to the Geldart powder classification [37] was identified as group B; in this case, the minimum fluidization velocity (u_{mf}) and the bubbling fluidization velocity coincide. Values of u_{mf} (Table 1), used as references to set devolatilization experimental conditions in this work (see Section 2.3), were determined according to the simplified Ergun equation (Equation (1)) [38], assuming the correlation constants C1 = 27.2 and C2 = 0.0408, as suggested by Grace [38].

$$Re_{mf} = \sqrt{C1^2 + C2Ar} - C1 \tag{1}$$

Equations (2) and (3) define the Reynolds number at minimum fluidization conditions (Re_{mf}) and the Archimedes number (Ar), both dimensionless: N_2 density (ρ_{N_2}) and dynamic viscosity (μ_{N_2}) were properly evaluated as a function of temperature, g is the gravity acceleration, ρ_p is the particle density, experimentally determined.

$$Re_{mf} = \frac{d_p \rho_{N_2} u_{mf}}{\mu_{N_2}} \tag{2}$$

$$Ar = \frac{\rho_{N_2}(\rho_p - \rho_{N_2})g d_p^3}{\mu_{N_2}} \tag{3}$$

2.5. Devolatilization Tests: Data Processing

From the ABB measurements, outlet molar flow rates ($F_{i,out}$, where i = CO_2, H_2, CH_4, CO) as functions of time (t) were determined assuming that N_2 inlet flow rate was the internal standard and that, during the tests with diluted air, O_2 reacted completely.

The outlet gas component flow rate ($F_{i,out}$) (NL/min), individual pellets weight (w_p), and masks composition allowed for the calculation of the integral-average values of:

- gas yield (η_{av}, Equation (4)) (mol/100 g sample);
- carbon conversion ($\chi_{av,C}$, Equation (5)) (g/g);
- outlet H_2/CO ratio (λ, Equation (6)) (mol/mol);
- percentage of i on dry and dilution free basis ($Y_{av,i}$, Equation (7)) (mol/mol).

The quantities defined in Equations (4)–(7) average out of the three repetitions were calculated for each set "material kind/fluidizing agent/bed temperature", provided with related standard deviations [39].

$$\eta = \frac{\int \Sigma_i F_{i,out}\, dt}{w_p} \cdot 100 \tag{4}$$

With i = H_2, CO, CO_2, CH_4

$$\chi_C = \frac{w_{C,in} - (12\,\text{g/mol} \Sigma_j \int F_{j,out} dt)}{w_{C,in}} \cdot 100 \tag{5}$$

With j = CO, CO$_2$, CH$_4$.

where:

$w_{C,in} = w_p \cdot \%wt_C$

w_p and $\%wt_C$ are the weight of the pellet (g) and the weight percentage of carbon;

$$\lambda = \frac{\int F_{H_2,out}dt}{\int F_{CO,out}dt} \tag{6}$$

$$Y_i = \frac{\int F_{i,out}dt}{\int \Sigma_i F_{i,out}dt} \tag{7}$$

With i = H$_2$, CO, CO$_2$, CH$_4$

2.6. Modeling Approach

The core simulation model was developed using Aspen Plus ®, which is a process simulation software often used to model chemical processes involving solid, liquid, and gaseous streams using mass and energy conservations equations and phase equilibrium databases [28]. The following assumptions were made: (i) steady-state and no kinetic reactions; (ii) autothermal system; (iii) the sulfur converts entirely to H$_2$S; (iv) the nitrogen converts entirely to NH$_3$ [20,40].

A simple schematic of the fluidized bed reactor is reported in Figure 2 (LHS). As stated in previous literature [41,42], in fluidized bed gasifiers, the devolatilization step occurs at the top of the reactor as soon as the feedstock enters the chamber. The remaining feedstock, namely the char, falls into the bed where the oxygen (air) is fed from the bottom and reacts with it. Given the sub-stoichiometric conditions, one can assume that most of the oxygen is consumed by the char transported by the bed. The modeling approach chosen to represent these physics is shown in Figure 2 (RHS).

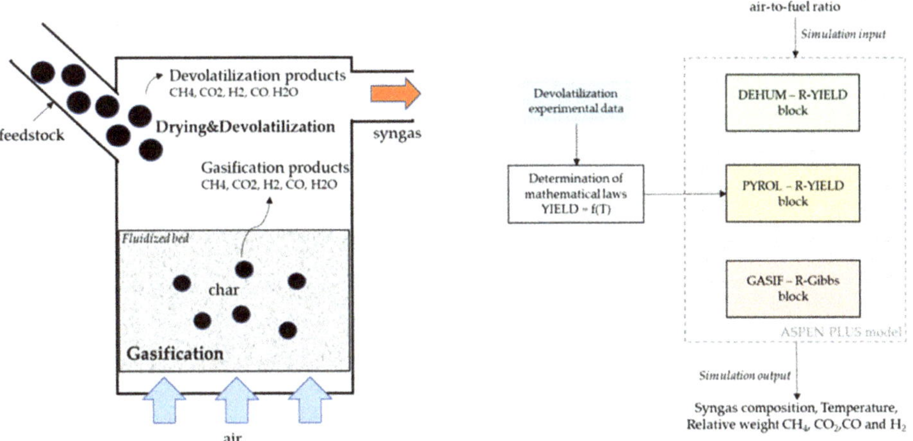

Figure 2. Schematics of the gasifier reactions main steps, on the left hand-side (LHS); overview of model implementation, on the right hand-side (RHS).

The devolatilization step was modeled using mathematical laws derived using the data from the experimental campaign (pure N$_2$), while the char gasification was modeled using a thermodynamic equilibrium reactor. The feedstock composition and the air-to-fuel ratio were used as input to the model. The syngas composition, temperature, and relative weights between the four components CH$_4$, CO$_2$, CO, and H$_2$ were obtained as the output of the model.

The inlet stream of feedstock is first sent to a dehumidification block (R-YIELD) named "DEHUM", where water (humidity) is separated from the feedstock. As the next step, the feedstock is sent to an R-YIELD block named "PYROL". This block is used to model the fast pyrolysis step using the devolatilization data obtained during the experimental campaign. The remaining material is sent to two subsequent reactor blocks: an R-YIELD block named "PREC", and an R-GIBBS block named "GASIF". In the former, the elemental composition of the remaining feedstock (after the completion of the devolatilization step) is determined. In the latter, based on the elemental composition obtained in the previous reactor, the gasification step is modeled using a non-stoichiometric equilibrium approach, thus finding the gas composition minimizing the Gibbs free energy. The gasification temperature is calculated by the ASPEN model as an output.

As stated above, the R-GIBBS equilibrium reactor is modeled as an autothermal system: from the energy balances of all the heat streams released/consumed by the reactions, the equilibrium temperature is calculated. This approach represents the physics of fluidized bed reactors, where the partial combustion of the feedstock supplies the heat for the endothermic reactions occurring during gasification and making, therefore, the system autothermal. The resulting model is thus able to represent an autothermal reactor. As the last step, the gas obtained in the GASIF block is merged with the gas obtained from the devolatilization step and the water vapor from the dehumidification block. The resulting gas thus represents the syngas obtained from the gasification process.

As already mentioned, data from devolatilization tests at different temperatures were used to obtain the dependence of the production of the main gases from the temperature. Based on the analysis of the data, a polynomial equation of the form of (Equation (8)) was used to model the CH_4 productivity, and an exponential equation of the form of (Equation (9)) was used to model the H_2, CO, and CO_2 productivity. In the equations, T is the gasification temperature, while $prodX$ represents the productivity of the generic gas X, expressed as moles of X for 100 g of sample. E_1 and E_2 represent the parameters for the exponential model (fitted based on the data), while P_1, P_2 and P_3 are the parameters for the polynomial model.

$$prodX = P_{1,X} \cdot T^2 + P_{2,X} \cdot T + P_{3,X} \tag{8}$$

$$prodX = E_{1,X} \cdot e^{E_{2,X} \cdot T} \tag{9}$$

The details of the derivations of the productivity laws are reported in Appendix A.

The results of the model were compared against the results of a simple thermodynamic-equilibrium model, based on the same assumptions, and developed in Aspen Plus ®. The thermodynamic model was based on an approach widely used in the literature [40] where pyrolysis and gasification are combined in a single step after decomposing the feedstock based on its elemental composition. The syngas composition is found as the one minimizing the Gibbs free energy. The comparison was carried out by fixing the feedstock composition and varying the equivalence ratio (ER) between air and fuel. The simulations were carried out with air with a lower oxygen content than normal air (10 %wt. O_2, 90 %wt. N_2), to keep the gasification temperature within the range 500–800 °C, which is the range of validity of the obtained devolatilization laws.

The comparison between the two models allowed estimating the difference between a pure thermodynamic model and the novel hybrid approach, where the kinetics effects related to the pyrolysis steps are considered. It must be stressed that both models are based on an R-GIBBS reactor to simulate the char gasification step. Therefore, in both cases, results represent a thermodynamic equilibrium limit and benchmark for the process; nevertheless, the results are significantly different, and the innovative algorithm of the model can better catch the real behavior in a fluidized bed fed with nonconventional feedstock.

3. Results and Discussion

3.1. Results of Physic-Chemical Characterization of FFP2 and Chirurgical Masks

Tables 2–4 report numerical results of proximate, compositional, and ultimate analyses performed through oven tests, muffle tests, TGA, and CHNS/O analyzer (see Section 2.2).

Table 2. Moisture and ash determination of FFP2 and chirurgical masks as received (EN 15414-3:2011).

	wt%	
	FFP2	**CHIRUGICAL**
Moisture	0.97	0.58
Ash	0.91	0.62

Table 3. Proximate and compositional analyses of FFP2 and chirurgical masks as received (ASTM E1131-20: average value of three tests $\pm$ standard deviation) [34].

	wt%	
	FFP2	**CHIRUGICAL**
Highly volatile matter	1.2 ± 0.5	ND [1]
Medium volatile matter	95.9 ± 2.1	97.2 ± 2.0
Combustible	1.4 ± 0.4	1.3 ± 0.5
Ash	ND [1]	ND [1]

[1] ND = No Detectable.

Table 4. Ultimate analysis of FFP2 and chirurgical masks on a dry basis.

	wt%	
	FFP2	**CHIRUGICAL**
C	82.34 ± 3.03	84.68 ± 0.29
H	7.44 ± 1.36	14.24 ± 2.76
N	0	0
S	0.81 ± 0.20	1.71 ± 0.18

Results obtained from moisture and ash determination tests are shown in Table 2.

The moisture content of the masks is low; this turns out to be advantageous for pyrolysis processes, which are more efficient using a material with reduced moisture content: it's commonly known that moisture increases energy consumption as it affects both the solid internal temperature history (due to endothermic evaporation) and the total energy required to bring the charge to the pyrolysis temperature [43].

Table 2 also shows low values of ash content. Its presence can determine possible problems in the plant, such as corrosion, deterioration of the heat exchange surfaces, obstruction, and equipment malfunction. In Table 3, the results of the proximate analysis are reported: it emerges a volatile matter (highly and medium volatile matter) content much relevant (>97%) against a very low fixed carbon (<1.4%). We cannot be confident of the ash determination via the technical standard E1131-20, because of a too high standard deviation compared to the corresponding average value.

Data obtained from CHNS elemental analyses conducted on the filter layers are summarized in Table 4. Oxygen content is the complement to 100 of the sum of the concentrations of other elements (C, H, N, S). From the literature analysis [30], it was deduced that chirurgical masks are mainly made with polypropylene and polyethylene, consisting of hydrocarbon chains. However, these are not the exclusive materials used for single-use disposable masks production [44] and the samples examined are not identical. This is the reason why that from CHNS analysis, further elements, including sulfur and oxygen, are contained in the filter layers. In particular, the estimated oxygen content in FFP2 masks (9.42%) is not negligible, but the presence of other possible oxidate compounds, such as cotton, nylon, etc. as reported in Medical Expo [29], and the experimental procedure allows justifying the abundant presence of oxidate species in the gas phase.

The following table (Table 5) summarizes the semi-quantitative elemental composition for FFP2 and chirurgical masks measured through XRF (see Section 2.2).

Table 5. Semi-quantitative elemental composition for FFP2 and chirurgical masks ashes via XRF.

Element	wt%	Abs Err. [2]	wt%	Abs Err. [2]
	FFP2		**CHIRURGICAL**	
Mg	2.479	0.011	6.013	0.019
Al	0.000	0.000	0.240	0.000
Si	1.267	0.002	4.145	0.004
P	0.321	0.001	0.293	0.001
S	0.276	0.000	0.407	0.001
K	0.253	0.001	0.428	0.002
Ca	16.860	0.010	11.860	0.010
Ti	9.153	0.007	14.180	0.010
Fe	0.214	0.000	4.358	0.004
Cu	0.015	0.000	0.519	0.000
Sb	0.322	0.006	0.282	0.003
Ba	0.000	0.000	0.270	0.000

[2] Abs err. = absolute error.

From Table 5 emerges that magnesium, silicon, calcium, phosphorous, sulphur, potassium, calcium, titanium, iron, copper, and antimony are the most abundant in both types of masks; in chirurgical masks, it is also possible to find traces of aluminum and barium. The other elements detectable by the instrument are not reported as their concentration was lower than 0.200 wt%.

3.2. Results of Devolatilization Tests

Figures 3–5 show six examples of devolatilization tests at 700 °C, for each combination of material kind and fluidizing agent. As already noted by Di Giuliano et al. [45], for similar experiments on biomass pellets, gaseous products (H_2, CO, CO_2, and CH_4) released by the unsteady state process determined the characteristic asymmetric shape of $F_{i,out}(t)$ peak.

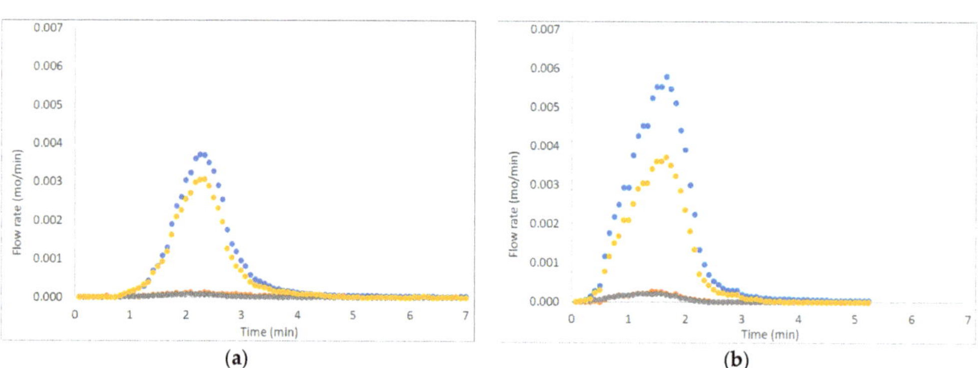

(a)　　　　　　　　　(b)

Figure 3. Example of H_2, CO, CO_2, and CH_4 outlet molar flow rates ($F_{i,out}$) as functions of time (t), produced by devolatilization tests of (**a**) chirurgical masks and (**b**) FFP2 mask, in a fluidized bed reactor at 700 °C, in a nitrogen atmosphere (• H_2, • CO, • CO_2; • CH_4).

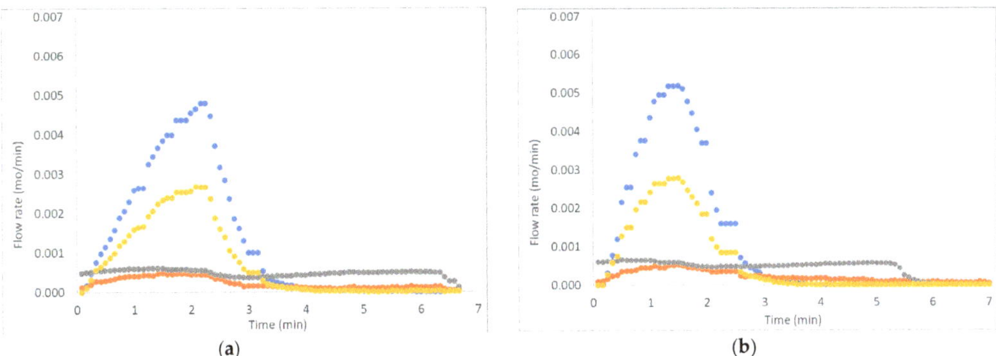

Figure 4. Example of H_2, CO, CO_2, and CH_4 outlet molar flow rates ($F_{i,out}$) as functions of time (t), produced by devolatilization tests of (**a**) chirurgical masks and (**b**) FFP2 mask, in a fluidized bed reactor at 700 °C, in nitrogen and air atmosphere. (• H_2, • CO, • CO_2; • CH_4).

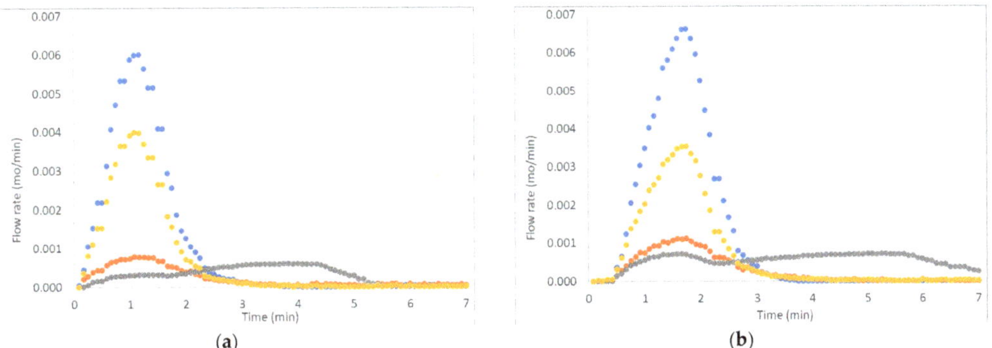

Figure 5. Example of H_2, CO, CO_2, and CH_4 outlet molar flow rates ($F_{i,out}$) as functions of time (t), produced by devolatilization tests of (**a**) wet chirurgical masks and (**b**) wet FFP2 mask, in a fluidized bed reactor at 700 °C, in nitrogen and air atmosphere. (• H_2, • CO, • CO_2; • CH_4).

Pyrolysis tests were carried out with two fluidizing agents: N_2 or N_2 plus air. To mime pseudo-steam gasification processes, pellets were immersed in distilled water before the devolatilization tests, flowing N_2 and air in the reactor, in order to take into account the presence of steam in the gasifier.

The outgoing gas molar flow rates calculated are plotted as a function of time: by way of example, results presented regard one of the three pellets examined at 700 °C under each operating condition as reported from Figure 3 to Figure 5. During pyrolysis/devolatilization, line charts representing the molar flow rate as a function of time are in the form of bell-shaped curves because the described process is not continuous but a semi-batch configuration.

In particular, it can be seen that:

1. In a nitrogen atmosphere, there is a small number of oxidized compounds compared to CH_4 and H_2. This is justified by the fact that, as discussed for the characterization test, disposable face masks contain mainly polypropylene and polyethylene consisting of long chains of hydrocarbons (HCs). On the other hand, the small amount of oxygen developed is mainly attributable to the presence of other unidentified components constituting the filter layers and the small amount (corresponding to the volume

between the two feeding valves) of air that inevitably enters the reactor during sample loading.

2. When air is added to the inlet gas mixture, there is a greater development of oxidized compounds (CO and CO_2) that increase with the increasing of temperature. A slow development of a CO_2 bell-shaped curve can be observed, in particular, at high temperatures when more gas is produced. The CO_2 curve takes more time than the others to return to zero.

3. For each operating condition, as the temperature increases, syngas composition changes: CH_4 develops mainly at low temperatures, whereas H_2 at high temperatures; this is due to the higher decomposition that takes place at a higher temperature.

The average values of the fundamental parameters in pyrolysis processes are reported using bar charts in which colored bars represent average values out of the three repetitions for each set "material kind/fluidizing agent/bed temperature"; errors bars refer to the related standard deviations. The gas yield η_{av} (Equation (4)) is reported in Figure 6a,b for chirurgical and FFP2 masks, respectively. H_2/CO ratio values are reported in Figure 7a–c for dry masks, N_2 and N_2+air of a fluidizing agent, wet masks with N_2+air flow. Carbon conversion $\chi_{av,C}$ (Equation (5)) is reported in Figure 8a,b while syngas composition $Y_{avi,out}$ (Equation (7)) is reported in Figure 9a–h. Numerical data are detailed in Appendix B of this work.

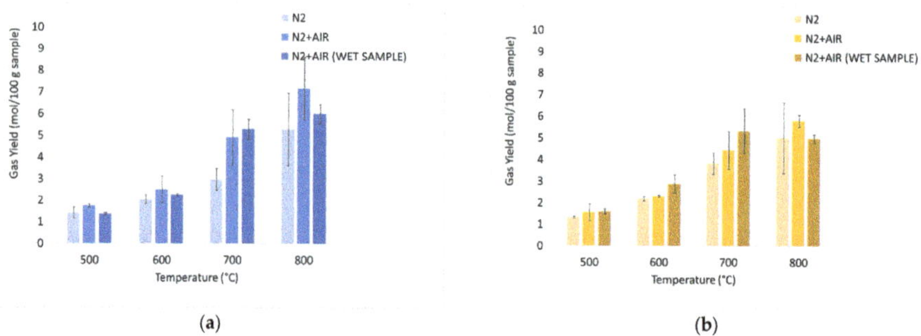

Figure 6. Experimental results of devolatilization tests as a function of temperature: integral-average gas yield (η_{av}, Equation (4)) of chirurgical (**a**) and FFP2 (**b**) masks.

In Figure 6, it is possible to notice high productivity values: this is due to the high content of volatile compounds, as already highlighted by TGA. Bar charts show an increase in gas productivity as the temperature rises for each operating condition; there are no significant differences between the two types of masks. The addition of air leads to an increase in productivity: the highest values are recorded for dry masks at 800 °C and are equal to 7.16 mol/100 g_{sample} for chirurgical masks, and 5.80 mol/100 g_{sample} for FFP2 masks. At 700°C there is an increase in gas productivity of about 7.6% for wet chirurgical masks and 20.3% for wet FFP2 masks compared to dry masks.

Product distributions (H_2/CO ratio and tar compounds) are influenced by the initial H/C ratio, temperature, gas feeding, residence time in the reactor. For this experimental campaign, Figure 7 reports that the H_2/CO ratio increases as temperature rises in each operating condition, except in the nitrogen atmosphere. At 800 °C, an unexpected result was obtained due to the presence of air that entered during the test run and led to the increase of CO molar productivity. So, in the N_2 atmosphere, the highest values are recorded at 700 °C (22.46 mol/mol for chirurgical masks, while FFP2 has a 23% lower value, which can be explained by the lower content of hydrogen in its elemental composition). The addition of air leads to an increase in the production of oxidized compounds, resulting in a reduction of H_2/CO values, especially for chirurgical masks. It is possible to notice how at 700 °C there is a reduction of the previously mentioned values of 76% for chirurgical masks and 67% for FFP2 masks.

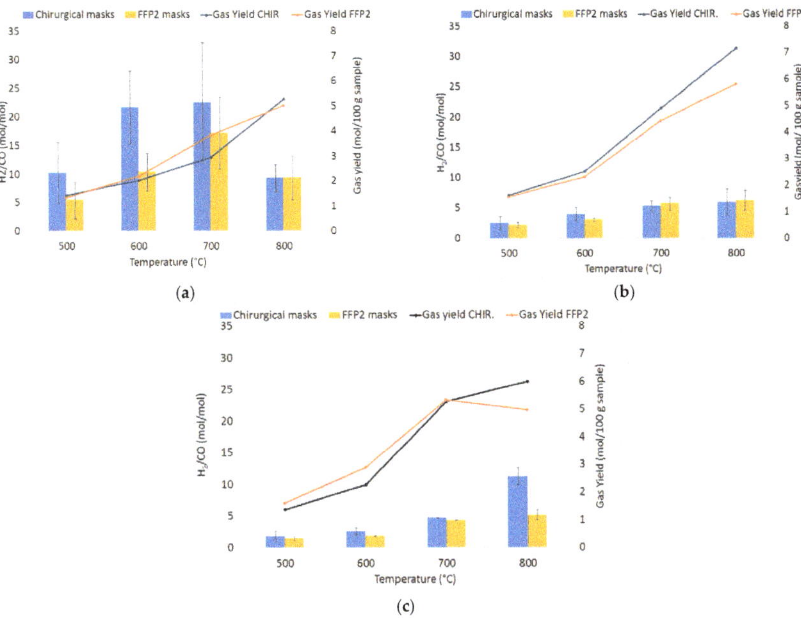

Figure 7. Experimental results of devolatilization tests as a function of temperature: H_2/CO ratio values in nitrogen (**a**), diluted air (**b**), and diluted air using wet pellets (**c**).

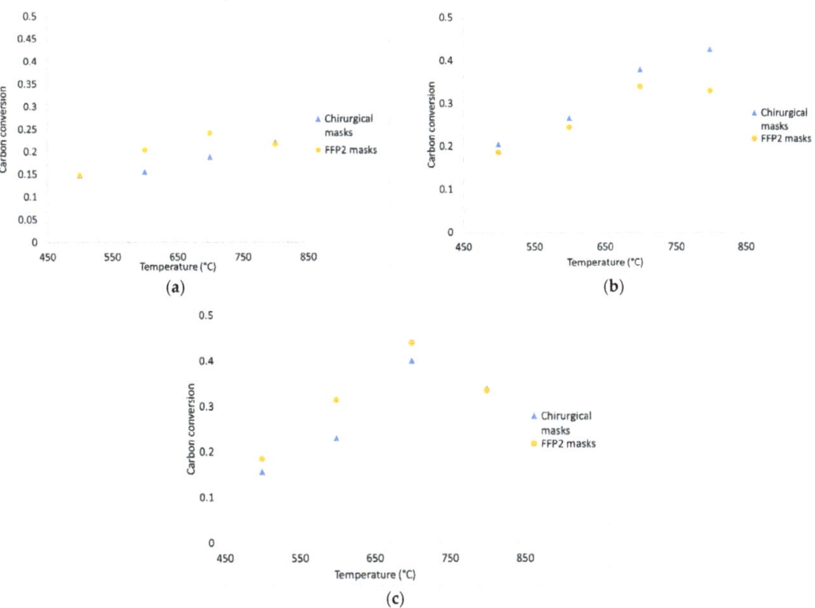

Figure 8. Experimental data of carbon conversion obtained from devolatilization tests in nitrogen atmosphere (**a**), diluted air (**b**), and diluted air using wet pellets (**c**).

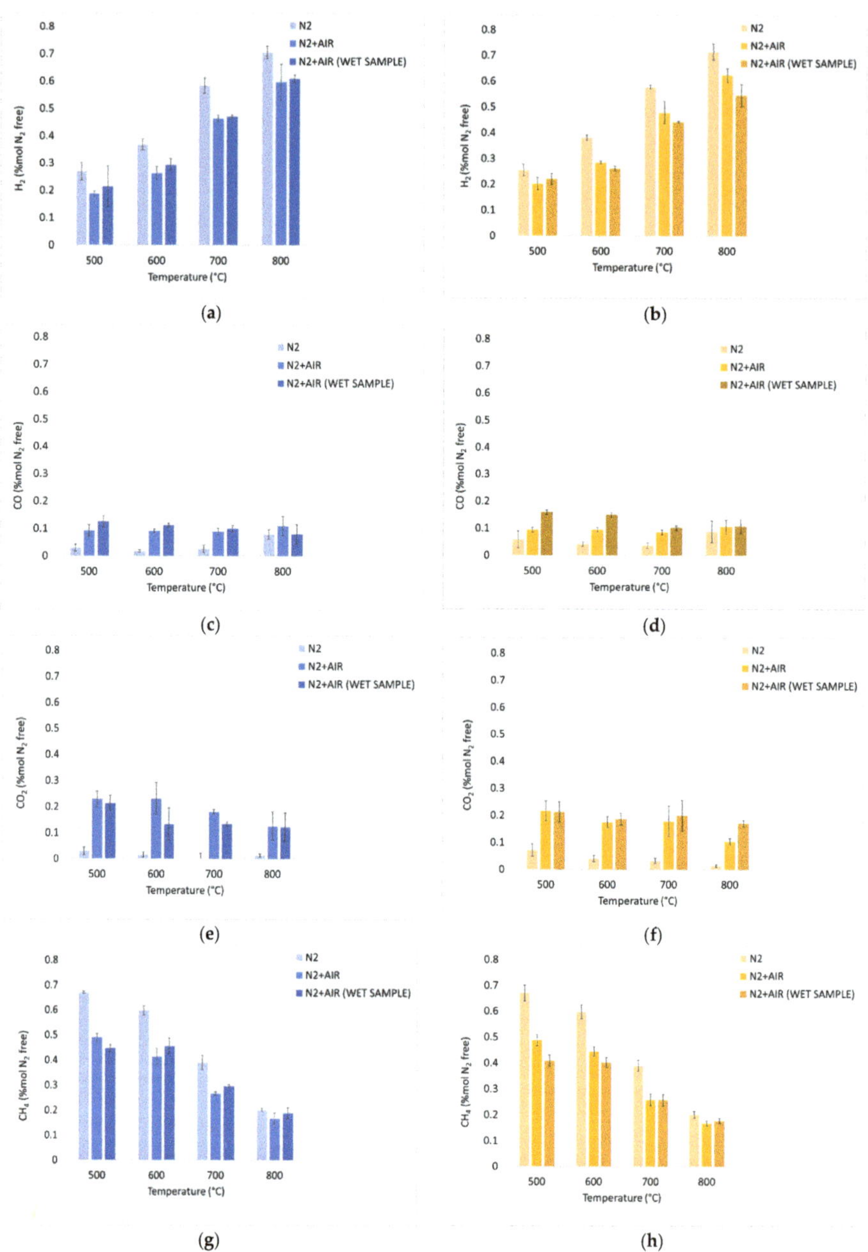

Figure 9. Experimental results of devolatilization tests in a nitrogen atmosphere, in a fluidized sand bed, as functions of temperature: integral average H_2 mol% dry, dilution-free $Y_{av,i}$ (Equation (7)) of chirurgical (**a**) and FFP2 (**b**) masks; CO mol% dry, dilution-free $Y_{av,i}$ (Equation (7)) of chirurgical (**c**) and FFP2 (**d**) masks; CO_2 mol% dry, dilution-free $Y_{av,i}$ (Equation (7)) of chirurgical (**e**) and FFP2 (**f**) masks; CH_4 mol% dry, dilution-free $Y_{av,i}$ (Equation (7)) of chirurgical (**g**) and FFP2 (**h**) masks. Figure 11 shows the aromatic hydrocarbon compounds detected and quantified in the liquid phase condensed downstream the devolatilization tests.

The H_2/CO ratio defines the use of syngas: it can be used to produce hydrogen for values $H_2/CO \geq 50$, methanol for $H_2/CO \sim 2$, ethanol for $H_2/CO \leq 1$ and Fischer–Tropsch synthesis (FTS) for $H_2/CO \geq 2$. The high values obtained for chirurgical and FPP2 masks may suggest their use in renewable fuels production.

From Figure 8a–c, it emerges that carbon conversion is always increasing for chirurgical and FFP2 masks from 500 °C to 700 °C. From the graphs, there is a greater carbon conversion when air is blown in; in particular, at 700 °C, there is an increase of 102% when air is added in the reactor and 123% when masks are wetted.

At 800 °C, the obtained conversion values are out of trend, this is justified by the inaccuracies that characterized the tests high temperatures: at high temperatures, the slow development of the CO_2 bell-shaped curves caused an overlap of the curves between each test which led to an underestimation of the quantity of gas produced. Moreover, conversion percentage does not exceed 45%; this happens because:

1. Some of the carbon is collected during the condensable capture process.
2. During the testing campaign, the rapid condensation of gaseous products, which are exposed to room temperature both on the reactor head and in the piping before the condensation zone, results in a very viscous residue in pipes, like TAR, which causes blackening and obstruction.

The presence of water determines an increase in productivity at 800 °C for chirurgical masks that show the highest average value equal to 7.65 mol/mol.

Figure 9 illustrates the percentage of each component on a nitrogen-free basis; Figure 9a,b show a growing trend which is related to the greater quantity of hydrocarbons that are decomposed at higher temperatures. In a nitrogen atmosphere, there is a high hydrogen content in syngas composition: at 800 °C, both chirurgical and FFP2 masks record a value of 71% for Y_{H_2} in N_2 atmosphere. This high hydrogen content is linked to the composition of the masks, which are mainly made of plastic material. Compared to inert atmosphere, values obtained at 800 °C are 15% and 14% lower for dry and wet chirurgical masks, while they are 13% and 23% lower for dry and wet FFP2 masks.

It is possible to say that for each temperature, the presence of oxygen in the incoming gas mixture leads to an increase in oxidized compounds in syngas. A demonstration of that is given by Figure 9c,d, which shows that air caused an increase in the CO produced. However, it is not possible to define a clear trend with regard to Y_{CO}. Complete oxidation reactions are not favored at high temperatures; as a consequence, there is a decrease in Y_{CO_2} at high temperatures. This trend is verifiable in Figure 9e,f, where the introduction of oxygen into the mixture increased the production of oxidized compounds.

In addition, Figure 9g,h Y_{CH_4} shows a decreasing trend as the temperature rises, and there are no differences between the two types of masks.

Based on the obtained results, it is possible to assert that the inert environment favors the production of hydrogen, allowing to obtain higher H_2/CO. There are no substantial differences between the introductions of dry or wet masks; this is because the masks are made of water-repellent material that does not absorb a well-defined amount of water directly involved in reactions such as methane and TAR reforming or water gas shift that would determine a different composition of the downstream syngas.

To the data exposed so far are added those of the μ-GC Agilent 490 that allows the identification of the components present in the syngas stream after passing the cooling bath section (see Section 2.3), also including gases that are detected by the ABB analyzers system. The data obtained show the presence of n-butane (C_4H_{10}) as air is introduced in the gas mixture, both for dry and wet disposable masks.

After the pseudo-pyro-gasification tests were carried out, sand contained in the reactor was extracted and visually examined. Figure 10 shows ash residues and bits of unburned masks: the char residue. No relevant agglomeration phenomena of bed particles are detected.

Figure 10. Post-test sand.

3.3. Results from GC-MS

Figure 11 shows the results of the analysis on the liquid solvent samples containing the condensable gases generated during the devolatilization tests, i.e., aromatic hydrocarbons. The compounds identified and quantified are acenaphthylene, fluorene, phenanthrene, anthracene, pyrene, phenol, naphthalene, styrene xylene, toluene, and benzene.

Figure 11 shows clearly that for all the test conditions and feedstocks, benzene has the highest content compared to the other aromatic hydrocarbons (HCs) detected, ranging between 50 and 100 mg/g_{sample}. Toluene is the second-most abundant hydrocarbon compound, up to 50 mg/g_{sample}. Light hydrocarbons (one-ring compounds: styrene, xylene, toluene, and benzene) in general are more abundant in all the tests (30–130 mg/g_{sample}), while heavy hydrocarbons are present in very low quantities, <10 mg/g_{sample}.

Comparing the contents of the aromatic HCs produced for tests carried out at different operating conditions, it is possible to observe that in general, increasing the devolatilization temperature from 500 °C to 800 °C, the HCs contents decrease, for almost all the tests, the temperature of 800 °C corresponds to the lowest HCs content.

Furthermore, wet samples treated with N_2+air (Figure 11e,f) gave in general considerably lower contents of HCs for almost all the temperatures and feedstocks, compared to dry samples treated with N_2 and with N_2+air. The moisture included in the wet samples turns into steam at the operating temperature and probably enhances steam reforming reactions that convert the HCs.

It is therefore observed that the lowest content of aromatic hydrocarbons in the set of devolatilization tests here reported are obtained for wet samples treated with N_2+air, at the highest operating temperature, 800 °C. In particular, for these conditions, both FFP2 and chirurgical masks samples gave a total HC content of approximately 33 mg/g_{sample} composed exclusively of light one-ring compounds.

In Figure 12a, a comparison between the total aromatic HCs is shown for different feedstocks, temperatures, and fluidizing agents.

Figure 12a shows the influence of temperature and fluidizing agents for the devolatilization of the chirurgical masks. It is shown that the higher quantities of HCs are produced at 500 and 600 °C, while for higher temperatures, their thermal decomposition is enhanced. From 600 °C to 700 °C, a reduction of tar is observed for all the fluidizing agents, which is further enhanced at 800 °C, for which the total HC content is the lowest. For the samples treated with N_2 and wet samples treated with N_2+air an increase of the HC content from 500 to 600 °C is observed; this phenomenon could be related to the formation of volatile compounds that occurs at temperatures not lower than 600 °C.

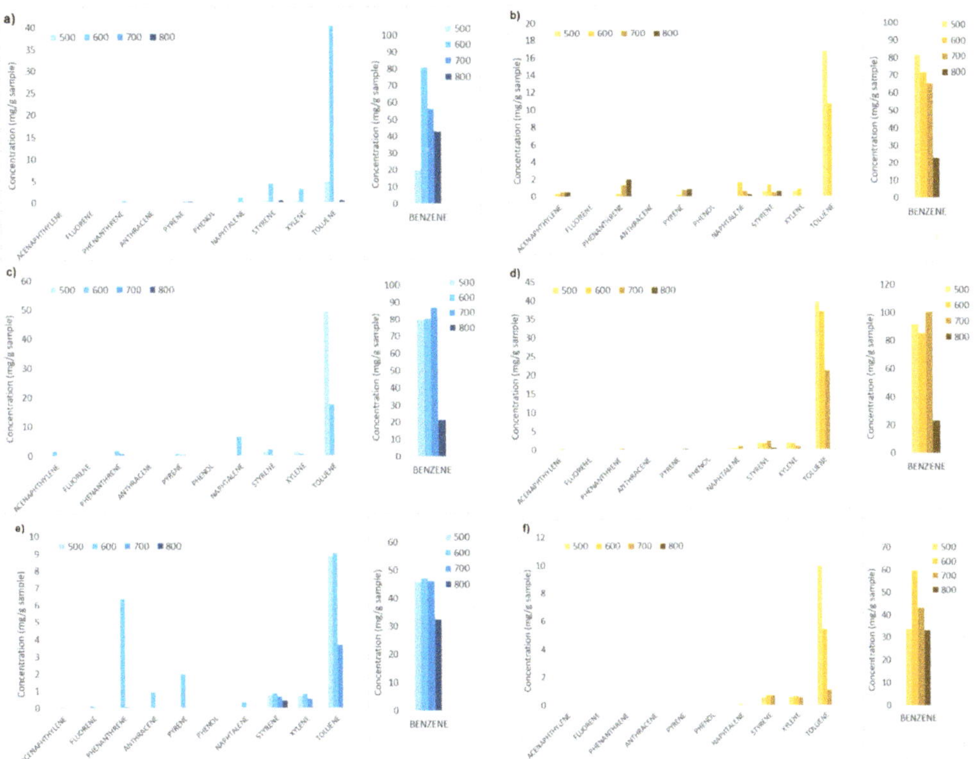

Figure 11. Aromatic hydrocarbon compounds and relative contents obtained in devolatilization tests carried out at different temperatures (500, 600, 700, 800 °C) and fluidizing agents (N_2, N_2+air, N_2+air-wet sample). (**a**,**b**) Tests with N_2, chirurgical and FFP2 masks respectively; (**c**,**d**) tests with N_2+air, chirurgical and FFP2 masks respectively; (**e**,**f**) tests of wet samples with N_2+air, chirurgical and FFP2 masks, respectively.

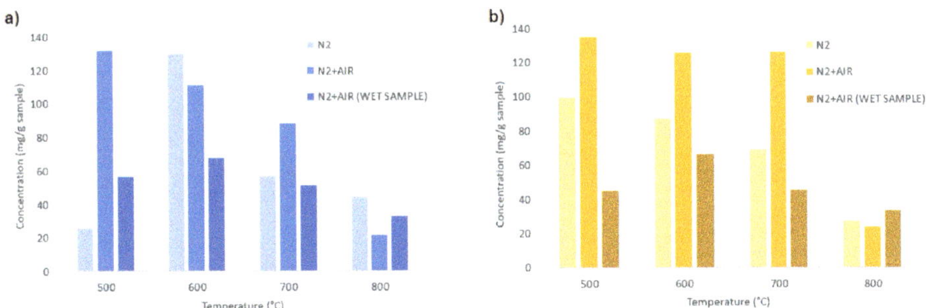

Figure 12. Total aromatic hydrocarbons for different temperatures (500, 600, 700, 800 °C) and fluidizing agent (N_2, N_2+air, N_2+air-wet sample); (**a**) chirurgical masks; (**b**) FFP2 masks.

For the FFP2 masks instead (Figure 12b), the total HC contents are similar at 500, 600, and 700 °C, for an equal fluidizing agent. A strong reduction of the aromatic HCs is instead observed when the devolatilization temperature is 800 °C, in which the HC contents are reduced to <40 mg/g$_{sample}$ for all the samples.

In general, samples treated with N_2+air give the higher content of HCs, while wet samples treated with N_2+air give the lower content for equal operating temperatures and feedstock. The total HC content produced at 800 °C is approximately 20–40 mg/g$_{sample}$, while for lower temperatures, the HC produced ranges between 40 and 140 mg/g$_{sample}$.

3.4. Simulations Results

Figure 13 shows the results of the model comparison between the new hybrid approach and a thermodynamic-equilibrium approach. The figure reports the relative weight of the four main components of the syngas CO, CO_2, H_2, and CH_4 for different values of ER.

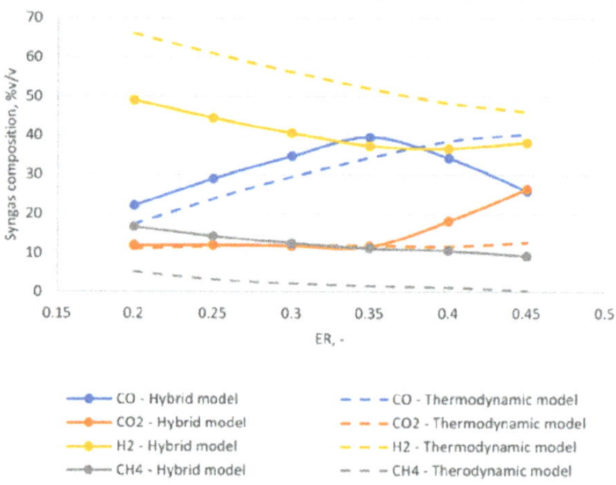

Figure 13. Comparison between new hybrid approach and thermodynamic model results: relative weights of CO, CO_2, CH_4, and H_2.

Figure 13 shows that the thermodynamic model estimates a higher molar fraction of hydrogen compared to the hybrid approach, while the opposite is true for methane. For the case of methane, the thermodynamic model predicts a very low content that decreases with increasing ER and leads to almost negligible content of the component, in agreement with the thermodynamic equilibrium conditions. However, this is not always true under real operation conditions.

Figure 13 additionally shows a good agreement between the two models for the trends of CO_2 and CO at low ERs. With increasing ERs, however, a decrease of the molar fraction of CO and a sharp increase of CO_2 are observed for the hybrid model. In particular, the CO content has a peak for an ER equal to 0.35. This trend can be explained by looking at the gasification temperature as a function of ER, reported in Figure 14. By increasing the value of ER, an increase of the gas produced in the devolatilization step is observed, as the temperature of the process increases. This implies a reduction of the feedstock content available for gasification. At the same time, higher oxygen is available for oxidation reactions, thus favoring the complete oxidation and thus the formation of carbon dioxide at the expense of CO production.

Lastly, Figure 13 highlights a general higher equilibrium temperature obtained using the hybrid approach compared to the thermodynamic equilibrium model, strictly connected to the higher concentration of CO_2 at high ERs, resulting from more exothermic oxidation reactions obtained using the hybrid approach.

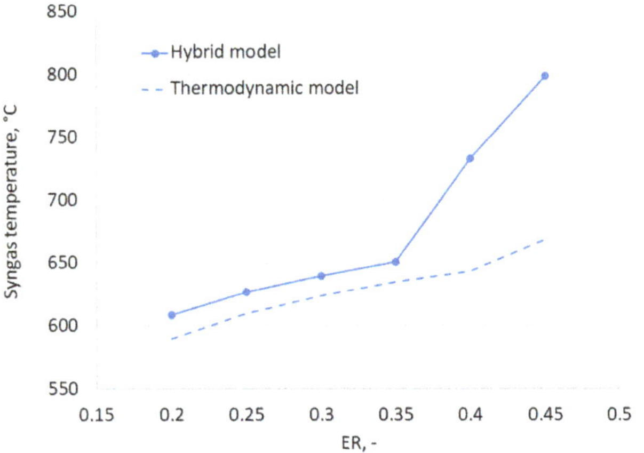

Figure 14. Syngas temperature at the outlet of the gasifier as a function of ERWe further conducted a qualitative verification against the experimental findings from Xiao et al. [46], who published experimental results of gasification tests carried out in a fluidized bed reactor for polypropylene feedstock, the main component of chirurgical masks.

The work by Xiao et al. [46] reports, for example, a constant and non-negligible amount of CH_4 in the syngas for different ER tests. By using empirical correlations for the devolatilization step, the hybrid approach allows introducing non-equilibrium effects and thus obtaining a similar non-negligible concentration of methane. The trend of methane justifies the decrease of hydrogen concentration because of the mass balance on hydrogen species.

The trend for CO_2 is also in agreement with experimental results reported by Xiao et al. [46]. This trend, resulting from kinetic regime of the investigated phenomena, cannot be observed in the thermodynamic model results.

It must be noted that it was not possible to quantitatively validate the suggested hybrid approach against experimental data due to a lack of experimental data available for gasification of the same plastic material in the temperature range for which the devolatilization empirical correlations were derived. However, the results in Figure 14 clearly show how the thermodynamic equilibrium model fails to predict non-equilibrium phenomena, affecting real gasifier operation. By including empirical laws for the devolatilization step, the hybrid approach allowed considering a few of these effects.

Future work can be devoted to the extension of the devolatilization experiments to a higher temperature, thus obtaining a larger range for simulating the process, a complete extension of the validation process against experimental data for continuous gasification processes.

4. Conclusions

This work had as its purpose the valorization of waste products, i.e., disposable masks, through their use as an alternative to fossil fuels. To do that, an experimental study was carried out through a physic-chemical characterization of the material. The parameters investigated during this last phase were: type of mask used, different temperature levels (500, 600, 700, and 800 °C), and gasifying agents (nitrogen and a mixture of air and nitrogen). Gas productivity increases with increasing temperature, especially in the presence of air. As the temperature increases, syngas composition changes: CH_4 develops mainly at 500 °C (68% in N_2 environment and about 45% in N_2 and air environment), while H_2 at 800 °C (71%

in N_2 atmosphere at and about 60% in N_2 and air); this is due to the higher decomposition that takes place at higher temperatures.

In a nitrogen atmosphere, there is a reduced amount of oxidized compounds compared to CH_4 and H_2, resulting in a high H_2/CO ratio (22.46 mol/mol for chirurgical masks). On the other hand, when air is added to the inlet gas mixture, oxidized compounds (CO and CO_2) are developed. For carbon conversion, the highest values are obtained at 700 °C and 800 °C, however, these do not exceed 50%, and this is attributable to the rapid condensation of vaporous products, which are exposed to room temperature both on the reactor head and in the piping before condensation zone, results in a very viscous residue in pipes, similar to tar, which causes blackening and obstruction; Aromatic HCs generated in the devolatilization tests were analyzed and quantified by GC-MS; the results show that benzene and toluene are the most abundant compounds generated (50 to 100 mg/g_{sample}), while heavy HCs are present in very low quantities (<10 mg/g_{sample}). In particular, tests carried out at higher temperatures (800 °C) and with wet samples presented the lowest contents of aromatic HCs produced. In particular, for these conditions, both FFP2 and chirurgical mask samples gave a total HC content of approximately 33 mg/g_{sample} composed exclusively of light 1-ring compounds.

The simulation study aimed at demonstrating the utilization of experimental data to model the devolatilization step of the gasification process. A hybrid model was developed and compared with the simple thermodynamic approach. The results showed that the hybrid model was able to predict a few non-equilibrium effects as a higher methane content at the expense of hydrogen content and a peak of CO production for an equivalence ratio equal to 0.35. The results shown in this work illustrated how the hybrid approach was able to qualitatively predict non-equilibrium phenomena while thermodynamic models failed to capture these effects.

Author Contributions: Conceptualization, K.G.; methodology, K.G., M.S. and R.M.; software, M.S. and R.M.; validation, M.S., R.M. and R.F.; formal analysis, R.F.; investigation, K.G., R.F. and E.S, resources, K.G.; data curation, M.S, R.M, R.F. and E.S.; writing—original draft preparation, R.F., M.S., R.M. and E.S.; writing—review and editing, R.F., M.S., R.M., E.S. and K.G.; visualization, R.F. and E.S.; supervision, K.G.; project administration, K.G.; funding acquisition, K.G. and M.S. All authors have read and agreed to the published version of the manuscript.

Funding: This research received no external funding.

Data Availability Statement: The data presented in this study are available on request from the corresponding author.

Acknowledgments: The authors warmly thank Fabiola Ferrante for FT-IR, XRF and CHNS/O analyses.

Conflicts of Interest: The authors declare no conflict of interest.

Appendix A

Table A1. Average values of pyrolysis parameters in N_2 atmosphere for dry samples.

Temperature	Sample	η^{av} (mol/100 g_{sample})	H_2/CO	$Y^{av}_{H_2}$ (%mol N_2 Free)	Y^{av}_{CO} (%mol N_2 Free)	$Y^{av}_{CO_2}$ (%mol N_2 Free)	$Y^{av}_{CH_4}$ (%mol N_2 Free)	χ^{av}
500 °C	Chirurgical	1.41	10.20	27%	3%	3%	67%	15%
	FFP2	1.33	5.39	26%	6%	7%	61%	15%
600 °C	Chirurgical	2.02	21.67	37%	2%	2%	60%	15%
	FFP2	2.20	10.34	38%	4%	4%	54%	20%
700 °C	Chirurgical	2.96	22.46	58%	3%	0%	39%	19%
	FFP2	3.83	17.15	58%	4%	3%	35%	24%
800 °C	Chirurgical	5.27	9.27	71%	8%	1%	20%	22%
	FFP2	5.02	9.28	71%	9%	1%	18%	22%

Table A2. Average values of pyrolysis parameters in N_2 and air atmosphere for dry samples.

Temperature	Sample	η^{av} (mol/100 g_{sample})	H_2/CO	$Y^{av}_{H_2}$ (%mol N_2 Free)	Y^{av}_{CO} (%mol N_2 Free)	$Y^{av}_{CO_2}$ (%mol N_2 Free)	$Y^{av}_{CH_4}$ (%mol N_2 Free)	χ^{av}
500 °C	Chirurgical	1.74	2.10	19%	9%	23%	49%	20%
	FFP2	1.56	2.18	20%	9%	22%	49%	19%
600 °C	Chirurgical	2.51	2.88	26%	9%	23%	41%	27%
	FFP2	2.31	3.03	29%	9%	18%	44%	24%
700 °C	Chirurgical	4.90	5.28	46%	9%	18%	27%	38%
	FFP2	4.43	5.68	48%	9%	18%	26%	34%
800 °C	Chirurgical	7.16	5.91	60%	11%	13%	17%	43%
	FFP2	5.80	6.20	62%	10%	10%	17%	33%

Table A3. Average values of pyrolysis parameters in N_2 and air atmosphere for wetted samples.

Temperature	Sample	η^{av} (mol/100 g_{sample})	H_2/CO	$Y^{av}_{H_2}$ (%mol N_2 Free)	Y^{av}_{CO} (%mol N_2 Free)	$Y^{av}_{CO_2}$ (%mol N_2 Free)	Y^{av}_{CH4} (%mol N_2 Free)	χ^{av}
500 °C	Chirurgical	1.37	1.77	21%	13%	21%	45%	16%
	FFP2	1.60	1.41	22%	16%	21%	41%	18%
600 °C	Chirurgical	2.27	2.56	29%	12%	13%	46%	23%
	FFP2	2.88	1.76	26%	15%	19%	40%	32%
700 °C	Chirurgical	5.28	4.73	47%	10%	13%	30%	42%
	FFP2	5.33	4.35	44%	10%	20%	26%	44%
800 °C	Chirurgical	5.99	7.65	61%	8%	12%	19%	34%
	FFP2	4.97	5.15	55%	11%	17%	18%	34%

Appendix B

Figure A1 reports the exponential and polynomial laws that were obtained based on the experimental data from the devolatilization tests. These expressions were used as input in the Aspen Plus ®model to simulate the devolatilization step in a R-YIELD block.

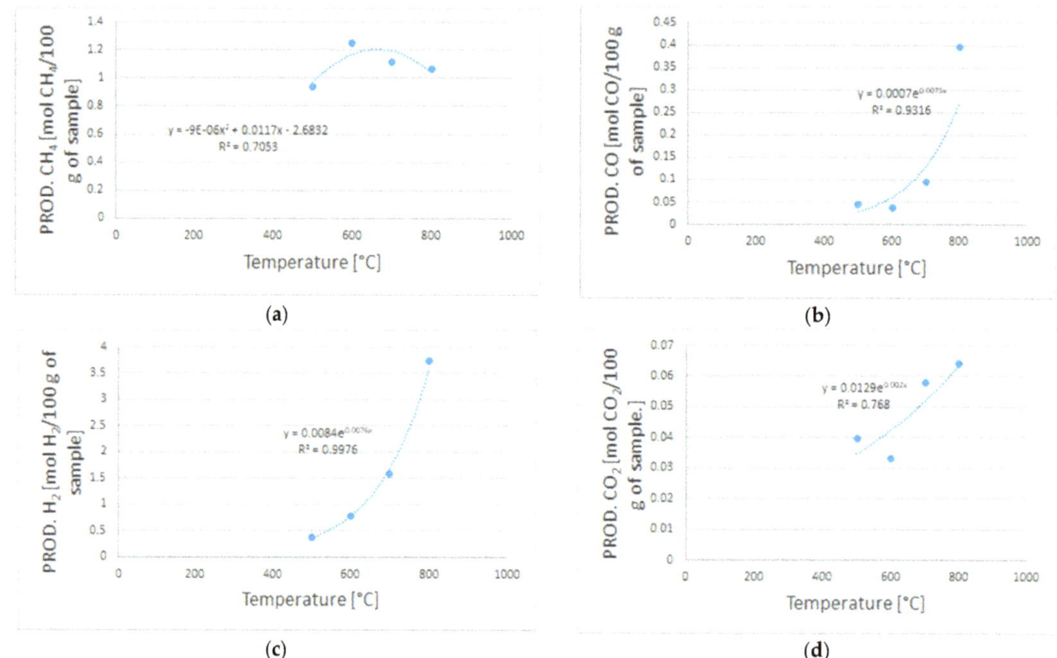

Figure A1. Polynomial equation (**a**) used to model the CH_4 productivity, and exponential equations (**b–d**) used to model the H_2, CO, and CO_2 productivity.

References

1. Malinauskaite, J.; Jouhara, H. The trilemma of waste-to-energy: A multi-purpose solution. *Energy Policy* **2019**, *129*, 636–645. [CrossRef]
2. Siwal, S.S.; Chaudhary, G.; Saini, A.K.; Kaur, H.; Saini, V.; Mokhta, S.K.; Chand, R.; Chandel, U.K.; Christie, G.; Thakur, V.K. Key ingredients and recycling strategy of personal protective equipment (PPE): Towards sustainable solution for the COVID-19 like pandemics. *J. Environ. Chem. Eng.* **2021**, *9*, 106284. [CrossRef] [PubMed]
3. Centi, G.; Perathoner, S. Chemistry and energy beyond fossil fuels. A perspective view on the role of syngas from waste sources. *Catal. Today* **2020**, *342*, 4–12. [CrossRef]
4. Sebastiani, A.; Macrì, D.; Gallucci, K.; Materazzi, M. Steam-oxygen gasification of refuse derived fuel in fluidized beds: Modelling and pilot plant testing. *Fuel Process. Technol.* **2021**, *216*, 106783. [CrossRef]
5. Siwal, S.S.; Zhang, Q.; Devi, N.; Saini, A.K.; Saini, V.; Pareek, B.; Gaidukovs, S.; Thakur, V.K. Recovery processes of sustainable energy using different biomass and wastes. *Renew. Sustain. Energy Rev.* **2021**, *150*, 111483. [CrossRef]
6. Green Energy: From Waste to Syngas Production—Instrumentation Monthly. Available online: https://www.instrumentation.co.uk/green-energy-from-waste-to-syngas-production/ (accessed on 28 June 2021).
7. De-la-Torre, G.E.; Aragaw, T.A. What we need to know about PPE associated with the COVID-19 pandemic in the marine environment. *Mar. Pollut. Bull.* **2021**, *163*, 111879. [CrossRef]
8. Parku, G.K.; Collard, F.; Görgens, J.F. Pyrolysis of waste polypropylene plastics for energy recovery: In fl uence of heating rate and vacuum conditions on composition of fuel product. *Fuel Process. Technol.* **2020**, *209*, 106522. [CrossRef]
9. Ahmad, I.; Khan, M.I.; Khan, H.; Ishaq, M.; Tariq, R.; Gul, K.; Ahmad, W. Pyrolysis study of polypropylene and polyethylene into premium oil products. *Int. J. Green Energy* **2015**, *12*, 663–671. [CrossRef]
10. Hou, Q.; Zhen, M.; Qian, H.; Nie, Y.; Bai, X.; Xia, T.; Laiq Ur Rehman, M.; Li, Q.; Ju, M. Upcycling and catalytic degradation of plastic wastes. *Cell Rep. Phys. Sci.* **2021**, *2*, 100514. [CrossRef]
11. Hou, Q.; Qi, X.; Zhen, M.; Qian, H.; Nie, Y.; Bai, C.; Zhang, S.; Bai, X.; Ju, M. Biorefinery roadmap based on catalytic production and upgrading 5-hydroxymethylfurfural. *Green Chem.* **2021**, *23*, 119–231. [CrossRef]
12. Huang, F.; Li, W.; Hou, Q.; Ju, M. Enhanced CH_4 production from corn-stalk pyrolysis using Ni-5CeO$_2$/MCM-41 as a catalyst. *Energies* **2019**, *12*, 774. [CrossRef]
13. GSTC. Waste to Energy Gasification; GSTC. Available online: https://globalsyngas.org/syngas-technology/syngas-production/waste-to-energy-gasification/ (accessed on 6 April 2021).

14. Puig-Arnavat, M.; Bruno, J.C.; Coronas, A. Review and analysis of biomass gasification models. *Renew. Sustain. Energy Rev.* **2010**, *14*, 2841–2851. [CrossRef]
15. González-Vázquez, M.P.; Rubiera, F.; Pevida, C.; Pio, D.T.; Tarelho, L.A.C. Thermodynamic analysis of biomass gasification using aspen plus: Comparison of stoichiometric and non-stoichiometric models. *Energies* **2021**, *14*, 189. [CrossRef]
16. Galvagno, A.; Prestipino, M.; Chiodo, V.; Maisano, S.; Brusca, S.; Lanzafame, R. Biomass blend effect on energy production in a co-gasification-CHP system. *AIP Conf. Proc.* **2019**, *2191*, 020082. [CrossRef]
17. Tauqir, W.; Zubair, M.; Nazir, H. Parametric analysis of a steady state equilibrium-based biomass gasification model for syngas and biochar production and heat generation. *Energy Convers. Manag.* **2019**, *199*, 111954. [CrossRef]
18. Kaushal, P.; Tyagi, R. Advanced simulation of biomass gasification in a fluidized bed reactor using ASPEN PLUS. *Renew. Energy* **2017**, *101*, 629–636. [CrossRef]
19. George, J.; Arun, P.; Muraleedharan, C. Stoichiometric Equilibrium Model Based Assessment of Hydrogen Generation through Biomass Gasification. *Procedia Technol.* **2016**, *25*, 982–989. [CrossRef]
20. De Kam, M.J.; Morey, R.V.; TIffany, D.G. Integrating biomass to produce heat and power at ethanol plants. *Appl. Eng. Agric.* **2009**, *25*, 227–244. [CrossRef]
21. Haydary, J. Aspen simulation of two-stage pyrolysis/gasification of carbon based solid waste. *Chem. Eng. Trans.* **2018**, *70*, 1033–1038. [CrossRef]
22. Khuriati, A.; Purwanto, P.; Setiyo Huboyo, H.; Suryono, S.; Bawono Putro, A. Application of aspen plus for municipal solid waste plasma gasification simulation: Case study of Jatibarang Landfill in Semarang Indonesia. *J. Phys. Conf. Ser.* **2018**, *1025*, 012006. [CrossRef]
23. Mutlu, Ö.Ç.; Zeng, T. Challenges and Opportunities of Modeling Biomass Gasification in Aspen Plus: A Review. *Chem. Eng. Technol.* **2020**, *43*, 1674–1689. [CrossRef]
24. Gagliano, A.; Nocera, F.; Bruno, M.; Cardillo, G. Development of an Equilibrium-based Model of Gasification of Biomass by Aspen Plus. *Energy Procedia* **2017**, *111*, 1010–1019. [CrossRef]
25. Salman, C.A.; Omer, C.B. Process modelling and simulation of waste gasification-based flexible polygeneration facilities for power, heat and biofuels production. *Energies* **2020**, *13*, 4264. [CrossRef]
26. Beheshti, S.M.; Ghassemi, H.; Shahsavan-Markadeh, R. Process simulation of biomass gasification in a bubbling fluidized bed reactor. *Energy Convers. Manag.* **2015**, *94*, 345–352. [CrossRef]
27. Pauls, J.H.; Mahinpey, N.; Mostafavi, E. Simulation of air-steam gasification of woody biomass in a bubbling fluidized bed using Aspen Plus: A comprehensive model including pyrolysis, hydrodynamics and tar production. *Biomass Bioenergy* **2016**, *95*, 157–166. [CrossRef]
28. Aspen Plus v 12.0. Available online: https://www.aspentech.com/en/-/media/aspentech/home/support-and-training/deployment-solutions/aspenone-v12-getting-started-guide.pdf?la=en (accessed on 8 February 2021).
29. Ffp2—All Medical Device Manufacturers. Available online: https://www.medicalexpo.com (accessed on 18 February 2022).
30. Jung, S.; Lee, S.; Dou, X.; Kwon, E.E. Valorization of disposable COVID-19 mask through the thermo-chemical process. *Chem. Eng. J.* **2021**, *405*, 126658. [CrossRef] [PubMed]
31. Abraham, J.P.; Plourde, B.D.; Cheng, L. Using heat to kill SARS-CoV-2. *Rev. Med. Virol.* **2020**, *30*, 8–10. [CrossRef]
32. Solid Recovered Fuels. Determination of Moisture Content Using The Oven Dry Method. Moisture in General Analysis Sample. 2011. Available online: https://www.iso.org/standard/71327.html (accessed on 18 February 2021).
33. The British Standards Institution Methods for Analysis and Testing of Coal and Coke (Multi-Part Document BS 1016) 2022. Available online: https://landingpage.bsigroup.com/LandingPage/Series?UPI=BS%201016 (accessed on 18 February 2022).
34. ASTM International Standard Test Method for Compositional Analysis by Thermogravimetry 2003. Available online: https://standards.globalspec.com/std/14182172/ASTM%20E1131-20 (accessed on 18 February 2021).
35. Gutierrez, A.C.G. Characterization of the Fuel Fraction of Solid Urban Solid Wastes of the Municipality of Santo André Aiming.
36. UNI CEN/TS 15439:2008 Biomass Gasification. Tar and Particles in Product Gases. Sampling and Analysis 2008. Available online: https://www.uni.com/ (accessed on 8 February 2021).
37. Gibilaro, L.G. Fluidization Dynamics. Butterworth Heinemann: Oxford, UK, 2011; ISBN 0750650036.
38. Yang, W. *Handbook of Fluidization and Fluid-Particle Systems*; CRC Press: Boca Raton, FL, USA, 2003.
39. Malsegna, B.; Di Giuliano, A.; Gallucci, K. Experimental study of absorbent hygiene product devolatilization in a bubbling fluidized bed. *Energies* **2021**, *14*, 2399. [CrossRef]
40. Ramzan, N.; Ashraf, A.; Naveed, S.; Malik, A. Simulation of hybrid biomass gasification using Aspen plus: A comparative performance analysis for food, municipal solid and poultry waste. *Biomass Bioenergy* **2011**, *35*, 3962–3969. [CrossRef]
41. Basu, P. *Biomass Gasification, Pyrolysis and Torrefaction: Practical Design and Theory*; Academic Press: Cambridge, MA, USA, 2013; ISBN 9780123964885.
42. Basu, P. *Combustion and Gasification in Fluidized Beds*; CRC Press: Boca Raton, FL, USA, 2006; ISBN 9780849333965.
43. Zaror, C.A.; Pyle, D.L. The pyrolysis of biomass: A general review. *Proc. Indian Acad. Sci. Sect. C Eng. Sci.* **1982**, *5*, 269–285. [CrossRef]
44. Aragaw, T.A. Surgical face masks as a potential source for microplastic pollution in the COVID-19 scenario. *Mar. Pollut. Bull.* **2020**, *159*, 111517. [CrossRef] [PubMed]

45. Di Giuliano, A.; Lucantonio, S.; Gallucci, K. Devolatilization of residual biomasses for chemical looping Gasification in Fluidized Beds Made up of Oxygen-Carriers. *Energies* **2021**, *14*, 311. [CrossRef]
46. Xiao, R.; Jin, B.; Zhou, H.; Zhong, Z.; Zhang, M. Air gasification of polypropylene plastic waste in fluidized bed gasifier. *Energy Convers. Manag.* **2007**, *48*, 778–786. [CrossRef]

Article

The Fractionation of Corn Stalk Components by Hydrothermal Treatment Followed by Ultrasonic Ethanol Extraction

Nianze Zhang [1,2], Chunyan Tian [1,*], Peng Fu [1], Qiaoxia Yuan [3], Yuchun Zhang [1], Zhiyu Li [1] and Weiming Yi [2]

1   School of Agricultural Engineering and Food Science, Shandong University of Technology, Zibo 255000, China; 19403010075@stumail.sdut.edu.cn (N.Z.); fupeng@sdut.edu.cn (P.F.); zhangyc@sdut.edu.cn (Y.Z.); lizhiyu@sdut.edu.cn (Z.L.)
2   Shandong Research Center of Engineering and Technology for Clean Energy, Zibo 255000, China; yiweiming@sdut.edu.cn
3   College of Engineering, Huazhong Agricultural University, Wuhan 430070, China; qxyuan@mail.hzau.edu.cn
*   Correspondence: tiancy@sdut.edu.cn

Abstract: The fractionation of components of lignocellulosic biomass is important to be able to take advantage of biomass resources. The hydrothermal–ethanol method has significant advantages for fraction separation. The first step of hydrothermal treatment can separate hemicellulose efficiently, but hydrothermal treatment affects the efficiency of ethanol treatment to delignify lignin. In this study, the efficiency of lignin removal was improved by an ultrasonic-assisted second-step ethanol treatment. The effects of ultrasonic time, ultrasonic temperature, and ultrasonic power on the ultrasonic ethanol treatment of hydrothermal straw were investigated. The separated lignin was characterized by solid product composition analysis, FT-IR, and XRD. The hydrolysate was characterized by GC-MS to investigate the advantage on the products obtained by ethanol treatment. The results showed that an appropriate sonication time (15 min) could improve the delignification efficiency. A proper sonication temperature (180 °C) can improve the lignin removal efficiency with a better retention of cellulose. However, a high sonication power 70% (840 W) favored the retention of cellulose and lignin removal.

Keywords: ultrasonic; components fractionation; lignocellulose; ethanol treatment

Citation: Zhang, N.; Tian, C.; Fu, P.; Yuan, Q.; Zhang, Y.; Li, Z.; Yi, W. The Fractionation of Corn Stalk Components by Hydrothermal Treatment Followed by Ultrasonic Ethanol Extraction. *Energies* **2022**, *15*, 2616. https://doi.org/10.3390/en15072616

Academic Editors: Andrea Di Carlo and Elisa Savuto

Received: 15 March 2022
Accepted: 30 March 2022
Published: 3 April 2022

Publisher's Note: MDPI stays neutral with regard to jurisdictional claims in published maps and institutional affiliations.

1. Introduction

Biomass use is becoming increasingly important with the growing concern related to the development of sustainable energy sources [1]. There are many types of biomass resource available, such as lignocellulosic biomass (LCB) consisting of cellulose, hemicellulose and lignin, food waste, and manure [2]. Corn stalk, a typical representative of lignocellulose, is often burned in an open environment, causing a great environmental pollution problem, and the wasting of this resource at the same time. Corn stalk is abundant in storage, accounting for about 32.46% of the total straw, and has great potential for application. Corn stalk can be converted to obtain biofuels or high value-added chemicals [3]. In recent years, the efficient conversion and utilization of corn stalk has become a research hotspot [4,5].

The complex structural features of LCB protect its carbohydrates from microbial or enzymatic degradation [6,7]. The recalcitrance of LCB stems mainly from the structural properties of the plant cell wall; it is a network mechanism formed by the cross-linking of xylans and lignin, so when utilizing LCB, it is first necessary to separate the composite tissue into each component to further enhance the value [8].

The pulp industry has developed several separation techniques that use strong acid or base reagents to separate lignin from the lignocellulosic biomass by decomposing and dissolving the lignin into the liquid [9]. However, this method of treatment only targets the utilization of cellulose and does not consider lignin and hemicellulose as resources, generating great waste. During the recovery of lignin, lignin condensation is produced, resulting in a limitation of recovered lignin, so this treatment is not applicable to the recovery of lignin.

Organic solvent treatment is a future component fractionation green technology [10]. The organic solvent treatment method allows lignin and hemicellulose to be dissolved in an aqueous organic solvent. In contrast to acid or alkali treatment, an organic solvent treatment prevents changes in the lignin structure due to its mild conditions. When treating lignocellulosic biomass by organic solvents, the lignin can be dissolved into the organic solvent solution and the cellulose is retained in the solid. Due to the aromatic nature of the lignin fragments, lignin is freely dissolved in organic solvents [11]. Organic solvent pretreatments usually include organic acids or alcohols, such as acetic acid, acetone and ethanol. Most organic solvents have low boiling points and can be easily separated. The usual conditions for organic solvent pretreatment are 1–3 h at 150 °C to obtain 60–90% lignin removal. Glucose yields in organic solvent pretreatment biomass saccharification range from 46% to 99% [12].

Already widely used in other areas, ultrasounds can be used to enhance the separation efficiency of components in lignocellulosic biomass. Ultrasound is a green technology that has the advantage of reducing energy consumption and causes less pollution [13]. Ultrasounds can break the complex structure of lignocellulose and remove hemicellulose and lignin, allowing various chemical reagents to better bind to the cellulose and facilitate the reaction. As a result, ultrasonic technology has been widely valued and developed for biomass pretreatment [14,15]. According to reports, the ultrasonic treatment of biomass increases the crystallinity of biomass and accomplishes the separation of the three major components. Cellulose is present in solid form and lignin and hemicellulose are dissolved into the solvent. However, the application of ultrasounds in LCB is still in its infancy. Nevertheless, there are relatively few studies on adding ultrasonic assistance to the process of treating biomass with organic solvents, so the application of ultrasound should be expanded.

The hydrothermal–ethanol method, which combines an organic solvent treatment and hydrothermal treatment, has gained importance [16,17]. The efficiency of the hydrothermal–ethanol method for the separation of lignin is lower than that of the organic solvent treatment. The biomass is hydrothermally treated and the lignin is removed from the cell walls, but with cooling, microspheres are formed that reattach to the cellulose surface. The hydrothermally treated biomass is then subjected to ethanol treatment, which causes this portion of the lignin microspheres to re-polymerize and form a lignin coating to attach to the cellulose surface, thus leading to a significant impact on the efficiency of lignin removal. Therefore, the focus of the study shifted to how to improve the low efficiency of lignin removal in the hydrothermal–ethanol method [18].

The characteristic point of this study is the combination of hydrothermal treatment, ethanol treatment, and ultrasonic treatment, which combines the advantages of the three treatments to achieve a stepwise separation of the three major components of LCB. By hydrothermal treatment, the hemicellulose component of corn stalk was removed. The lignin component was removed from the corn stalk by ultrasound-assisted ethanol. Finally, cellulose-rich solids were obtained. The problem of the low efficiency of lignin removal by the hydrothermal–ethanol method was solved with the aid of an ultrasound.

In this study, the goal was set to explore the problem and to improve the efficiency of lignin removal by the hydrothermal–ethanol method. This study used corn stalks as a feedstock. By studying the ultrasonic time, ultrasonic temperature, and ultrasonic power, the aim was to obtain an efficient and clean fraction separation technique. With the aid of an ultrasound, the problem of the low efficiency of lignin removal by the hydrothermal–ethanol method can be effectively solved, which is beneficial for the separation and utilization of cellulose, lignin, and hemicellulose.

2. Materials and Methods

2.1. Raw Materials and Analytical Methods

The corn stalk was obtained from Jiangsu province, China, crushed through a grinder, and then the obtained corn stalk powder was sieved through an environmentally friendly vibrating sieve machine (SDNS 300t, Hubei, China) to obtain 40–60 mesh corn stalk. The obtained corn stalk was subjected to hydrothermal treatment at 180 °C and solid–liquid separated in an autoclave (Parr 4575a, Moline, IL, USA) with water as a solvent, and the resulting solid product was dried in an electric thermostatic drying oven (WGL-858, Hebei, China) at 105 °C, which is the hydrothermal straw (HT) required for this experiment.

2.2. Ultrasonic-Assisted Ethanol Treatment of Hydrothermal Straw

The separation of the components was completed according to the hydrothermal–ethanol method [17,18]. The problem of low lignin removal efficiency was solved by attaching ultrasonic assistance to the second ethanol treatment step. Fixed factors: solids content 5 ω%, ethanol concentration 50 φ%, material as HT, pulse 2 s/8 s. The effects of ultrasonic time, ultrasonic temperature, and ultrasonic power on lignin removal efficiency were investigated in a single-factor test. The ultrasonic generator was selected from an ultrasonic processor (Jy98-222DN, Beijing, China) with 1200 W and φ20 variable amplitude rod, and the ultrasonic power was output in a percentage. The ultrasonic temperature was set at 140 °C, ultrasonic power at 40%, and the experiment was set at the ultrasonic times of 0, 15, 30, 45, 60 min. The ultrasonic time was set to 30 min, ultrasonic power was set to 40%, and the experiment was set with ultrasonic temperatures of 100, 120, 140, 160, 180 °C. The sonication time was set to 30 min, sonication temperature was set to 140 °C, and the experiment was set with sonication powers of 10, 25, 40, 55, 70%. After the treatment was completed, the mixture was filtered through a Brinell funnel, the solid was dried in a blast drying oven at 105 °C to a constant weight, and the liquid was centrifuged to obtain ethanolic lignin. Figure 1 shows the overview of the hydrothermal–ethanol treatment.

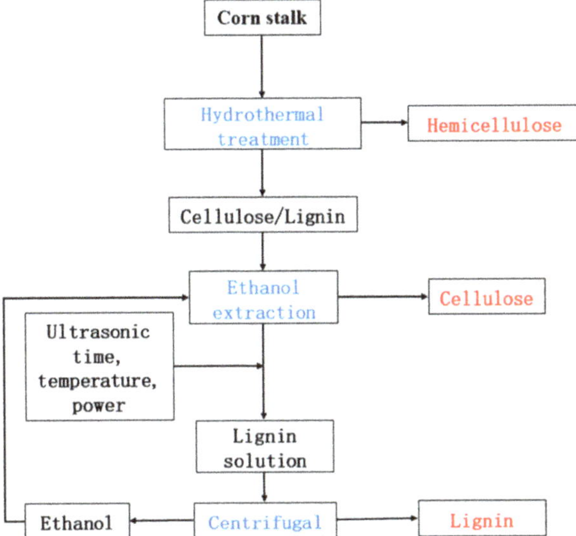

Figure 1. The overview of hydrothermal—ultrasonic ethanol treatment.

2.3. Analysis and Calculation Methods

2.3.1. Hydrolysate Composition Analysis

The compositional changes in the liquid phase products were identified using a gas chromatography–mass spectrometry (GC–MS) analyzer (Agilent 5973-6080, Santa Clara, CA, USA). The column used for the test samples was DB-1701. The aqueous phase was extracted with Dichloromethane, filtered with a 0.22 μm filter membrane, and determined on the machine. The initial temperature was 40 °C, and the temperature was increased at 5 °C/min to 240 °C for 5 min; the shunt ratio was 30:1; and the solvent was delayed for 2 min [19]. The NIST17.L mass spectrometry library was used to compare with the compounds and to select the identified components.

2.3.2. Fourier Transform Infrared Spectrometry (FT-IR) Analysis

A Fourier infrared spectrometer (Thermo Nicolet 5700, Thermo Electron, Waltham, MA, USA) was used to qualitatively analyze the functional groups and the chemical bonds of the solid products to determine the changes in the components during the separation process [20,21]. The wavelength range was 400–4000 cm^{-1} with a resolution better than 0.09 cm^{-1} and a wavenumber accuracy of 0.01 cm^{-1}.

2.3.3. Content of Cellulose, Lignin, and Hemicellulose Fractions in Solid Products

The content of cellulose, lignin, and hemicellulose fractions in solid products based on the Van Soest principle was measured using a cellulose tester (ANKOPM A220i, Macedon, NY, USA) [22–24]. The sample was boiled and treated by neutral detergent, and the undissolved residue was neutral detergent fiber, which is mainly cell wall components, including hemicellulose, cellulose, acid insoluble lignin and silicate. The sample was treated with acid detergent, and the remaining residue was acid detergent fiber, which included cellulose, acid insoluble lignin and silicate, and the difference between neutral detergent fiber and acid detergent fiber was the hemicellulose content. The residue of acid detergent fiber after 72% sulfuric acid treatment was acid insoluble lignin and silicate, and the cellulose content of the sample was obtained by subtracting the residue after 72% sulfuric acid treatment from the acid detergent fiber value. The residue after 72% sulfuric acid treatment was dried and then ashed, and the part that escaped during the ashing process was the acid-washed lignin content.

2.3.4. X-ray Diffractometry (XRD) Analysis

The fiber crystal structure in the obtained solid products was characterized by an X-ray diffractometer (D8-02, Karlsruhe, Germany) [25,26]. The dried sample was put into the test piece, the glass press was used to flatten the test piece to keep the surface flat, then the test piece was put into the sample holder. The test conditions: Diffraction angle 2 min scans were performed in the range 5–60°, angular resolution: FWHM(Full Width at Half Maxima) $\leq \pm0.1$; angular reproducibility: $\pm0.0001°$.

2.3.5. Calculation Methods of Evaluation Indicators

Calculation of yields of solid yield, cellulose retention rate, and lignin extraction rate [16]. The calculation method in the text is as follows:

$$\text{solid yield (d\%)} = \frac{\text{Solid mass}}{\text{Stalk mass}} \times 100\% \tag{1}$$

$$\text{Cellulose retention rate (d\%)} = \frac{\text{Solid cellulose mass}}{\text{Stalk cellulose mass}} \times 100\% \tag{2}$$

$$\text{Lignin extraction rate (d\%)} = 1 - \frac{\text{solid ligin mass}}{\text{Stalk ligin mass}} \times 100\% \tag{3}$$

Stalk mass is the mass of reactants, solid mass is the mass of the solid product obtained by the reaction, the solid cellulose mass is the mass of cellulose in the solid product, the stalk cellulose mass is the mass of cellulose in the stalk, the solid lignin mass is the mass of lignin in the solid product, and the stalk lignin mass is the mass of lignin in the stalk.

3. Results and Discussion

3.1. Characteristics of Raw Materials

Table 1 shows the characteristics of the corn stalk and hydrothermal stalk. From the table, the contents of the three components of the hydrothermal stalk and corn stalk can be seen.

Table 1. Characteristics of corn stalk and hydrothermal stalk.

Components	Corn Stalk	Hydrothermal Stalk
Chemical composition (daf%)		
Hemicellulose	31.31 ± 0.69	8.22 ± 0.14
Cellulose	40.13 ± 0.40	57.06 ± 1.11
Lignin	9.74 ± 0.87	23.15 ± 0.56

The number after $\pm$ denotes Standard Deviation.

The hydrothermal–ethanol method is a good method for the separation of lignocellulosic biomass fractions. The hydrothermal treatment of corn stalk released 90% of hemicellulose and had a high hemicellulose separation efficiency.

3.2. Effect of Different Conditions of Ultrasonic Ethanol Treatment on Solid Products

3.2.1. Effect of Different Times of Ultrasonic Ethanol Treatment on a Solid Product

Figure 2 shows the effect of the time factor on the three main components and yields of the solid products. Compared with the control sample, the ultrasonic ethanol treatment had a significant effect on the yield of the solid product, and the mass loss was more than 10%, which may have been partially caused by the loss of various components of the corn stalk. With the increase in the ultrasonic time, the change in the mass decreased and then increased, and reached an inflection point at 15 min of ultrasonic time. Before 15 min, the mass kept reducing with the increase in time, and the lowest point reached 87.65%. The mass kept increasing again when the time exceeded 15 min. From the experimental results, it is clear that the hydrothermal straw was treated with ethanol by ultrasonication, and many substances were precipitated from the solid, which may partly be the presence of cellulose, lignin, hemicellulose, and soluble substances present in the straw, so the composition of the solid product needs to be analyzed to elucidate the effect of ultrasonication time on the dissolution of hydrothermal stalk [2,20].

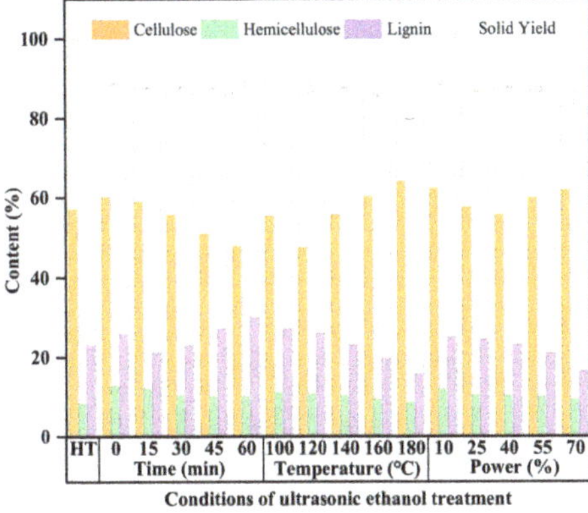

Figure 2. Effect of different ultrasonic ethanol treatments on solid product yields and fractions.

Figure 3 shows the effect of the time factor on the cellulose retention and lignin removal of the solid products. Cellulose retention and lignin removal are important indicators used to judge the effectiveness of ultrasonic ethanol treatment [27,28]. In the process of exploring the ultrasonic time on ethanol treatment, the retention of cellulose decreased continuously with the increase in time and showed a linear relationship with R2 reaching 0.9831 and the equation $y = -4.9836x + 99.414$, the retention of cellulose decreased from 93.34% at 0 min to 74.18% at 60 min. Moreover, the lignin removal efficiency was the opposite of the mass change, which rose and then decreased, the inflection point appeared at 15 min, the lignin removal efficiency was 20.20%, and with the increase in time, the lignin removal efficiency gradually decreased. At 60 min, it reached −15.08%, indicating that with the increase in time, the condensation of the removed lignin occurred, resulting in more lignin content and more lignin in the solid product of the attached lignin.

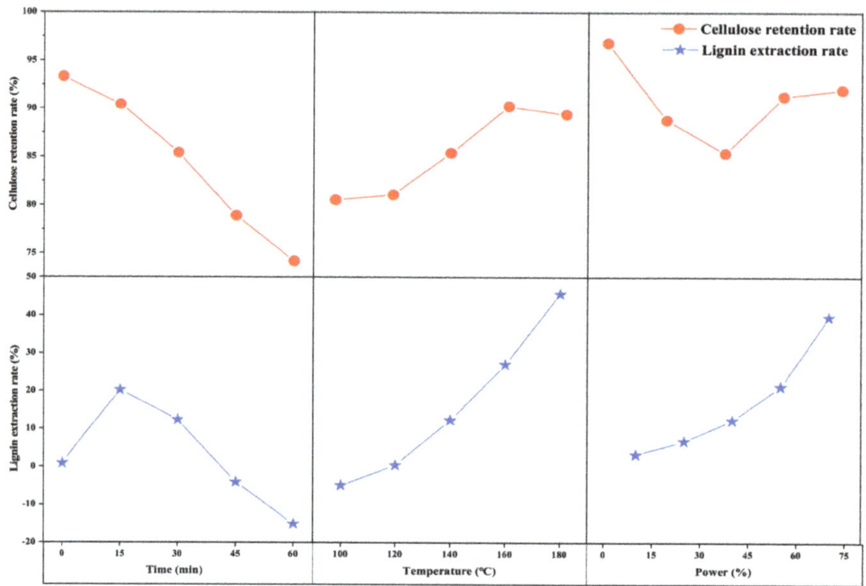

Figure 3. Lignin removal and cellulose retention of solid products.

The ultrasonic treatment destroys wax-containing and silicon-containing components deposited on the lignocellulosic surface [29,30] which is the main reason for the change in mass. Hydrothermal straw was treated with ultrasound-assisted ethanol, which allows this part of the fraction to be dissolved, as seen in the 0 min group, which was effectively removed under the action of ultrasound [15].

The effect on the loss of mass as the time increased, as can be seen from Figures 1 and 2, was due to mass changes in cellulose and lignin. When ultrasound acts on cellulose, the penetration of heat into the fiber structure is enhanced due to the sonar effect generated by ultrasound, and the high temperature and pressure provide conditions for the unbinding of the hydrogen bonds of cellulose molecules [31]. These effects cause damage to the cellulose surface and interior, and the morphological structure is disrupted so that part of the cellulose is hydrolyzed during the treatment. The presence of ultrasound can also destroy the lignin microspheres deposited on the surface of lignocellulose [16]. At the same time, it can also act internally to break the joints between lignin and cellulose, allowing the lignin to break away from the entire structure and dissolve into the organic solvent ethanol [32,33]. As the treatment time increases, the ultrasonic treatment causes condensation and the degradation of lignin, which dissolves from hydrothermal straw into the solution while releasing low molecular weight phenols that clear the free radicals released during the instantaneous collapse. This leads to the polymerization of lignin and deposition on the biomass surface, resulting in an increase in lignin mass and a negative lignin removal efficiency [34–36].

3.2.2. Effect of Ultrasonic Ethanol Treatment at Different Temperatures on a Solid Product

Figure 2 shows the effect of the temperature factor on the three main components and yields of the solid products. As the temperature increased, the yield of the solid product obtained gradually decreased. The analysis of the first three points and the linear fit revealed that R2 was equal to 0.9989, obtaining the function $y = -0.89x + 90.387$, which indicates that at low temperatures, the mass loss increases linearly with temperature, and when the temperature is higher than 140 °C, the rate of mass loss increases and more material undergoes hydrolysis and is removed from the solid, which may be part of hemicellulose, lignin, and part of cellulose [37,38].

Figure 3 shows the effect of the temperature factor on the cellulose retention and lignin removal of the solid products. The changes of each component in the hydrothermal straw with temperature changes can be better analyzed [39]. The cellulose retention showed rose and then fell, the inflection point was at 160 °C, the retention rate reached the highest at 90.29%, and the lignin removal rate was a linear rising trend. When the point of 100 °C was excluded, a linear fit to the remaining temperature points could be obtained as a linear function $y = 15.032x - 6.176$. R^2 was equal to 0.9901, which is consistent with the linear law, indicating that the lignin removal rate was linearly and positively correlated with the temperature. This indicates that after hydrothermal treatment, hemicellulose was removed from the straw and the three components of lignocellulose caused the dense structure to be destroyed. Cellulose can only be hydrolyzed at low temperatures. With the increase in temperature, the lignin removal rate increased, the detached lignin in solution was constantly polymerized, attached, and removed, and the increase in lignin molecules and the increase in the active degree of cellulose played a good protective role. In the investigation of the effect of temperature on the hydrothermal straw fraction, the loss of mass is directly related to the rate of lignin removal [40,41].

3.2.3. Effect of the Different Powers of Ultrasonic Ethanol Treatment on a Solid Product

Figure 2 shows the effect of the ultrasonic power factor on the three main components and yields of the solid products. As the ultrasound power increases, the quality of the solid product obtained after treatment constantly decreases. The increase in ultrasonic power and the cavitation bubbles lead to the fragmentation of the cell wall and an increase in the permeability of the solvent, which leads to a continuous loss of mass quality [42].

Figure 3 shows the effect of the ultrasonic power factor on the cellulose retention and lignin removal rate of the solid products. As can be seen in Figure 2, the lignin removal rate increased with the increase in ultrasonic power, and the efficiency of lignin removal rate increased, indicating that the ultrasonic power had a significant effect on the release of lignin. The cellulose retention rate decreased and then increased with the increase in power, and the cellulose retention rate reached the lowest value at 40% power. This is mainly due to the cavitation effect of ultrasound, which leads to the breaking of aryl ether bonds in lignin, and the breaking of hydrogen bonds between lignin and cellulose [43,44]. This promotes the release and dissolution of more lignin into the organic solvent. Thus, an increase in ultrasonic power can enhance the cavitation effect, which leads to an increase in the lignin removal efficiency [14,45].

3.3. Effect of Different Conditions of Ultrasonic Ethanol Treatment on the Properties of Solid Products

The FT-IR spectra of the solid products obtained with different sonication factors are shown in Figures 4–6. Their FT-IR spectra are very similar and agree with reports found in the literature [46–49]. The infrared absorption wave numbers of cellulose OH and C-O are 3200~3400 and ~1050 cm^{-1}, and those of hemicellulose C=O (from ketone and carboxyl groups) and C-C (from sugar skeleton) are 1715 ~ 1765 and 615 cm^{-1}, respectively; lignin is rich in O-CH$_3$ (from methoxy), C-O-C (from the aromatic-alcohol ester bond), C=C (from the aromatic ring) and C-OH (from phenolic hydroxyl group), whose wave numbers are located at 1430~1270, 1270, 1450 and 1613, 1215 cm^{-1}.

For the solid products treated with different ultrasonic factors of ethanol, the solid products were compared according to the characteristic peaks of infrared spectra. It was found that the characteristic peaks of hemicellulose were undetectable, indicating the removal effect of hemicellulose in the treatment process. In addition, comparing the characteristic peaks of cellulose and lignin, the signals of these two components were obvious, which, combined with the retention rate of cellulose and the removal rate of lignin, proved that cellulose and lignin were still present in the solid products obtained after the ethanol treatment with ultrasound.

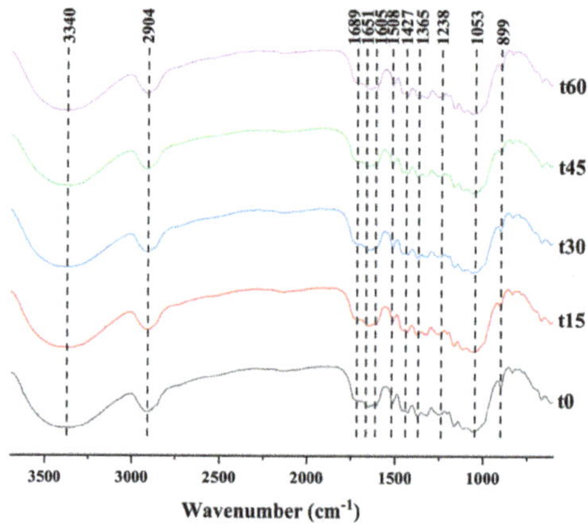

Figure 4. Infrared spectra of corn stalk treated with different ultrasound times.

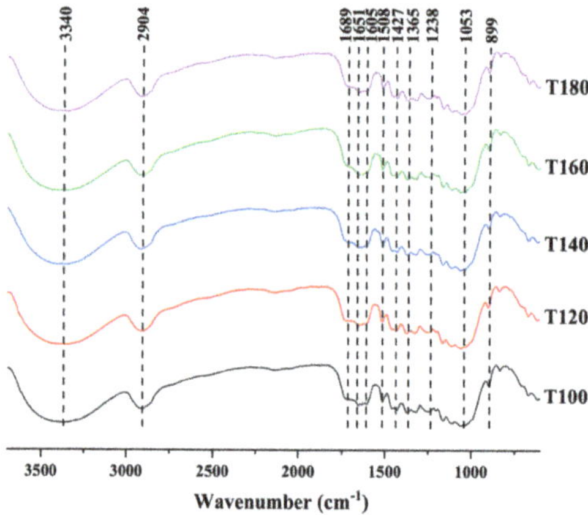

Figure 5. Infrared spectra of corn stalk treated with different ultrasound temperatures.

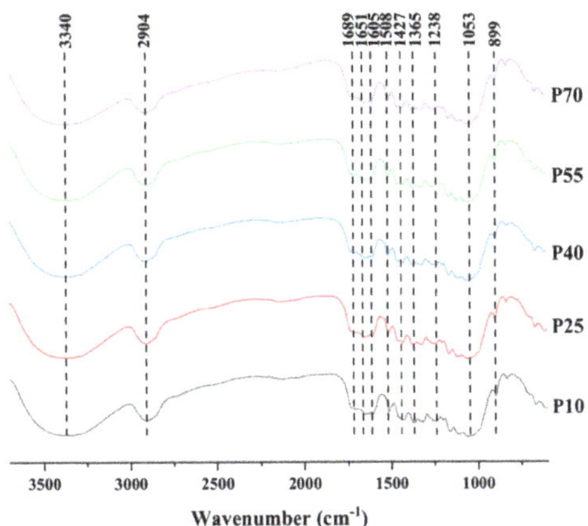

Figure 6. Infrared spectra of corn stalk treated with different ultrasonic power.

The solid products obtained were analyzed by XRD and the results of the calculated cellulose crystallinity are listed in Table 2. From Table 2, we can see that after sonication, the crystallinity of type I and type II cellulose changed. With the increase in sonication time, type I cellulose decreased, then increased, and then decreased, and the inflection points existed at 15 min and 45 min, with a maximum value of 40.43% and a minimum value of 34.06%. Type II cellulose rose and then fell, and the inflection point existed at 45 min, reaching a maximum value of 44.06%. With the increase in the ultrasonic temperature, type I cellulose decreased and then increased, the inflection point appeared at 120 °C, reached a minimum value of 29.49%, and then with the increase in temperature until 180 °C, reached a maximum value of 40.38%. Type II cellulose decreased and then increased, with the inflection point appearing at 100 °C, reaching a minimum value of 30.70%, and then reaching a maximum value of 42.67% with the increase in temperature up to 180 °C. With the increase in ultrasonic power, type I cellulose rose and then fell, and then rose and then decreased, reaching an inflection point at 25%, reaching a maximum value of 37.28% at 55% power, and a minimum value of 28.92% at 70% power. Type II cellulose showed the same trend as type I cellulose, reaching a maximum value of 39.93% at 55% power and a minimum value of 35.03% at 70% power.

Table 2. Crystallinity of solid products.

		CrI-I	CrI-II
	HT	35.54	37.64
	0	34.47	38.22
Time	15	34.06	39.05
(min)	30	36.48	39.79
	45	40.43	44.06
	60	39.93	39.56
	100	30.01	30.70
	120	29.49	32.24
Temperature	140	36.48	39.79
(°C)	160	38.05	41.16
	180	40.39	42.67
	10	35.54	38.62
	25	33.44	37.74
Power	40	36.48	39.79
(%)	55	37.28	39.94
	70	28.92	35.03

With the enhancement of ultrasonic conditions, type I cellulose transforms to type II cellulose, and cellulose crystallinity increases, which is mainly due to the removal of lignin and hemicellulose from lignocellulose and the hydrolysis of the amorphous zone. Ultrasonic waves promote the removal of lignin and hemicellulose, which makes the cellulose content increase relatively, thus leading to the increase in crystallinity [49,50]. When it reaches 45 min, the crystallinity starts to decrease, which is mainly due to the destruction of cellulose crystals caused by the ultrasonic waves on the hydrogen bonds between cellulose molecules, in addition to the re-polymerization of lignin, which leads to the reduction in cellulose content, both of which lead to the decrease in crystallinity [51].

3.4. Effect of Different Conditions of Ultrasonic Ethanol Treatment on the Characteristics of Liquid Products

The composition of the liquid products obtained by different ultrasonic ethanol treatments was analyzed, and the specific composition distribution is listed in Table 3. For the liquid products obtained by ultrasonic ethanol treatment, excluding the two added solvents of water and ethanol, the composition of other components was the same. Ethyl formate was the main product, which is mainly the hemicellulose and cellulose hydrolysis of five-carbon sugar and six-carbon sugar. After the reaction, they were first converted to furfural, then furfural in the ethanol H^+ ionized in the process of treatment into furfuryl alcohol, and hydrolysis occurred under the catalytic effect of H^+ to obtain ethyl formate [52,53].

Table 3. GC-MS composition analysis of ultrasonic ethanol-treated liquid products.

RT (min)		3.85	4.74	5.77	7.37	27.62
		Formic acid, ethenyl ester $C_3H_4O_2$	Ethyl formate $C_3H_6O_2$	Ethyl Acetate $C_4H_8O_2$	1,1-diethoxy-Ethane, $C_6H_{14}O_2$	2-methyl-Benzaldehyde, C_8H_8O
Time (min)	0	4.65	87.58	1.14	3.16	3.48
	15	6.38	88.62	1.23	3.77	-
	30	5.90	90.69	-	3.41	-
	45	4.72	85.47	1.28	3.69	4.84
	60	5.50	88.33	0.92	4.11	1.13
Temperature (°C)	100	3.69	92.56	0.93	2.82	-
	120	3.54	93.28	1.11	2.07	-
	140	5.90	90.69	-	3.41	-
	160	4.97	85.58	1.82	5.34	2.29
	180	20.96	46.26	3.66	19.18	9.95
Power (%)	10	19.28	58.37	12.19	10.16	-
	25	36.00	30.77	20.37	12.87	-
	40	5.90	90.69	-	3.41	-
	55	8.87	74.86	8.50	7.76	-
	70	30.25	22.17	15.45	7.78	24.35

3.5. Characterization of Ethanol Lignin

Figures 7–9 show the fingerprint regions of the lignin FT-IR spectra. According to the literature, the plots were analyzed to obtain the types of various characteristic peaks [54–58]. The peaks at 1601, 1509, and 1423 cm^{-1} belong to the benzene ring vibration of the benzene propane skeleton; the absorption peak at 1367 cm^{-1} demonstrates the presence of S-type and condensed cross-linked G-type lignin structural units; 1238 cm^{-1} is characteristic of G-type lignin structural units; 1329 cm^{-1} is characteristic of S-type lignin structural units; the 1121 cm^{-1} peak indicates the C-H bond deformation vibration in S-type lignin; and 1033 cm^{-1} indicates C-H bond deformation vibration in G-type lignin. The 832 cm^{-1} peak is the out-of-plane bending in $C_{2,6}$ of the S-type lignin unit. The C=O stretching vibration was at 1720 cm^{-1}, the conjugated aryl carbonyl group stretching at was at 1657 cm^{-1}, the C-H deformation and methoxy bending was at 1455 cm^{-1}, and the methyl acetate

C-H stretching was at 1367 cm^{-1}. The FT-IR spectra were very similar, and the relative intensities of the characteristic peaks were almost unchanged, indicating similar changes in the chemical structure of the lignin obtained by the ultrasonic ethanol treatment.

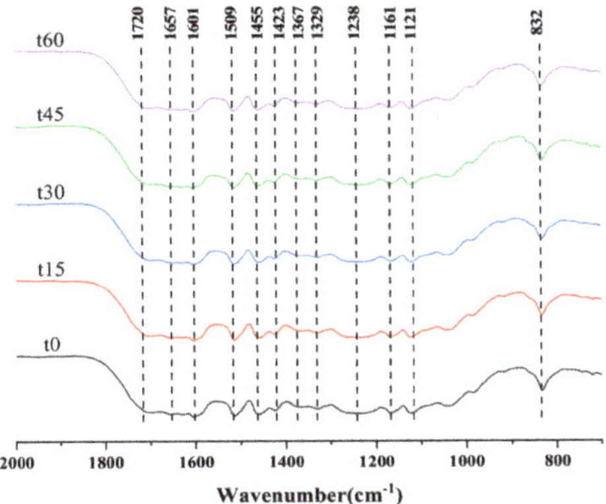

Figure 7. Infrared spectra of lignin obtained by the different sonication time treatments.

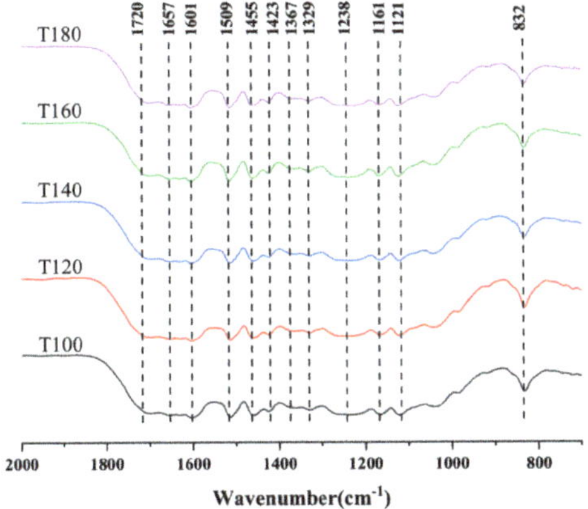

Figure 8. Infrared spectra of lignin obtained by different ultrasonic temperature treatments.

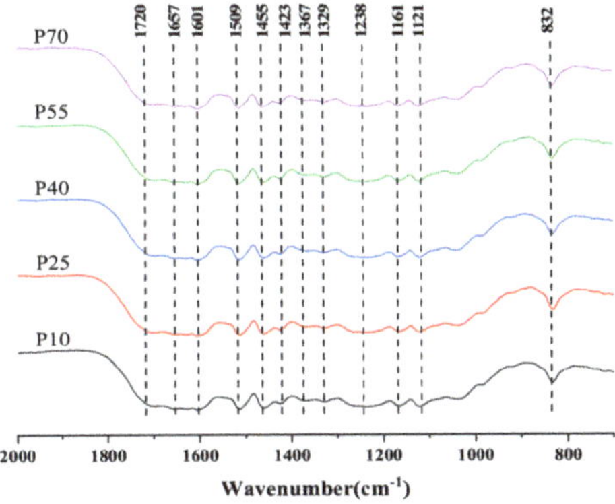

Figure 9. Infrared spectra of lignin obtained by different ultrasonic power treatments.

3.6. Hydrothermal–Ultrasonic Ethanol Method for Mass Flow Analysis

In order to investigate the role of ultrasound in the hydrothermal–ethanol method, the mass flow of the three major components of hydrothermal straw was analyzed. Figure 10a shows the hydrothermal–ethanol method, with the experimental conditions of 180 °C and 30 min, and Figure 10b shows the hydrothermal–ultrasonic ethanol method with the experimental conditions of 180 °C, 40%, and 30 min. Figure 10a,b have the same temperature and time; the only difference is the influence of ultrasound. It can be seen in the Figures that the retention of cellulose, and the removal of lignin can be favorably increased in the solid product obtained with the effect of ultrasound. When compared with the hydrothermal–ethanol method, the cellulose content of 51.03% was much greater than that of 42.27% in the hydrothermal–ultrasonic ethanol method, and the lignin content of 12.58% was lower than that of 14.84% in the hydrothermal–ethanol method.

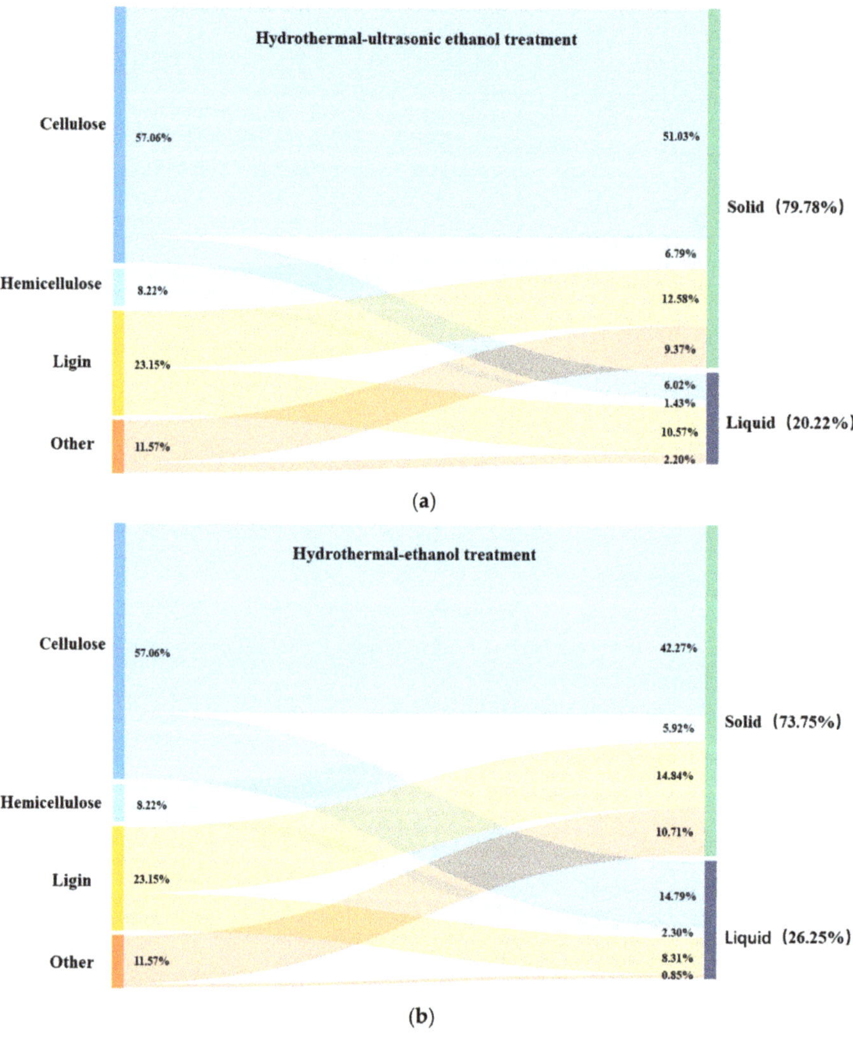

Figure 10. Diagram of the three-component mass flow of hydrothermal straw.

In the process of ultrasound-assisted hydrothermal–ethanol treatment, most of the cellulose was retained and existed in the solid, and only a small portion of small vimentin was hydrolyzed and converted into monosaccharides and furfuryl alcohol, and then hydrolyzed to obtain ethyl formate under the catalytic effect of H^+. A part of hemicellulose was also present in the solid, which was mainly stabilized by hydrogen bonding with cellulose, leading to difficulty in detachment; thus, it remained in the solid product. A portion of lignin fused to ethanol to produce ethanolic lignin, and a portion of lignin continued to participate in the solid product, which required higher processing conditions for the purpose of removal. Ultrasonic ethanol treatment can better retain the cellulose fraction and remove the lignin fraction.

4. Conclusions

In this study, ultrasound was found to enhance the efficiency of the hydrothermal–ethanol method for the separation of lignocellulosic biomass fractions. The hydrothermal–ultrasonic ethanol method can effectively improve cellulose retention and lignin removal, and the analysis of the obtained organic lignin revealed that there was little effect on the lignin structure. The effects of ultrasonic time, ultrasonic temperature, and ultrasonic power on lignin were also investigated, and a comprehensive analysis of the factors yielded the optimal process as follows: the optimal sonication time was 15 min, the optimal sonication temperature was 180 °C, and the sonication power was 70% (840 W). A longer sonication time will be accompanied by the aggregation of lignin and affect the lignin removal rate. A higher sonication temperature will reduce the retention of cellulose. However, a high sonication power is beneficial for cellulose retention and lignin removal.

Author Contributions: Conceptualization: C.T., analysis and investigation: N.Z.; writing and preparation of manuscript: N.Z., C.T.; funding acquisition: C.T., Q.Y., P.F., Y.Z., Z.L. and W.Y. All authors have read and agreed to the published version of the manuscript.

Funding: This research was supported by the National Natural Science Foundation of China (51706126, and 31872400), National key research and development program of China (2019YFD1100600), Shandong University of Technology and Zibo City Integration Development Project (2019ZBXC380), and the Youth Innovation Support Program of Shandong Colleges and Universities (2019KJD013).

Institutional Review Board Statement: Not applicable.

Informed Consent Statement: Not applicable.

Data Availability Statement: No new data were created or analyzed in this study. Data sharing is not applicable to this article.

Acknowledgments: We gratefully thank to the Analysis and Testing Center in Shandong University of Technology for XRD and FT–IR analyses and Haifeng Zhao.

Conflicts of Interest: The authors declare no conflict of interest.

References

1. Bhatia, S.K.; Jagtap, S.S.; Bedekar, A.A.; Bhatia, R.K.; Patel, A.K.; Pant, D.; Rajesh Banu, J.; Rao, C.V.; Kim, Y.; Yang, Y. Recent developments in pretreatment technologies on lignocellulosic biomass: Effect of key parameters, technological improvements, and challenges. *Bioresour. Technol.* **2020**, *300*, 122724. [CrossRef] [PubMed]
2. Haldar, D.; Purkait, M.K. A review on the environment-friendly emerging techniques for pretreatment of lignocellulosic biomass: Mechanistic insight and advancements. *Chemosphere* **2021**, *264*, 128523. [CrossRef] [PubMed]
3. Kumar, B.; Bhardwaj, N.; Agrawal, K.; Chaturvedi, V.; Verma, P. Current perspective on pretreatment technologies using lignocellulosic biomass: An emerging biorefinery concept. *Fuel Process. Technol.* **2020**, *199*, 106244. [CrossRef]
4. Soltanian, S.; Aghbashlo, M.; Almasi, F.; Hosseinzadeh-Bandbafha, H.; Nizami, A.; Ok, Y.S.; Lam, S.S.; Tabatabaei, M. A critical review of the effects of pretreatment methods on the exergetic aspects of lignocellulosic biofuels. *Energy Convers. Manag.* **2020**, *212*, 112792. [CrossRef]
5. Vu, H.P.; Nguyen, L.N.; Vu, M.T.; Johir MA, H.; Mclaughlan, R.; Nghiem, L.D. A comprehensive review on the framework to valorise lignocellulosic biomass as biorefinery feedstocks. *Sci. Total Environ.* **2020**, *743*, 140630. [CrossRef] [PubMed]

6. Alonso, D.M.; Hakim, S.H.; Zhou, S.; Won, W.; Hosseinaei, O.; Tao, J.; Garcia-Negron, V.; Motagamwala, A.H.; Mellmer, M.A.; Huang, K.; et al. Increasing the revenue from lignocellulosic biomass: Maximizing feedstock utilization. *Sci. Adv.* **2017**, *3*, 1603301. [CrossRef]

7. Sankaran, R.; Parra Cruz, R.A.; Pakalapati, H.; Show, P.L.; Ling, T.C.; Chen, W.; Tao, Y. Recent advances in the pretreatment of microalgal and lignocellulosic biomass: A comprehensive review. *Bioresour. Technol.* **2020**, *298*, 122476. [CrossRef]

8. Sheng, Y.; Lam, S.S.; Wu, Y.; Ge, S.; Wu, J.; Cai, L.; Huang, Z.; Le, Q.V.; Sonne, C.; Xia, C. Enzymatic conversion of pretreated lignocellulosic biomass: A review on influence of structural changes of lignin. *Bioresour. Technol.* **2021**, *324*, 124631. [CrossRef]

9. Hoang, A.T.; Nizetic, S.; Ong, H.C.; Chong, C.T.; Atabani, A.E.; Pham, V.V. Acid-based lignocellulosic biomass biorefinery for bioenergy production: Advantages, application constraints, and perspectives. *J. Environ. Manag.* **2021**, *296*, 113194. [CrossRef]

10. Ab Rasid, N.S.; Shamjuddin, A.; Abdul Rahman, A.Z.; Amin NA, S. Recent advances in green pre-treatment methods of lignocellulosic biomass for enhanced biofuel production. *J. Clean. Prod.* **2021**, *321*, 129038. [CrossRef]

11. Haldar, D.; Purkait, M.K. Lignocellulosic conversion into value-added products: A review. *Process Biochem.* **2020**, *89*, 110–133. [CrossRef]

12. Meng, X.; Bhagia, S.; Wang, Y.; Zhou, Y.; Pu, Y.; Dunlap, J.R.; Shuai, L.; Ragauskas, A.J.; Yoo, C.G. Effects of the advanced organosolv pretreatment strategies on structural properties of woody biomass. *Ind. Crops Prod.* **2020**, *146*, 112144. [CrossRef]

13. Ong, V.Z.; Wu, T.Y. An application of ultrasonication in lignocellulosic biomass valorisation into bio-energy and bio-based products. *Renew. Sustain. Energy Rev.* **2020**, *132*, 109924. [CrossRef]

14. Huerta, R.R.; Silva, E.K.; Ekaette, I.; El-Bialy, T.; Saldaña MD, A. High-intensity ultrasound-assisted formation of cellulose nanofiber scaffold with low and high lignin content and their cytocompatibility with gingival fibroblast cells. *Ultrason. Sonochem.* **2020**, *64*, 104759. [CrossRef]

15. Bundhoo, Z.M.A.; Mohee, R. Ultrasound-assisted biological conversion of biomass and waste materials to biofuels: A review. *Ultrason. Sonochem.* **2018**, *40*, 298–313. [CrossRef]

16. Li, J.; Feng, P.; Xiu, H.; Zhang, M.; Li, J.; Du, M.; Zhang, X.; Kozliak, E.; Ji, Y. Wheat straw components fractionation, with efficient delignification, by hydrothermal treatment followed by facilitated ethanol extraction. *Bioresour. Technol.* **2020**, *316*, 123882. [CrossRef]

17. Li, J.; Feng, P.; Xiu, H.; Li, J.; Yang, X.; Ma, F.; Li, X.; Zhang, X.; Kozliak, E.; Ji, Y. Morphological changes of lignin during separation of wheat straw components by the hydrothermal-ethanol method. *Bioresour. Technol.* **2019**, *294*, 122157. [CrossRef]

18. Choi, J.; Park, S.; Kim, J.; Cho, S.; Jang, S.; Hong, C.; Choi, I. Selective deconstruction of hemicellulose and lignin with producing derivatives by sequential pretreatment process for biorefining concept. *Bioresour. Technol.* **2019**, *291*, 121913. [CrossRef]

19. Yin, S.; Zhang, N.; Tian, C.; Yi, W.; Yuan, Q.; Fu, P.; Zhang, Y.; Li, Z. Effect of Accumulative Recycling of Aqueous Phase on the Properties of Hydrothermal Degradation of Dry Biomass and Bio-Crude Oil Formation. *Energies* **2021**, *14*, 285. [CrossRef]

20. Guo, X.; Liu, S.; Wang, Z.; Zhang, G. Ultrasonic-assisted extraction of polysaccharide from Dendrobium officinale: Kinetics, thermodynamics and optimization. *Biochem. Eng. J.* **2022**, *177*, 108227. [CrossRef]

21. Akhlisah, Z.N.; Yunus, R.; Abidin, Z.Z.; Lim, B.Y.; Kania, D. Pretreatment methods for an effective conversion of oil palm biomass into sugars and high-value chemicals. *Biomass Bioenergy* **2021**, *144*, 105901. [CrossRef]

22. Xu, J.; Zhang, S.; Shi, Y.; Zhang, P.; Huang, D.; Lin, C.; Wu, Y. Upgrading the wood vinegar prepared from the pyrolysis of biomass wastes by hydrothermal pretreatment. *Energy* **2021**, *244*, 122631. [CrossRef]

23. Silva, A.C.; Barbosa, M.S.; Barral, U.M.; Silva BP, C.; Fernandes JS, C.; Viana AJ, S.; Filho CV, M.; Bispo DF, A.; Christófaro, C.; Ragonezi, C.; et al. Organic matter composition and paleoclimatic changes in tropical mountain peatlands currently under grasslands and forest clusters. *CATENA* **2019**, *180*, 69–82. [CrossRef]

24. Xu, Q.; Zhao, M.; Yu, Z.; Yin, J.; Li, G.; Zhen, M.; Zhang, Q. Enhancing enzymatic hydrolysis of corn cob, corn stalk and sorghum stalk by dilute aqueous ammonia combined with ultrasonic pretreatment. *Ind. Crops Prod.* **2017**, *109*, 220–226. [CrossRef]

25. Manzato, L.; Rabelo LC, A.; De Souza, S.M.; Da Silva, C.G.; Sanches, E.A.; Rabelo, D.; Mariuba LA, M.; Simonsen, J. New approach for extraction of cellulose from tucumã's endocarp and its structural characterization. *J. Mol. Struct.* **2017**, *1143*, 229–234. [CrossRef]

26. Rizwan, M.; Gilani, S.R.; Durrani, A.I.; Naseem, S. Cellulose extraction of Alstonia scholaris: A comparative study on efficiency of different bleaching reagents for its isolation and characterization. *Int. J. Biol. Macromol.* **2021**, *191*, 964–972. [CrossRef]

27. Chen, X.; Li, H.; Sun, S.; Cao, X.; Sun, R. Co-production of oligosaccharides and fermentable sugar from wheat straw by hydrothermal pretreatment combined with alkaline ethanol extraction. *Ind. Crops Prod.* **2018**, *111*, 78–85. [CrossRef]

28. Rao, M.V.; Sengar, A.S.; CK, S.; Rawson, A. Ultrasonication—A green technology extraction technique for spices: A review. *Trends Food Sci. Technol.* **2021**, *116*, 975–991. [CrossRef]

29. Muthuvelu, K.S.; Rajarathinam, R.; Kanagaraj, L.P.; Ranganathan, R.V.; Dhanasekaran, K.; Manickam, N.K. Evaluation and characterization of novel sources of sustainable lignocellulosic residues for bioethanol production using ultrasound-assisted alkaline pre-treatment. *Waste Manag.* **2019**, *87*, 368–374. [CrossRef] [PubMed]

30. Luo, J.; Fang, Z.; Smith, R.L. Ultrasound-enhanced conversion of biomass to biofuels. *Prog. Energy Combust. Sci.* **2014**, *41*, 56–93. [CrossRef]

31. Louis AC, F.; Venkatachalam, S. Energy efficient process for valorization of corn cob as a source for nanocrystalline cellulose and hemicellulose production. *Int. J. Biol. Macromol.* **2020**, *163*, 260–269. [CrossRef]

32. Koutsianitis, D.; Mitani, C.; Giagli, K.; Tsalagkas, D.; Halász, K.; Kolonics, O.; Gallis, C.; Csóka, L. Properties of ultrasound extracted bicomponent lignocellulose thin films. *Ultrason. Sonochem.* **2015**, *23*, 148–155. [CrossRef]
33. Wang, Q.; Niu, L.; Jiao, J.; Guo, N.; Zang, Y.; Gai, Q.; Fu, Y. Degradation of lignin in birch sawdust treated by a novel Myrothecium verrucaria coupled with ultrasound assistance. *Bioresour. Technol.* **2017**, *244*, 969–974. [CrossRef]
34. Wang, S.; Li, F.; Zhang, P.; Jin, S.; Tao, X.; Tang, X.; Ye, J.; Nabi, M.; Wang, H. Ultrasound assisted alkaline pretreatment to enhance enzymatic saccharification of grass clipping. *Energy Convers. Manag.* **2017**, *149*, 409–415. [CrossRef]
35. Singh, S.; Agarwal, M.; Bhatt, A.; Goyal, A.; Moholkar, V.S. Ultrasound enhanced enzymatic hydrolysis of Parthenium hysterophorus: A mechanistic investigation. *Bioresour. Technol.* **2015**, *192*, 636–645. [CrossRef]
36. Bussemaker, M.J.; Xu, F.; Zhang, D. Manipulation of ultrasonic effects on lignocellulose by varying the frequency, particle size, loading and stirring. *Bioresour. Technol.* **2013**, *148*, 15–23. [CrossRef]
37. Zhou, Z.; Liu, D.; Zhao, X. Conversion of lignocellulose to biofuels and chemicals via sugar platform: An updated review on chemistry and mechanisms of acid hydrolysis of lignocellulose. *Renew. Sustain. Energy Rev.* **2021**, *146*, 111169. [CrossRef]
38. Zhou, M.; Tian, X. Development of different pretreatments and related technologies for efficient biomass conversion of lignocellulose. *Int. J. Biol. Macromol.* **2022**, *10*, 2741. [CrossRef]
39. Xu, Y.; Li, M. Hydrothermal liquefaction of lignocellulose for value-added products: Mechanism, parameter and production application. *Bioresour. Technol.* **2021**, *342*, 126035. [CrossRef]
40. Sun, Q.; Chen, W.; Pang, B.; Sun, Z.; Lam, S.S.; Sonne, C.; Yuan, T. Ultrastructural change in lignocellulosic biomass during hydrothermal pretreatment. *Bioresour. Technol.* **2021**, *341*, 125807. [CrossRef]
41. Sarker, T.R.; Pattnaik, F.; Nanda, S.; Dalai, A.K.; Meda, V.; Naik, S. Hydrothermal pretreatment technologies for lignocellulosic biomass: A review of steam explosion and subcritical water hydrolysis. *Chemosphere* **2021**, *284*, 131372. [CrossRef]
42. Shchukin, D.G.; Skorb, E.; Belova, V.; Möhwald, H. Ultrasonic Cavitation at Solid Surfaces. *Adv. Mater.* **2011**, *23*, 1922–1934. [CrossRef]
43. García, A.; Erdocia, X.; González Alriols, M.; Labidi, J. Effect of ultrasound treatment on the physicochemical properties of alkaline lignin. *Chem. Eng. Process. Process Intensif.* **2012**, *62*, 150–158. [CrossRef]
44. Sun, R.; Tomkinson, J. Comparative study of lignins isolated by alkali and ultrasound-assisted alkali extractions from wheat straw. *Ultrason. Sonochem.* **2002**, *9*, 85–93. [CrossRef]
45. García, A.; Alriols, M.G.; Llano-Ponte, R.; Labidi, J. Ultrasound-assisted fractionation of the lignocellulosic material. *Bioresour. Technol.* **2011**, *102*, 6326–6330. [CrossRef]
46. Liu, X.; Jiang, Z.; Feng, S.; Zhang, H.; Li, J.; Hu, C. Catalytic depolymerization of organosolv lignin to phenolic monomers and low molecular weight oligomers. *Fuel* **2019**, *244*, 247–257. [CrossRef]
47. Wang, K.; Wang, B.; Hu, R.; Zhao, X.; Li, H.; Zhou, G.; Song, L.; Wu, A. Characterization of hemicelluloses in Phyllostachys edulis (moso bamboo) culm during xylogenesis. *Carbohydr. Polym.* **2019**, *221*, 127–136. [CrossRef]
48. Choi, J.; Jang, S.; Kim, J.; Park, S.; Kim, J.; Jeong, H.; Kim, H.; Choi, I. Simultaneous production of glucose, furfural, and ethanol organosolv lignin for total utilization of high recalcitrant biomass by organosolv pretreatment. *Renew. Energy* **2019**, *130*, 952–960. [CrossRef]
49. Kawamata, Y.; Yoshikawa, T.; Aoki, H.; Koyama, Y.; Nakasaka, Y.; Yoshida, M.; Masuda, T. Kinetic analysis of delignification of cedar wood during organosolv treatment with a two-phase solvent using the unreacted-core model. *Chem. Eng. J.* **2019**, *368*, 71–78. [CrossRef]
50. Utekar, P.G.; Kininge, M.M.; Gogate, P.R. Intensification of delignification and enzymatic hydrolysis of orange peel waste using ultrasound for enhanced fermentable sugar production. *Chem. Eng. Process. Process Intensif.* **2021**, *168*, 108556. [CrossRef]
51. Wu, H.; Dai, X.; Zhou, S.L.; Gan, Y.Y.; Xiong, Z.Y.; Qin, Y.H.; Ma, J.; Yang, L.; Wu, Z.K.; Wang, T.L. Ultrasound-assisted alkaline pretreatment for enhancing the enzymatic hydrolysis of rice straw by using the heat energy dissipated from ultrasonication. *Bioresour. Technol.* **2017**, *241*, 70–74. [CrossRef] [PubMed]
52. Mongkolpichayarak, I.; Jiraroj, D.; Anutrasakda, W.; Ngamcharussrivichai, C.; Samec JS, M.; Tungasmita, D.N. Cr/MCM-22 catalyst for the synthesis of levulinic acid from green hydrothermolysis of renewable biomass resources. *J. Catal.* **2022**, *405*, 373–384. [CrossRef]
53. Fang, W.; Riisager, A. Efficient valorization of biomass-derived furfural to fuel bio-additive over aluminum phosphate. *Appl. Catal. B Environ.* **2021**, *298*, 120575. [CrossRef]
54. El Hage, R.; Brosse, N.; Chrusciel, L.; Sanchez, C.; Sannigrahi, P.; Ragauskas, A. Characterization of milled wood lignin and ethanol organosolv lignin from miscanthus. *Polym. Degrad. Stab.* **2009**, *94*, 1632–1638. [CrossRef]
55. Liu, L.; Patankar, S.C.; Chandra, R.P.; Sathitsuksanoh, N.; Saddler, J.N.; Renneckar, S. Valorization of Bark Using Ethanol–Water Organosolv Treatment: Isolation and Characterization of Crude Lignin. *ACS Sustain. Chem. Eng.* **2020**, *8*, 4745–4754. [CrossRef]
56. Cheng, X.; Guo, X.; Qin, Z.; Wang, X.; Liu, H.; Liu, Y. Structural features and antioxidant activities of Chinese quince (Chaenomeles sinensis) fruits lignin during auto-catalyzed ethanol organosolv pretreatment. *Int. J. Biol. Macromol.* **2020**, *164*, 4348–4358. [CrossRef] [PubMed]
57. Zevallos Torres, L.A.; Lorenci Woiciechowski, A.; De Andrade Tanobe, V.O.; Karp, S.G.; Guimarães Lorenci, L.C.; Faulds, C.; Soccol, C.R. Lignin as a potential source of high-added value compounds: A review. *J. Clean. Prod.* **2020**, *263*, 121499. [CrossRef]
58. Sheng, Y.; Ma, Z.; Wang, X.; Han, Y. Ethanol organosolv lignin from different agricultural residues: Toward basic structural units and antioxidant activity. *Food Chem.* **2022**, *376*, 131895. [CrossRef]

Article

Preliminary Results of Biomass Gasification Obtained at Pilot Scale with an Innovative 100 kWth Dual Bubbling Fluidized Bed Gasifier

Andrea Di Carlo [1,*], Elisa Savuto [1], Pier Ugo Foscolo [1], Alessandro Antonio Papa [1], Alessandra Tacconi [1], Luca Del Zotto [2], Bora Aydin [3] and Enrico Bocci [4]

[1] Industrial Engineering Department, University of L'Aquila, Piazzale E. Pontieri 1, Monteluco di Roio, 67100 L'Aquila, Italy; elisa.savuto@univaq.it (E.S.); pierugo.foscolo@univaq.it (P.U.F.); alessandroantonio.papa@univaq.it (A.A.P.); alessandra.tacconi@graduate.univaq.it (A.T.)

[2] CREAT, Centro di Ricerca su Energia, Ambiente e Territorio, Università Telematica eCampus, 22060 Novedrate, Italy; luca.delzotto@uniecampus.it

[3] Walter Tosto SpA, Via Erasmo Piaggio 62, 66100 Chieti, Italy; b.aydin@waltertosto.it

[4] Department of Nuclear, Subnuclear and Radiation Physics, Marconi University, 00193 Rome, Italy; e.bocci@unimarconi.it

* Correspondence: andrea.dicarlo1@univaq.it

Abstract: Biomass gasification is a favourable process to produce a H_2-rich fuel gas from biogenic waste materials. In particular, the dual bubbling fluidized bed (DBFB) technology consists of the separation of the combustion chamber, fed with air, from the gasification chamber, fed with steam, allowing to obtain a concentrated syngas stream without N_2 dilution. In a previous work, an innovative design of a DBFB reactor was developed and its hydrodynamics tested in a cold model; in this work, the novel gasifier was realized at pilot scale (100 kWth) and operated for preliminary biomass gasification tests. The results showed a high-quality syngas, composed of H_2 = 35%, CO = 23%, CO_2 = 20%, and CH_4 = 11%, as a confirmation of the design efficacy in the separation of the reaction chambers. The dry gas yield obtained was 1.33 Nm^3/kg of biomass feedstock and the carbon conversion was 73%. Tars were sampled and measured both in the raw syngas, giving a content of 12 g/Nm^3, and downstream from a traditional conditioning system composed of a cyclone and a water scrubber, showing a residual tar content of 3 g/Nm^3, mainly toluene. The preliminary tests showed promising results; further gasification tests are foreseen to optimize the main process parameters.

Keywords: biomass gasification; dual bubbling bed gasifier; innovative pilot scale gasifier; H_2-rich syngas

Citation: Di Carlo, A.; Savuto, E.; Foscolo, P.U.; Papa, A.A.; Tacconi, A.; Del Zotto, L.; Aydin, B.; Bocci, E. Preliminary Results of Biomass Gasification Obtained at Pilot Scale with an Innovative 100 kWth Dual Bubbling Fluidized Bed Gasifier. *Energies* 2022, 15, 4369. https://doi.org/10.3390/en15124369

Academic Editors: Artur Blaszczuk and Albert Ratner

Received: 6 May 2022
Accepted: 14 June 2022
Published: 15 June 2022

Publisher's Note: MDPI stays neutral with regard to jurisdictional claims in published maps and institutional affiliations.

1. Introduction

Gasification of a biogenic solid fuel is a thermo-chemical process able to convert biomass into a fuel product gas. Dual fluidized bed (DFB) steam gasification is a consolidated process to produce a hydrogen-rich syngas from biomass wastes [1–6] suitable for the synthesis of chemicals and liquid biofuels or for high efficiency power generation, using, for example, high-temperature fuel cells.

The major goals of these technologies are: (i) to maintain the temperature of the endothermic steam gasification process higher than 1073 K to obtain a high conversion efficiency, and (ii) to obtain a product gas almost free of nitrogen, while using air to burn char. This is ensured by the heat transport from an exothermic combustion chamber to the gasification chamber. The contact wall between them usually does not provide enough surface area to transfer indirectly (i.e., without a simultaneous mass transfer) to the gasifier the (very high) amount of heat required there, when the temperature gap with the combustor is limited.

In practical applications, the heat is mainly transferred by circulation of the granular solid bed material [7]. As demonstrated by Hofbauer et al. [8], this circulation is sufficient to transport heat from the combustor to the steam gasifier, with reference to a dual fluidized bed composed of two separated reactors (combustor and gasifier), thus without additional heat transfer through a contact wall. In their system, a circulation rate of 50 kg bed material per 1 kg dry wood chips feedstock is necessary, with a temperature difference below 50 K between the gasification zone (at 1073 K) and the combustor.

Kuramoto et al. [9] studied a different reactor geometry: the two reaction chambers are housed in a single vessel allowing for a contact wall surface area, which contributes to improve the heat transfer rate. In fact, the heat could be transferred indirectly through the contact wall surface, as well as directly by the solid circulation, reducing the intensity of the latter one required for steady state operation.

In order to counteract gas leakages between the two reaction chambers, through the interconnecting solid loops, siphons/loop seals must be considered and properly designed. For instance, if substantial N_2 leaks occur from the combustor to the gasifier, the product syngas would be diluted, undermining the effort in the realization of the DFB system.

A suggested way to evaluate these critical aspects is to perform tests at laboratory scale with reactor models able to faithfully reproduce the fluid-dynamic behaviour of the real DFB gasifier (cold flow models). Di Carlo et al. [10] realized a cold model of an innovative, pilot-scale dual bubbling fluidized bed gasifier (up to 100 kWth as biomass input), within the activity of the Research Project HBF 2.0, funded by the Italian Ministry of Economic Development [11]. In that work, Di Carlo et al., investigated bed material circulation between gasification and combustion chambers using Lagrangian Particle Tracking. In addition, the gas leakage between gasification and combustion chambers was evaluated by means of tracer gas injection and analysis. The results showed that, with this new reactor concept, the bed material circulation rate can be high enough to assure effective heat exchange between the two chambers, as required by an allothermal gasification process, with negligible gas leakages.

The aim of this work is to show the preliminary results obtained during gasification tests carried out with a pilot-scale gasifier, by the application of this new concept. The activities presented in this work are part of the Research Project HBF 2.0, funded by the Italian Ministry of Economic Development [11]. The main objective of the project is to integrate the new dual fluidized bed gasifier tested in this work with cold gas cleaning to feed a CHP based on an internal combustion engine to produce 25 kWe and 55 kWth. The next steps will be to integrate the dual fluidized bed with hot gas cleaning and conditioning, to convert tar and remove particulates at a temperature as high as the gasification temperature. With this solution it will be possible to produce a clean syngas, at high temperature, which could be exploited (i) to produce high-efficiency electricity by means of high-temperature fuel cells, (ii) for the catalytic synthesis of biofuel (Bio-SNG) or biochemicals, and (iii) for the production of pure hydrogen

2. Materials and Methods

2.1. HBF 2.0 Gasifier Concept

The new reactor concept consists of two concentric fluidized beds of olivine sand particles, arranged inside a single vessel (see Figure 1):

1. The gasification zone (external cylinder) fluidized by steam.
2. The combustion zone (internal cylinder) fluidized by air.

The two chambers are connected with orifices at a given vertical distance to allow bed material circulation. The circulation system considered here is based on the application of the physical concept originally studied experimentally by Kuramoto et al. [9].

The two fluidized beds operate at different temperatures and superficial velocities (us):

i.   The fast fluidized bed (combustor) at T~1173 K and us = 5–10 umf.
ii. The slow fluidized bed (gasifier) at T~1073 K and us = 2–3 umf.

The biomass is fed in the slow fluidized bed (gasifier).

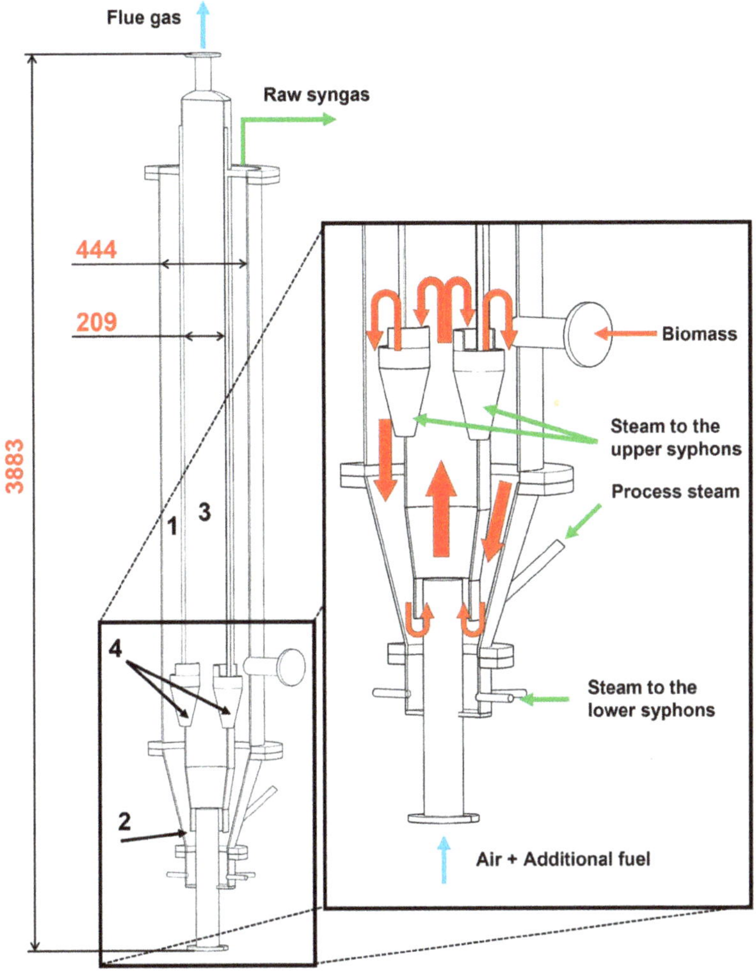

Figure 1. A 3D sketch of the new dual bubbling fluidized bed gasifier: (1) gasification chamber; (2) lower siphon; (3) combustion chamber; (4) upper siphon.

The different fluidization conditions imply different bed densities (i.e., pressure gradients) and a solid circulation between the two fluidized beds. As a result, heat is exchanged between the combustor and the gasifier, due to the circulation of bed material established within the device: (i) olivine sand and residual char in the slow fluidized bed (gasifier) flow into the fast bed through the lower orifice and (ii) hot sand is recycled back into the slow fluidized bed through the upper orifice. Combustion of char and additional fuel taking place in the fast fluidized bed supplies the heat to be transported to the gasification chamber. To avoid flue gas leakages, loop seals fluidized with steam are included in the reactor design (see Figure 1).

The main novelties of this design are:

- The system compactness, which makes it suitable for small-scale applications (0.1–1 MW as biomass input), due to integration of both reaction chambers (gasification and combustion) in the same cylindrical body;

- The heat exchange between the two chambers occurs through bed material circulation and additionally by conduction/convection through the lateral wall of the internal cylinder;
- The higher temperature chamber (combustor), operating at 1173–1223 K, is thermally well insulated; this reduces the drawback of thermal losses in small scale applications;
- Longer residence time in the combustor (bubbling bed) improves the combustion efficiency of char particles.

The main difference between this new concept and the well-known dual fluidized bed gasifier concepts such as that developed by Battelle Columbus [12], or the Fast Internal Circulating Fluidized Bed developed by Hofbauer et al. [8], is that in this case the two reactors are both bubbling bed (for the reason explained above) and they are arranged in a single vessel with the aim to reduce the thermal losses that occur in the small-scale applications.

2.2. Pilot Plant

The experience with the cold flow model [10] was used to realize the 100 kWth gasification pilot plant. The preliminary test and results reported in this work were obtained with a state-of-art configuration of the gasification installation, composed mainly of the gasifier, a cyclone for abatement of particulate content of the product gas, and a water scrubber for tar removal. The so-purified gaseous fuel is available for heat and power cogeneration by means of an internal combustion engine. The flowsheet and a picture of the pilot plant are shown in Figures 2 and 3, respectively.

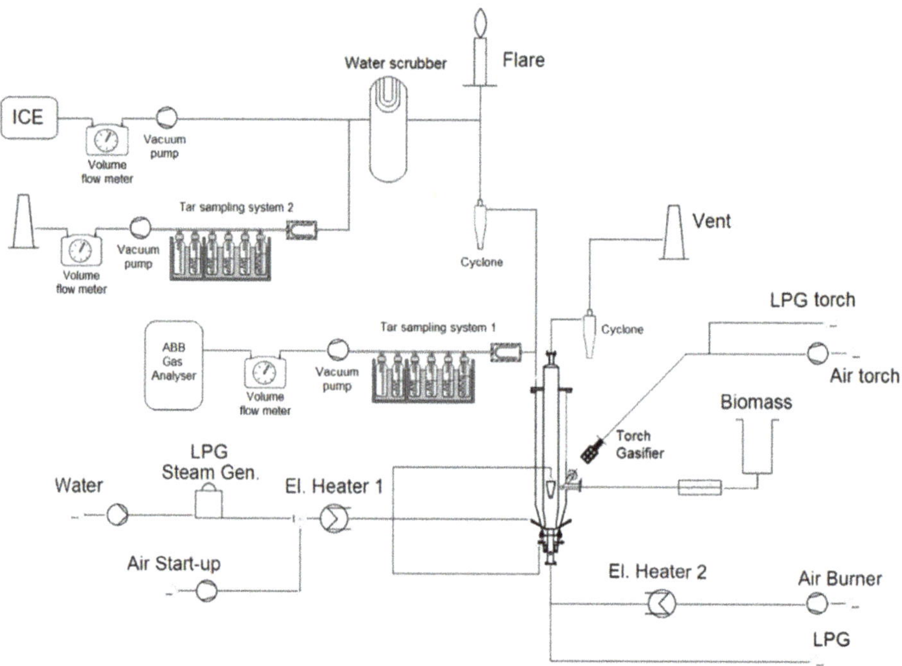

Figure 2. Flowsheet of the gasification pilot plant.

Figure 3. Picture of pilot plant.

The biomass feeding system is composed of a hopper and a two screws conveyor. Air is fed to the combustor by a blower and during the start-up phase it is also fed to the gasification zone. The incoming air is pre-heated to 200 °C by means of an electric heater before entering the reactors. Steam is produced by an LPG steam generator and overheated up to 200 °C by an electrical heater. The raw gas from the gasifier passes through a cyclone, to reduce the particulate content, and then is fed to one of the following two lines. Through the former stream, the gas goes directly to the flare for complete combustion, before being evacuated to the chimney. In the latter stream, the gas passes through a water scrubber equipped with a bundle of heat exchange tubes to help the residual steam condensation in order to remove most of the tars and water. A liquid ring vacuum pump is used to suck the entire gas flow from this stream. The volumetric flow of the cold gas exiting the liquid ring pump is finally measured and then the gas is sent to the flare. This line is used for a limited period, after steady state operation of the gasifier is established, to measure the fuel gas yield. The flue gas from the combustor is depulverized in an additional cyclone and then sent to the chimney.

A slipstream of the fuel gas, approximately 1% of the total flow (0.18 Nm^3/h), is sampled for the analysis of syngas composition and tar content. The slipstream is sucked through a porous element to separate the solid particles, and then it flows through a series of impingement bottles filled with 2-propanol to collect tar, according to the technical specification dictated by protocol CEN/TS 15439. The collected tar samples are analysed offline with a Gas Chromatograph with Mass Spectrometer (GC-MS) in order to identify and quantify tar compounds, and to calculate the tar content in the syngas. The tar concentration is calculated as the ratio between the mass of tar sampled to the total volume of dry gas that passes throw the impingement bottles. The clean, dry gas downstream from the impingement bottles is finally analysed by means of an online ABB analyser for the measurement of H_2, CO, CO_2, and CH_4 volume fractions. Gas bags are also collected for the offline GC measurement of H_2S, lighter hydrocarbons, and other permanent gases (i.e., N_2). Two of these sampling systems are present in the plant, one just downstream from the gasifier and another downstream from the water scrubber, in order to analyse both the raw syngas and the syngas after the conditioning system.

In the start-up procedure, the gasifier (and the combustor) fluidized bed is heated up by means of the incoming hot air (200 °C) and a 20 kW start-up LPG burner placed at the gasifier bed surface. When the bed temperature reaches approximately 300 °C, biomass feeding is started, together with its combustion in the gasification zone, until the desired gasification temperature is reached. The whole warm-up procedure lasts about 4 h.

An LPG injection system was also installed into the combustor, to allow changing the temperature level of the system without varying other operating parameters. With this installation, parametric studies can be carried out easily.

Temperatures, pressures at different points of the pilot plant, and the composition of the product gas were measured and recorded continuously.

3. Results and Discussion

3.1. Analysis of Materials

The bed material is composed of 90 kg of olivine sand supplied by Nuova Cives Srl with a particle average diameter $d_{3,2}$ = 557 μm. The particle size distribution (PSD) of the fresh batch of olivine inserted in the reactor was obtained by a Malvern Mastersizer 2000 laser diffraction particle size analyser. This PSD is shown in Figure 4.

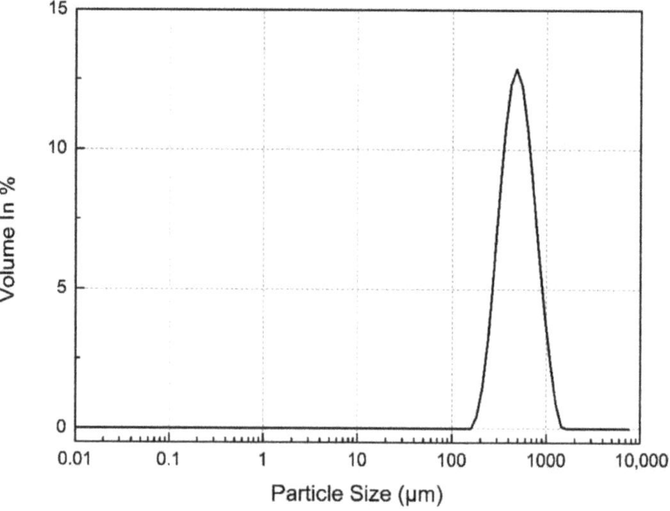

Figure 4. Particle size distribution of olivine.

The Sauter mean diameter of particles, defined as the diameter of a hypothetical particle having the same averaged specific surface area per unit volume as that of the given sample, was calculated by the following equation:

$$d_{3,2} = \frac{1}{\left(\sum_{i=1}^{n} \frac{x_i}{d_i} \right)} \tag{1}$$

x_i is the volume fraction of particles with an average diameter equal to d_i. Both x_i and d_i are known from the particle size distribution.

The biomass used for the tests was hazelnut shells. The proximate and ultimate analysis is shown in Table 1.

Table 1. Proximate and ultimate analyses of hazelnut shells (moisture content = 8.00%.).

Ash (wt%$_{db}$)	1.2
Volatile matter (wt%$_{db}$)	75.5
Fixed Carbon (wt%$_{db}$)	23.2
C (wt%$_{db}$)	50.96
H (wt%$_{db}$)	5.72
N (wt%$_{db}$)	0.42
S (wt%$_{db}$)	0.03
O (wt%$_{db}$)	41.60
HHV (MJ/kg)$_{db}$	19.93
LHV (MJ/kg)$_{db}$	18.12

Table 2 shows a list of the protocols adopted for the analyses of Table 1.

Table 2. Methods of reference for the characterizations of the hazelnut shells.

Characterization	Parameter	Ref. Method
Humidity	Amount of water in the as-received sample	UNI EN 14774-1 (ASTM E203)
Proximate Analysis	Ash Content	UNI EN 14775-TAPPI Standard T 211 om-93
	Volatile Matter (VM) Fixed Carbon (FC)	UNI EN 15148, mod. ASTM modif. D3175
Ultimate Analysis	Elemental analysis (C, H, N, O) Sulphur (S), Chlorine (Cl)	UNI EN 15104 UNI EN 15289
Calorific value	Higher Heating Value (HHV) Lower Heating Value (LHV)	UNI EN 14918, ISO 1928 DIN 51900–TAPPI Test T684

3.2. Gasification Tests

Figure 5 shows the temperatures measured during a typical start-up phase and test.

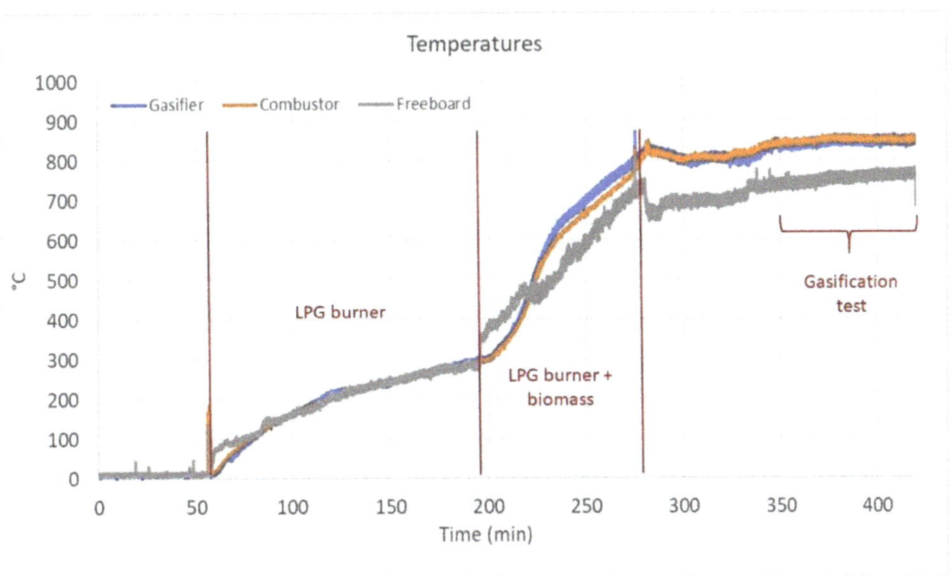

Figure 5. Temperature values measured during the start-up phase and during the gasification test.

As previously mentioned, the heating up of the reactor is carried out with a flow rate of pre-heated air injected to both chambers and by means of a 20 kW LPG burner facing the gasifier bed surface; the corresponding temperature profiles are reported in Figure 5 between minute 60 and 200, when the gasifier, combustor and freeboard temperatures increase simultaneously up to 300 °C. In this phase, the heating rate is approximately 2 °C/min.

Once the temperature level of 300 °C was reached, biomass is also added to the gasifier, in order to contribute to the temperature increase by combustion. Biomass is fed at a flow rate of 10 kg/h, corresponding to about 50 kW of thermal power input. This phase is reported in Figure 5, between minute 200 and 280, when it is possible to observe a considerably higher slope of the temperature increase versus time, compared to the previous phase. In this phase, the heating rate is about 6.5 °C/min. While the temperature in the reactor increases, the air flow rate to the gasifier is gradually decreased, in order to reduce the production rate of heat required for the reactor pre-heating, however, still keeping the bed in bubbling fluidization regime.

When the temperature of approximately 830 °C is reached, the LPG burner in the gasifier is turned off, the biomass flow is stopped, and the input gas to the gasifier is switched from air to steam, keeping the air input flow to the combustion chamber. Both flow rates are set according to the operating conditions chosen for the test. After a few minutes for the stabilization of the steam flow, the biomass is introduced again in the gasifier at the desired mass flow rate and the gasification test is started.

A steady flow of LPG is also introduced in the combustion reactor chamber as auxiliary fuel, in addition to the recirculated char, in order to produce the whole heat flow necessary to sustain the endothermic gasification reactions.

Figure 5 shows that, during the gasification test starting approximately at minute 350, the temperatures are approximately constant, meaning that the flow of auxiliary fuel (LPG) and the recirculated char in the combustor are enough to supply the heat necessary for the endothermic gasification reactions. The stable temperature regime in the reactor assures that, with these process parameters, the gasification is auto-thermal.

Furthermore, the average temperature in the upper part of the freeboard is about 750 °C, which is only 100 °C lower than the temperature measured in the gasifier bed; this confirms that the thermal insulation around the reactor vessel is efficient. As shown in Figure 5, the freeboard temperature seems to keep increasing slightly during the gasification test; it is foreseen that a higher temperature in the freeboard could be reached in a longer test run. This could improve further the quality of the syngas by enhancing the conversion of tars in the freeboard.

Biomass gasification tests were carried out with the input conditions reported in Table 3.

Table 3. Input conditions for gasification test.

Biomass feeding rate (kg/h)	15
Steam—central (kg/h)	10
Steam—lower orifice (kg/h)	2.5
Steam—upper orifice (kg/h)	1.5
Air combustor (kg/h)	39
LPG combustor (L/min)	10

The raw product gas composition in terms of H_2, CO, CO_2, and CH_4 volume fractions was measured online during the test. Detected values are reported in Figure 6 for a typical gasification test.

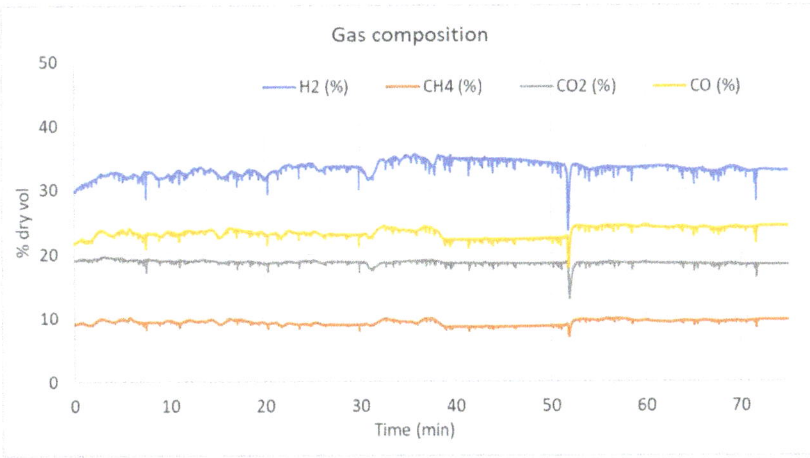

Figure 6. Raw fuel gas composition measured online during gasification test period.

Figure 6 shows that the gas composition is approximately constant during the 75 min of the test, meaning that the gasification process reached the steady state. H_2 content in the fuel gas is around 35%.

The results of the gasification tests were evaluated also in terms of some performance parameters, such as:

- The dry gas yield, η_{gas}, calculated as:

$$\eta_{gas} = \frac{\dot{Q}_{syngas,dry}}{\dot{m}_{bio}} \left(\frac{Nm^3}{kg_{bio}} \right) \tag{2}$$

- The CGE % (Cold Gas Efficiency), calculated as:

$$CGE\ (\%) = \frac{\dot{Q}_{dry,syngas}\ LHV_{syngas}}{\left(\dot{m}_{bio}LHV_{bio} + \dot{m}_{LPG}LHV_{LPG} \right)} 100 \tag{3}$$

- The carbon conversion, calculated as:

$$C\ conversion\ (\%) = \frac{\Sigma_{i=CO,\ CO2,\ CH4,}\ mol_i}{mol_{C\ bio}} \tag{4}$$

The main outputs of the gasification tests carried out in different days are summarised in Table 4.

The data reported in Table 4 show that the volume fraction of N_2 in the syngas is maximum 10%, so it can be deduced that the gas leakage from the combustor to the gasifier is minimum, and the separation of the gas streams between the two chambers is efficient. The relatively high value of the syngas LHV confirms this result.

The sum of the gas volume fractions detected is 95.6%, meaning that only approximately 4% of the gas composition is unknown, probably consisting of light hydrocarbons (C2–C6). These results are comparable with those obtained with similar gasification technologies as described in the work of Larsson et al. [13]. In fact, depending on the gasification technologies. the data points from operation with woody biomass and olivine for six gasifiers (GoBiGas, Senden, Chalmers, Gussing, Oberwart, Villach) show that the hydrogen concentration varies between 30 to 45%, the CO between 15 to 30%, the CO_2 between 20 to 30%, and the CH_4 between 5 to 12%.

Table 4. Results of the gasification test.

T gasifier (°C)	810–846
T combustor (°C)	810–872
Duration (min)	75
η_{gas} (Nm^3/kg)	1.33 ± 0.11
H_2 (%vol dry)	34.8 ± 1.0
CO (%vol dry)	23.2 ± 1.2
CO_2 (%vol dry)	19.7 ± 1.2
CH_4 (%vol dry)	10.6 ± 0.7
N_2 (%vol dry)	7.3 ± 2.3
H_2S (ppm dry)	55–120
Tar content (g/Nm^3)	12.4 ± 6.0
LHV (MJ/Nm^3)	10.5 ± 0.4
CGE (%)	65.8 ± 1.2
C conversion (%)	72.8± 2.5

There are relevant differences instead comparing the results of this work with those of the Milena gasifier [14]. The CH_4 and CO content in the syngas of the Milena gasifier is slightly higher (12% and 29%), while H_2 is lower (25%). These differences can be explained by the low S/B ratio used in the test of the Milena gasifier (0.28): the carbon is only partially gasified at such a low S/B ratio.

The measurement of the syngas flow rate allowed to calculate the η_{gas} obtained in the test, 1.33 Nm^3/kg, which is very similar to the values obtained in non-catalytic biomass gasification tests on laboratory scale [15].

The H_2S content measured in the syngas is of the order of 100 ppm, which is in the same order of magnitude of the values reported in literature for woody biomass [16,17].

The analysis of the tar samples collected downstream from the gasifier during the test gave as a result 12.4 g/Nm^3 on dry basis, coherently with the level of 10 g/Nm^3 generally expected for fluidized bed gasifiers [18,19]. Tar content of the raw syngas is too high for direct use in downstream systems for the production of energy; however, tars can be significantly reduced with conditioning systems.

3.3. Gas Conditioning Systems

A water scrubber was used as the conditioning system for the product gas.

Inside the scrubber, the temperature of the gas is strongly reduced thanks to fresh water flowing in the bundle of tubes; in this way, the humidity content in the gas condensates and generates water droplets that remove tar from the syngas stream.

The tar content at the outlet of the gasifier, obtained in the test described in the previous section, was compared with the tar content obtained downstream the water scrubber. The comparison of tar contents is reported in Figure 7.

It is possible to observe that the conditioning system led to a considerable reduction in tar content, compared to values measured at the exit of the gasifier (12.4 g/Nm^3). The syngas sampled downstream of the water scrubber showed a tar content of 3.2 g/Nm^3, resulting in an efficiency of more than 74% of the scrubber in tar removal. The residual tars are mainly toluene, which is a lighter hydrocarbon; the residual tar composition is compatible with the use of the syngas in the internal combustion engine (ICE). In fact, as reported in [18], some of the "light tars" (such as benzene and toluene) are not considered as harmful components in the gas, since these compounds are found in important quantities in gasoline.

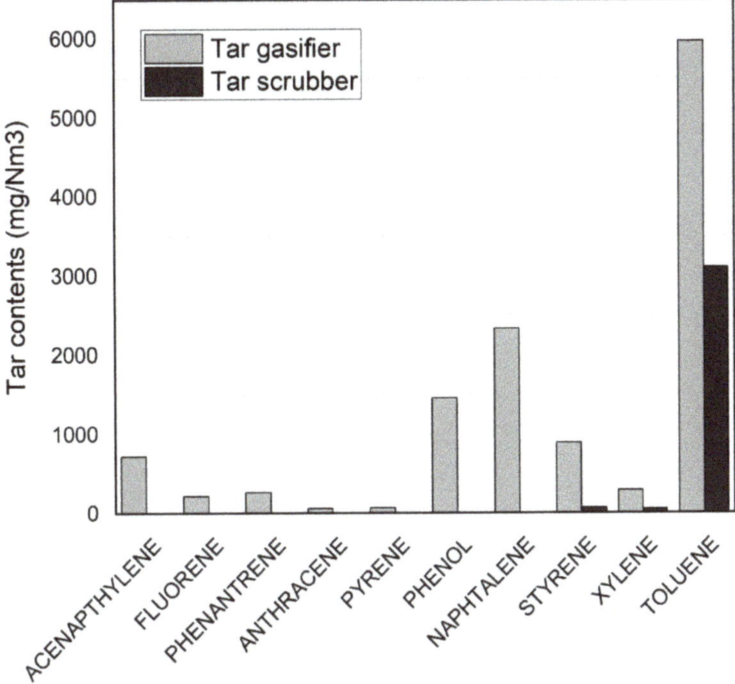

Figure 7. Tar contents measured in the raw syngas and downstream from the water scrubber.

3.4. Mass and Energy Balances

In order to check the reliability of the results obtained, the mass and energy balances were calculated for the gasification test. The ash produced was estimated based on the biomass analysis reported in Table 1. The char produced, which is circulated from the gasifier to the combustor, was assumed as 10% of the input biomass, as reported in [8]. The H_2O conversion was estimated around 30% based on the experimental findings reported in [8]. It was assumed that the nitrogen in the syngas is due to air leakage from the combustor. The flow rate of air was estimated from the knowledge of nitrogen found in the syngas.

The results of mass balance on the gasification chamber and energy balance on the whole dual-bed reactor are reported in Figures 8 and 9, respectively.

Figure 8 shows that the mass balance on the gasifier is closed with a very small error (around 4%), with the above assumptions dictated by previous experimental evidence [8]. This error can be further reduced if the light hydrocarbons would be also considered in the mass balance. These light hydrocarbons are certainly in the syngas, with a concentration between 2 to 5%, as reported in the work of Larsson et al. [13]. It is possible to observe that humid syngas represents the most abundant of the outputs, accounting for 93% of the total products. Tar and ash account for less than 1% of the total products; char is 5.5% of the total output of the gasifier and it is assumed as a mass and energy input for the combustion reactor.

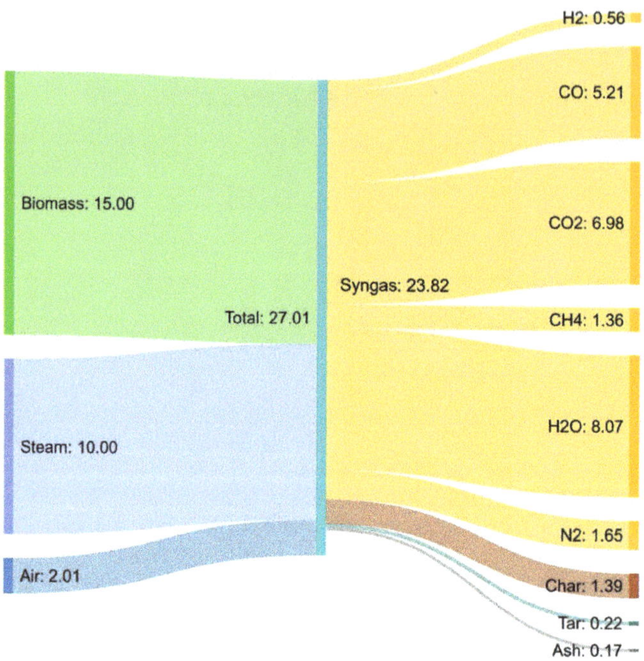

Figure 8. Gasifier mass balance (units in kg/h).

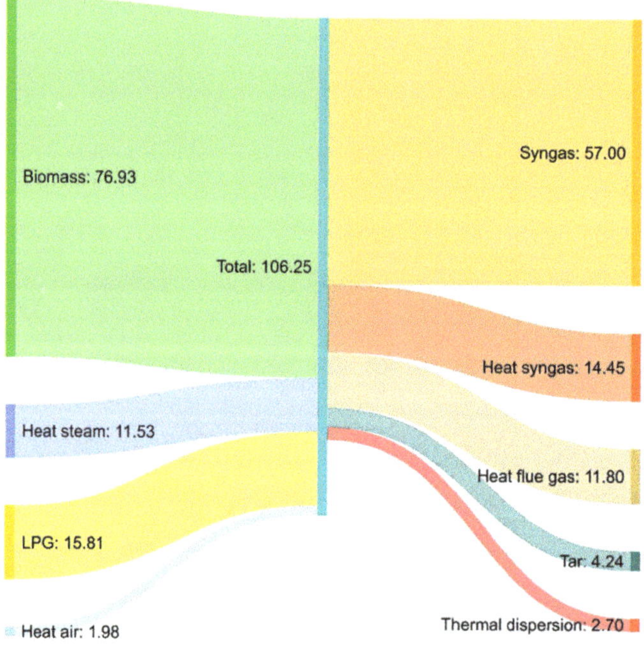

Figure 9. Dual-bed reactor energy balance (units in kW).

For the energy balance, except for the energy contents of biomass, LPG, syngas, and tar, the energies/time required for pre-heating the input flows (air and steam) and the residual heat content of the output flows were considered. In particular, air and steam were introduced in the gasifier at approximately 200 °C; the syngas exits the gasifier at around 800 °C, as measured in the freeboard, and the flue gas exits the combustion chamber at 850 °C. For the energy balance, the higher heating values (HHV) were considered, and the residual heat of the output streams were calculated, reporting the products to 25 °C.

Figure 9 shows the energy balance calculated on the whole dual-bed gasification reactor, which closes with an error of about 14%. In addition, in this case the error can be reduced if the light hydrocarbons would be taken into account in the energy balance. In fact, adding C_2H_4 [13] in the energy balance, and considering that its concentration, on a dry basis, can vary between 2 to 5%, the error would be reduced to less than 1%. The chemical energy of the syngas flow represents 63% of the total output.

The thermal dispersion of the reactor was calculated based on the temperature measured on the external surface of the reactor during the operation, equal to 70 °C; the convective heat transfer coefficient of air was considered as 5 W/m^2 K.

The sensible heat in the flow rate of syngas and that of flue gas account for approximately 16% and 12% of the total energy outputs, respectively. These energy outputs could be recovered internally in the process in practical applications, for instance used in heat exchangers for pre-heating the input air, evaporating water, and superheating steam to the gasifier. This heat recovery would allow to increase the temperatures of the input streams to the gasifier from 200 °C to 450 °C, and therefore to reduce the auxiliary fuel flow rate needed for the thermal requirements of the process, from about 15 kW in the test described here to about 11 kW.

4. Conclusions

In a previous work [10], an innovative design of a dual bubbling fluidized bed gasifier was developed, and cold model experiments were carried out to check the performance of the reactor geometry from a fluid-dynamic point of view. Subsequently, the reactor was realized at pilot scale (up to 100 kWth as thermal input), and in this work the results of the preliminary gasification tests are reported.

Temperature monitoring during the start-up phase shows that it is possible to reach the desired operating temperature in approximately 4 h, by heating the reactor by means of LPG and biomass combustion.

Gasification tests were carried out using hazelnut shells as biomass feedstock, feeding steam to the gasifier and air to the combustor. The heat produced in the combustor by burning the residual char and the auxiliary LPG fuel was found to be enough for the gasification process, as confirmed by the stability of the temperature values measured during the test.

The syngas composition analysed online also showed fair stability during a typical test that lasted approximately 70 min. The average volumetric composition of the syngas was: $H_2 = 35\%$, $CO = 23\%$, $CO_2 = 20\%$, and $CH_4 = 11\%$. The N_2 content in the gas was as low as 7%, which confirms the efficient design of the gasifier as far as the separation of the gas flows in the two chambers is concerned. Mass and energy balances, closed with a small error, add reliability to the results obtained.

The tar content of the raw syngas, obtained in the plain biomass gasification test, was 12 g/Nm^3. In order to reduce the tar content, a water scrubber downstream from the gasifier was used as the conditioning system. The water scrubber showed good efficiency in reducing the tar content from the syngas stream, with a residual tar content of about 3 g/Nm^3.

Future test campaigns are planned on the pilot plant, also changing the main process parameters. Further tests will allow to identify the best operating conditions to obtain a high-quality syngas with low content of pollutants.

Furthermore, new tests are planned to evaluate the behaviour of the gasifier using different biomass feedstocks: in particular, olive pits, olive pomace, and in the future also wood chips and wood pellets. For these tests, a new calibration of the two screw conveyors will be necessary to evaluate the exact flow rate of the biomass fed, with the density of the feedstocks being different than the one used in this work. Attention will also be given to the velocity of the second screw conveyor (the fast one) to reduce the residence time of the fuel and thus avoid the fuel starting to pyrolyze before entering the reactor. Injections of CO_2 or N_2 in the screw feeder and in the hopper will be considered to help with this problem.

5. Patents

The patent 102022000007613 that describe the gasification unit has been submitted to the Italian Patent Office.

Author Contributions: Conceptualization, A.D.C., E.S., P.U.F. and E.B.; methodology, A.D.C., E.S., A.A.P. and A.T.; investigation, A.D.C., E.S., A.A.P., A.T., L.D.Z. and E.B.; data curation, A.D.C. and E.S.; writing—original draft preparation, A.D.C. and E.S.; writing—review and editing, A.D.C., E.S., P.U.F., L.D.Z., A.A.P., A.T., B.A. and E.B.; supervision E.B. and P.U.F. All authors have read and agreed to the published version of the manuscript.

Funding: This research was funded by the Italian Ministry of Economic Development, under the research project HBF 2.0, proposal number: CCSEB_00224.

Acknowledgments: The authors affiliated kindly acknowledge the financial support of R&D Project HBF 2.0, funded by the Italian Ministry of Economic Development.

Conflicts of Interest: The authors declare no conflict of interest.

Nomenclature

u_s	Superficial velocity (m/s)
u_{mf}	Minimum fluidization velocity (m/s)
η_{gas}	Gas yield (Nm^3/kg)
$Q_{syngas,dry}$	Volumetric flow of dry syngas (Nm^3/h)
$\dot{m}_{bio}$	Input biomass flow (kg/h)
CGE	Cold Gas Efficiency (%)
Q_{syngas}	Syngas total volumetric flow (Nm^3/h)
LHV_{syngas}	Syngas lower heating value (MJ/Nm^3)
LHV_{bio}	Biomass lower heating value (MJ/kg)
$\dot{m}_{LPG}$	LPG mass flow (kg/h)
LHV_{LPG}	LPG lower heating value (MJ/kg)
$mol_{C\ bio}$	Total moles of C in the input biomass (mol)
T	Average temperature (°C)

References

1. Nikoo, M.B.; Mahinpey, N. Simulation of biomass gasification in fluidized bed reactor using ASPEN PLUS. *Biomass Bioenergy* **2008**, *32*, 1245–1254. [CrossRef]
2. Li, X.; Grace, J.; Lim, C.; Watkinson, A.; Chen, H.; Kim, J. Biomass gasification in a circulating fluidized bed. *Biomass- Bioenergy* **2004**, *26*, 171–193. [CrossRef]
3. Hofbauer, H.; Rauch, R.; Loeffler, G.; Kaiser, S.; Fercher, E.; Tremmel, H. Six years experience with the FICFB-gasification process. In Proceedings of the 12th European Conference and Technology Exhibition on Biomass for Energy, Industry and Climate Protection, Amsterdam, The Netherlands, 17–21 June 2002.
4. Igarashi, M.; Hayafune, Y.; Sugamiya, R.; Nakagawa, Y.; Makishima, K. Pyrolysis of Municipal Solid Waste in Japan. *J. Energy Resour. Technol.* **1984**, *106*, 377. [CrossRef]
5. Van der Drift, A.; Van der Meijden, C.M.; Boerrigter, H. MILENA gasification technology for high efficient SNG production from biomass. In Proceedings of the 14th European Conference and Technology Exhibition on Biomass for Energy, Industry and Climate Protection, Paris, France, 17–21 October 2005.
6. Matsuoka, K.; Kuramoto, K.; Murakami, T.; Suzuki, Y. Steam Gasification of Woody Biomass in a Circulating Dual Bubbling Fluidized Bed System. *Energy Fuels* **2008**, *22*, 1980–1985. [CrossRef]
7. Corella, J.; Toledo, A.J.M.; Molina, G. A Review on Dual Fluidized-Bed Biomass Gasifiers. *Ind. Eng. Chem. Res.* **2007**, *46*, 6831–6839. [CrossRef]

8. Hofbauer, H.; Fercher, E.; Fleck, T.; Rauch, R.; Veronik, G. Two years experience with the FICFB-gasification process. In Proceedings of the 10th European Conference and Technology Exhibition on "Biomass for Energy And Industry", Wurzburg, Germany, 8–11 June 1998.
9. Kuramoto, M.; Kunii, D.; Furusawa, T. Flow of dense fluidized particles through an opening in a circulation system. *Powder Technol.* **1986**, *47*, 141–149. [CrossRef]
10. Di Carlo, A.; Moroni, M.; Savuto, E.; Pallozzi, V.; Bocci, E.; Di Lillo, P. Cold model testing of an innovative dual bubbling fluidized bed steam gasifier. *Chem. Eng. J.* **2019**, *377*, 119689. [CrossRef]
11. HBF High Performance Flexible Small Scale Biomass Gasifier 2.0. Project Funded by the Italian Ministry of Economic Development. 2017. Available online: https://www.hbf2-0.it/ (accessed on 20 May 2022).
12. Spath, P.; Aden, A.; Eggeman, T.; Ringer, M.; Wallace, B.; Jechura, J. Biomass to Hydrogen Production Detailed Design and Economics Utilizing the Battelle Columbus Laboratory Indirectly-Heated Gasifier. 2005. Available online: http://www.osti.gov/bridge (accessed on 20 May 2022).
13. Larsson, A.; Kuba, M.; Vilches, T.B.; Seemann, M.; Hofbauer, H.; Thunman, H. Steam gasification of biomass—Typical gas quality and operational strategies derived from industrial-scale plants. *Fuel Process. Technol.* **2021**, *212*, 106609. [CrossRef]
14. Van Der Meijden, C.M.; Van Der Drift, A.; Vreugdenhil, B.J. Experimental results from the allothermal biomass gasifier Milena Experimental results from the allothermal biomass gasifier Milena Experimental Results From The Allothermal Biomass Gasifier Milena. In Proceedings of the 15th European Biomass Conference, Berlin, Germany, 7–11 May 2007; Available online: https://www.researchgate.net/publication/264092142 (accessed on 20 May 2022).
15. Savuto, E.; Di Carlo, A.; Steele, A.; Heidenreich, S.; Gallucci, K.; Rapagnà, S. Syngas conditioning by ceramic filter candles filled with catalyst pellets and placed inside the freeboard of a fluidized bed steam gasifier. *Fuel Process. Technol.* **2019**, *191*, 44–53. [CrossRef]
16. Abdoulmoumine, N.; Adhikari, S.; Kulkarni, A.; Chattanathan, S. A review on biomass gasification syngas cleanup. *Appl. Energy* **2015**, *155*, 294–307. [CrossRef]
17. Cui, H.; Turn, S.Q.; Keffer, V.; Evans, D.; Tran, T.; Foley, M. Study on the fate of metal elements from biomass in a bench-scale fluidized bed gasifier. *Fuel* **2013**, *108*, 1–12. [CrossRef]
18. Milne, T.A.; Evans, R.J.; Abatzoglou, N. *Biomass Gasifier "Tars": Their Nature, Formation, and Conversion*; National Renewable Energy Laboratory: Golden, CO, USA, 1998. [CrossRef]
19. Basu, P. *Biomass Gasification and Pyrolysis*; Elsevier Inc.: Amsterdam, The Netherlands, 2010.

MDPI
St. Alban-Anlage 66
4052 Basel
Switzerland
Tel. +41 61 683 77 34
Fax +41 61 302 89 18
www.mdpi.com

Energies Editorial Office
E-mail: energies@mdpi.com
www.mdpi.com/journal/energies